THEORY AND PRACTICE IN EXPERIMENTAL BACTERIOLOGY

THEORY AND PRACTICE IN EXPERIMENTAL BACTERIOLOGY

G. G. MEYNELL
M.D.
Guinness Professor of Microbiology
Lister Institute of Preventive Medicine

ELINOR MEYNELL
M.D., DIP. BACT.
Department of Microbiology
Lister Institute of Preventive Medicine

SECOND EDITION

CAMBRIDGE
AT THE UNIVERSITY PRESS
1970

Published by the Syndics of the Cambridge University Press
Bentley House, 200 Euston Road, London N.W.1
American Branch: 32 East 57th Street, New York, N.Y.10022

Library of Congress Catalogue Card number: 72-85729

Standard Book Number: 521 07682 X

First published 1965
Second edition 1970

Printed in Great Britain
at the University Printing House, Cambridge
(Brooke Crutchley, University Printer)

CONTENTS

CONTENTS

4. STERILIZATION

5. EXAMINATION OF BACTERIA BY MICROSCOPY

6. QUANTITATIVE ASPECTS OF MICROBIOLOGICAL EXPERIMENTS

TABLES

FIGURES

FIGURES

PLATES

PREFACE TO THE SECOND EDITION

The enormous advantages of bacteria and phage as models for experimental biology were already well known in 1961, when we first discussed the writing of this book, and have become even more widely recognized in the interval. The fact remains, however, that to use bacteria successfully still requires a background of bacteriological technique, if only to avoid wasting time and effort. So the book was primarily intended for biologists turning to work with bacteria and we particularly tried to bring out the rational basis of the subject, rather than writing down arbitrary lists of instructions. This edition follows the same plan. The major changes are a new chapter on Genetic Technique and sections on photodynamic inactivation, fluorescence microscopy and autoradiography, although a large part of the old text has also been re-written: notably, the sections on bacterial growth; the thermodynamics of disinfection; microscopy and staining methods, now linked more firmly to bacterial physiology; and the interpretation of exponential processes in terms of birth-death models. About one-third of the references have been replaced and are again chosen less on priority than because they illustrate a particular point or provide a useful key to the literature. It may seem surprising to have kept so many of the older papers dating back to the 1900s, considering the mass of later work on the same topics, but more often than not they have a freshness and originality of lasting value.

We should like to thank Professor Peter Armitage, Dr J. M. Creeth, Dr Dmitri Karamata and Dr Alan Stone for their most helpful comments on the draft, and also Dr Denis Herbert and the Editors of the *Journal of General Microbiology* for permission to reproduce Fig. 1.2, Degenhardt & Co. for Fig. 5.6, and Dr G. L. Ercolani for Fig. 6.6.

G. G. M.
E. W. M.

February 1969

PREFACE TO THE FIRST EDITION

In the past, the techniques of experimental bacteriology have, perhaps, received less emphasis in textbooks than those used in medical and public health laboratories. Although all branches of bacteriology overlap to some extent, the day-to-day problems encountered in experiments are not of a kind that usually disturb routine tests, and we have therefore tried to provide a documented guide to the basic bacteriological techniques, including some theory as well as working rules. We have also included a number of tables that greatly lessen the computation involved in standard procedures like viable counts by the dilution method. All the techniques described here are relatively simple, but others can be traced from the references which have been chosen largely for their bibliographies and not on grounds of priority. Much of this material is applicable to micro-organisms of all kinds, and, although the main emphasis is on bacteria, many of the examples concern viruses and yeasts.

A book of this sort inevitably relies heavily on methods devised by others, and we would like to express our gratitude to the following authors and manufacturers, and to the publishers and editors of the journals cited below, for giving us permission to reproduce their work:

Dr D. J. Finney and the *Journal of General Microbiology* (Table 1.1); Professor B. D. Davis and the *Journal of Bacteriology* (p. 37); Dr A. D. Hershey and *Virology* (p. 39); Professor Joshua Lederberg and the *Proceedings of the National Academy of Sciences* (p. 63), the *Journal of Bacteriology* (p. 65) and Year Book Medical Publishers Inc. (p. 40); Messrs Hilger and Watts Ltd (Fig. 1.3); Evans Electroselenium Ltd (Fig. 1.5); Professor E. L. Gaden and *Biotechnology and Bioengineering* (Fig. 3.2); Drayton Castle Ltd (Fig. 4.8); Dr J. C. Kelsey and the *Lancet* (Fig. 4.9); Albert Browne Ltd (Table 4.5); Dr P. A. P. Moran and the *Journal of Hygiene* (Table 6.2); Professor W. G. Cochrane and *Biometrics* (Fig. 6.5, Table 6.4); Dr D. J. Finney and the Cambridge University Press (Table 6.6); Dr I. A. DeArmon and the *Journal of Bacteriology* (Table 6.8); Mr J. Taylor and the *Journal of Applied Bacteriology* (Table 6.10); Professor L. L. Kempe and the *Journal of Biotechnical and Microbiological Technology and Engineering* (Table 6.11); Dr A. L. Fernelius and the *Journal of Bacteriology* (Fig. 6.13); Professor H. Orin Halvorson and the *Journal of Bacteriology* (Table 6.12); Dr C. S. Weil and *Biometrics*

(Table 6.13); the Editors of *Nature* (Fig. 6.14); Mr M. R. Young (Plate 5.1); Carl Zeiss (Plate 5.2); and Dr M. G. Macfarlane and Dr C. M. Gray for the method quoted on p. 10. The draft was read by Dr Janice Taverne, Professor R. E. O. Williams and Professor Peter Armitage, and we should like to express our thanks to them for their criticisms and suggestions.

G.G.M.
E.W.M.

January 1964

& Marr, 1968), since the formation of sister cells and their nuclei and even the mass of individual cells is measurable by light microscopy (p. 146, 156: Mitchison, 1961). Thus, it may be asked if the individual generation times of two daughters or cousins are correlated, and whether such differences arise during growth proper or solely from delays in cell separation (Powell & Errington, 1963*a*). Or the observed distribution of cell sizes can be compared with prediction, e.g. on the assumption that individuals grow exponentially and divide once they reach a critical size (Koch, 1966*a*). Although many of these details are still in dispute, such studies show the distribution of individual division times to be markedly skewed to the left, with a minority of organisms dividing after relatively long periods. The corresponding distributions of individual cell ages are also extremely skew with newly formed cells predominating (Powell, 1956*a*). Most of these observations concern cell lengths and divisions, but nuclear division times are equally scattered (Schaechter *et al.* 1962).

Bacterial populations

The asynchronous growth of individual bacteria is reflected in the exponential growth of their cultures, which increase evenly in number without giving any indication of the stepwise changes that characterize the growth of small clones. The only natural instance of synchronous growth is found immediately after diluting certain stationary phase cultures in fresh medium (Cutler & Evans, 1966). Numerous methods have been devised for obtaining synchronous division (see Campbell, 1957; Maaløe, 1962), many of which entail fairly drastic treatment of the culture, like thymine-starvation or temperature shifts, which may produce artefacts. The current preference is to separate cells of a given age from an asynchronously-dividing culture maintained in a constant environment. Two such methods depend on collecting the youngest and smallest cells by filtration (Anderson & Pettijohn, 1960) or by their spontaneous separation from a population adhering to the undersurface of a membrane filter (Cooper & Helmstetter, 1968). Even with the most successful methods, asynchrony reappears in 2–3 generations as innate differences between individual cells quickly become manifest. No method is known which will maintain synchrony indefinitely.

The use of closed cultures using a finite volume of medium is well recognized to be unsuitable for many purposes because the cultural conditions are continually changing as the cell concentration increases and the medium is progressively depleted, with consequent changes in the composition and properties of the cells. The alternative is some form of open

1

BACTERIAL GROWTH

It must be admitted, whatever a bacteriological purist may hop vast majority of experiments are carried out perfectly satisfacto conditions which are almost completely undefined. Most people a culture growing in a closed system like a tube of broth is chan generation to generation, just as they know that the cells would mentally different in glucose-salts medium. Yet, more often th these differences can be set aside, simply because experience te when it is safe to do so. To do this successfully, however, one nee something of bacterial growth under different conditions and h be measured and controlled.

KINETICS OF GROWTH

An outstanding feature of bacterial growth is the manner in dimensions of each cell and the relative proportions of its constit DNA, RNA and protein, are co-ordinated in different cultural (see Maaløe, 1960; Kjeldgaard, 1967; Powell *et al.* 1967). On su a stationary phase culture in fresh medium, cell mass increases numbers so that the cells enlarge (Hershey & Bronfenbrenner, analogous changes are seen when a culture growing expon glucose-salts medium is transferred to broth ('shift up'). The reve when cells growing in broth reach the stationary phase or are t to a simple medium ('shift down'). Only when exponential g persisted for several generations do the concentrations of dif constituents increase at the same rate, as judged by measureme culture as a whole. Even then, the culture is likely to be grossly eous at the cellular level (pp. 3, 9, 197, 288).

Individual bacteria

For many years, it has been known that, although bacterial p may increase exponentially at a constant rate, the rates of divisi vidual bacteria invariably show gross differences. These are i the nature of microbial growth itself and have therefore pro starting-point for numerous studies of growth and cell division (s system, exemplified by the chemostat, in which new medium is continually added to the culture vessel and an equivalent volume of culture removed. In this way, the composition of the medium and the growth rate of the culture are maintained constant, so avoiding the major drawback of closed cultures. It should be appreciated, however, that although continuous culture provides a constant environment, it cannot produce a cell population that is uniform—for example, in individual division times. Quite apart from the differences arising between individuals at cell division that are intrinsic in bacterial growth (see Wilson & Miles, 1964, p. 384), there are long-term changes apparent in bacterial populations caused by mutation and selection—the phenomenon of 'periodic selection'—which leads to successive shifts in the composition of the culture (Atwood, Schneider & Ryan, 1951).

Closed cultures. An inoculum of non-growing cells does not begin growing immediately on transfer to fresh medium, and the duration of this *lag* is measured arbitrarily as in Fig. 1.1 (see Finn, 1955).

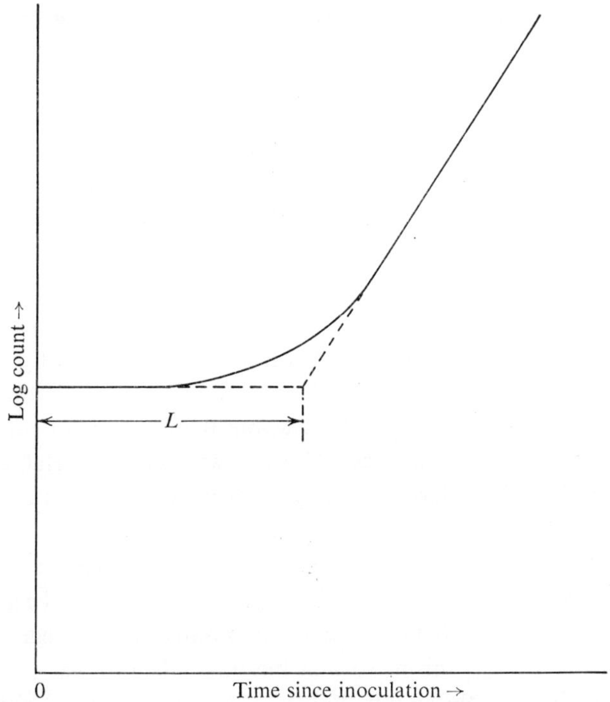

Fig. 1.1. Measurement of lag. Since the growth curve is usually concave upwards, as shown here by the continuous line, the duration of the lag phase is arbitrarily taken as the period, *L*.

It is also apparent from Fig. 1.1 that, once the lag is past, the count increases at a constant rate. Since bacteria increase by binary fission, the count increases as follows:

Number of generations	0	1	2	3	4	...
Total number of organisms	N	$2N$	$4N$	$8N$	$16N$	...
or	2^0N	2^1N	2^2N	2^3N	2^4N	...

The total number of organisms, N_g, present after g generations is therefore

$$N_g = 2^g N \tag{1.1}$$

and

$$\log_2 N_g = g + \log_2 N.$$

The number of generations, g, that occur in a given time, t, is μt, where μ is the number of generations per unit time and may be called the *specific growth rate* (Herbert, Elsworth & Telling, 1956). Equation 1.1 then becomes

$$N_t = 2^{\mu t} N. \tag{1.2}$$

An alternative form of equation 1.2 is derived from the constancy of the growth rate during the exponential phase. Since growth is asynchronous

$$dN/dt = \mu' N.$$

On integration, $N_t = Ne^{\mu' t}$, where e is the base of natural logarithms. On taking logarithms, this equation becomes: $\log_e N_t = \log_e N + \mu' t$. Similarly, $\log_{10} N_t = \log_{10} N + \mu'' t$. A plot of log N_t, using logarithms to any base, against t must therefore always give a straight line when growth is exponential, and exponential growth is also termed 'logarithmic' for this reason. Exponential growth is synonymous with 'balanced growth' where 'every extensive property of the growing system increases by the same factor' (Campbell, 1957), provided the term is applied to microbial populations as distinct from individuals in whom the synthesis of components like DNA may be discontinuous at low growth rates (see Helmstetter & Cooper, 1968). Neither term implies that growth is not restricted by deficiency of nutrients, for exponential balanced growth occurs in the chemostat where the concentration of one nutrient is intentionally limited.

Although N_t is usually plotted in $\log_{10}$, plotting in $\log_2$ has the advantage that each unit on the ordinate then corresponds to a twofold increase in the number or mass of organisms so that the time taken for each generation is read directly from the graph. Values of $\log_2$ are obtained either from $\log_{10}$ ($\log_2 = 3{\cdot}3219 \log_{10}$) or from the values tabulated on p. 32. Alternatively, the data can be plotted in $\log_{10}$ and the ordinate divided in $\log_2$

simply by marking off divisions at intervals equivalent to 0·3 on the $\log_{10}$ scale ($\log_{10} 2 = 0{\cdot}301$).

The numerical values of the growth rate constants, μ, μ' and μ'', depend on the time unit (hours, minutes, etc.) chosen for the experiment, but although their values differ, they all have the same general significance. Equation 1.2 shows that for logarithms to any base (2, *e*, 10, ...) the specific growth rate equals

$$\frac{\log N_t - \log N}{t}.$$

In one generation, N increases to $2N$, the simplest case being $N = 1$ with $N_t = 2$. As $\log 1 = 0$, the above equation shows that the growth rate constant equals

$$\log 2/G,$$

where G is the time taken for the culture to double in mass or number as the case may be. Therefore, for logarithms to base 2, *e* and 10,

$$\begin{aligned} \mu &= \log_2 2/G = 1/G, \\ \mu' &= \log_e 2/G = 0{\cdot}693/G \\ \mu'' &= \log_{10} 2/G = 0{\cdot}301/G. \end{aligned}$$

It is evident that when G is expressed in hours, the values of the specific growth rates are 60 times greater than when the time unit is 1 min.

The time taken for a culture to double in concentration or mass, symbolized above by G, has been referred to in the past as the *mean generation time*, since the generation times of individual cells are known to be distributed. However, the mean of the distribution of individual division times does not correspond to G, which is therefore better called the *doubling time* of the culture (Powell, 1956*a*; Painter & Marr, 1967).

Every closed culture ultimately reaches the point at which the growth rate falls to zero to exhaustion of one or more nutrients. When the energy source is limiting, such as glucose in glucose-salts medium, its initial concentration was once thought to be proportional to the final mass, N, of bacteria

$$N = YC,$$

where Y is the *yield coefficient* (see Monod, 1949). If this relationship held, then no energy would be required by the cell other than for growth; i.e. the *energy of maintenance* would be zero. That this is not so has become clear (Pirt, 1965). Furthermore, Y varies for a given strain grown at different rates (Powell, 1967). As originally expressed, the above relationship concerned the dry weight of organisms and of substrate, but it is now usual to choose more fundamental parameters such as the weight of cells formed per μM ATP generated per μM of substrate (see Gunsalus & Shuster, 1961; Senez, 1962).

Eventually, in a closed culture, the viable count falls. Nevertheless, in many cultures division is probably continuing, albeit slowly, and the viable count falls because the death rate of the organisms exceeds their division rate. This also occurs in infected animals. Conventional viable counts measure only the net result of death and division and so cannot distinguish this situation from that in which deaths but no divisions occur. However, the true division rate of bacteria is now measurable by genetic techniques, as by superinfecting lysogenic bacteria with a phage related to their prophage, a method which, used in conjunction with viable counts, also enables the true death rate of the organisms to be measured (Meynell, 1959; Maw & Meynell, 1968).

Continuous culture. The principle of a simple continuous culture system can readily be understood without giving the argument in detail (see Herbert, Elsworth & Telling, 1956; Herbert, 1959, 1961*b*; Powell, 1965). Consider a perfectly stirred vessel containing a constant volume of culture medium to which fresh sterile medium is added at a constant rate while organisms and medium are removed at the same rate. When the concentration of organisms is small, they divide at their maximum rate because nutrients are in excess. Eventually, however, the organisms increase to a point at which they consume one of the nutrients to such an extent that its concentration in the medium falls and begins to restrict the microbial growth rate. The organisms therefore divide more slowly and so come to adjust their growth rate to the concentration of the limiting nutrient (which is in turn determined by the rate at which fresh medium is added and by the size of the culture vessel) until the number of new organisms produced balances that removed from the vessel: the microbial concentration then becomes constant and the system has reached a steady state. The only conditions in which this can fail to come about is if medium is added faster than the organisms can divide, or if the machine is set for extremely small doubling times when the organisms may cease to divide at all (Tempest, Herbert & Phipps, 1967).

The specific growth rate, μ, is generally considered as related to the concentration of the limiting nutrient in the vessel, s, by

$$\mu = \mu_m \frac{s}{K_s + s},$$

where μ_m is the *growth rate constant*, the maximum possible growth rate of the organisms in that medium, and K_s is a constant. This relation resembles the Michaelis–Menton equation and a plot of μ against s gives a curve rising to an asymptotic value, μ_m. Although this relation is clearly an oversimplification, because μ would fall if s were increased indefinitely (e.g. because of plasmoptysis), it has proved useful in practice. The behaviour of an idealized system of this sort is predictable (Herbert, Elsworth & Telling, 1956) as shown in Fig. 1.2 where various properties of the culture are plotted against D, the dilution rate (D = the flow-rate of fresh

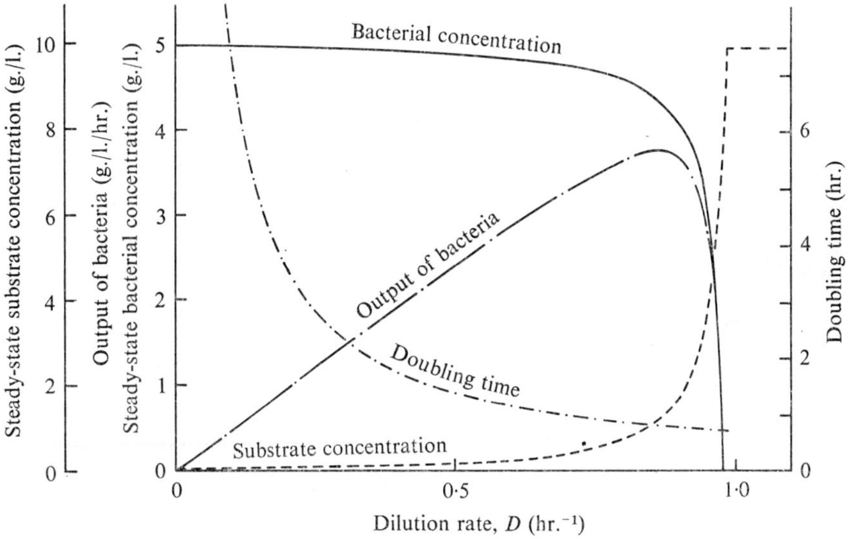

Fig. 1.2. Steady-state behaviour of an idealized continuous culture system. From Herbert, Elsworth & Telling (1956).

medium/the volume of medium in the vessel). When $D > \mu_m$, the organisms are washed out of the vessel faster than they can divide and their ultimate concentration is therefore zero. When $D = 0$, the system is closed, like a tube of broth, the ultimate bacterial concentration is maximal for the system, the limiting nutrient is exhausted and the doubling time is infinite. At intermediate points, as D decreases, the steady state bacterial concentration increases, while the concentration of limiting nutrient falls and the doubling time increases. What is perhaps less to be expected is that the output of bacteria per unit volume of medium in the vessel passes through a maximum at a relatively high dilution rate, and then falls.

In practice, these predictions are not completely fulfilled. As D increases, the bacterial concentration may decrease more slowly than expected, due to organisms adhering to the wall of the vessel where they grow in a film, parts of which enter the bulk of the culture (Herbert, Elsworth & Telling, 1956). Although such a film may be almost invisible, it contributes sufficient organisms to disturb the behaviour of the culture (Larsen & Dimmick, 1964). A second common discrepancy is that bacterial mass does not tend to constancy at small D: in carbon-limited cultures of *Aerobacter aerogenes*, the mass decreases; whereas, when growth is limited by N, S, P, K or Mg, it rises. Nitrogen-limited cells, for example, store progressively more glycogen as D falls, so increasing their mass (Tempest & Dicks, 1967).

This account of continuous culture applies to any single stage stirred fermentor, like the chemostat and also the 'turbidostat' in which the bacterial concentration is fixed at a predetermined level by turbidimetry while D is left to find its own level. The difference between the two instruments is that the cell population is maintained most stably by a chemostat operating at small D, because small fluctuations in D do not then greatly affect the cell concentration, whereas the changes in cell concentration to which the turbidostat responds are maximal when D approximates to μ_m (Herbert, 1959). Many more complex systems have been devised, as by using fermentors arranged in series and recycling part of the terminal effluent (see Herbert, 1961*b*). The principles of continuous culture are continually being extended, as by taking into account the transient states that intervene when the culture is changed from one steady state to another and by considering the dependence of Y on μ (Powell, 1965, 1967).

The construction of chemostats and other continuous systems has often been described in detail, particularly in the literature concerned with industrial microbiology, and extend from relatively straightforward apparatus for aerobic (Herbert, Elsworth & Telling, 1956; Baker, 1968) and anaerobic culture (Hobson & Summers, 1967) to others that are highly complex, with continuous monitoring of pH, dissolved O_2, foaming, E_h and CO_2 (see Herbert, Phipps & Tempest, 1965).

METHODS FOR MEASURING BACTERIAL MASS AND CONCENTRATION

The following sections describe various methods for determining bacterial mass, total and viable counts, their precision, and the effect of technical error. Sampling errors arising, for example, from the random distribution of organisms in suspension, are discussed more fully in Chapter 6.

METHODS OF MEASUREMENT

Most methods tend to obscure the fact that bacterial cultures are grossly heterogeneous. The estimates they provide are averages, derived from individuals which, although they may well be genetically identical, nevertheless differ in many characteristics like length and breadth (Henrici, 1923), antibiotic resistance (Hughes, 1956), cell wall composition (Pennington, 1950) and age (Powell, 1955, 1956*a*). The extent of these differences which almost certainly arise from non-genetic causes can sometimes be determined directly, as in measuring the size of bacteria by microscopy or the amounts of enzyme formed by individual cells (Rotman, Zderic & Edelstein, 1963; Collins, 1964). Other differences can only be determined retrospectively. One example is antibiotic-resistance. Another is the varying number of phage particles released by individual infected bacteria which differ in a way that cannot be determined before lysis, for example from the distribution of cell volumes in the culture (Delbrück, 1945). Some differences, as in size, are eliminated if cell division is synchronized, but others, like differences in composition, may be inherent in the nature of bacterial growth.

BACTERIAL MASS

The relation between dry weight, chemical composition, and light-scattering power of a culture necessarily depends considerably on the species of organism and the conditions in which it is grown. However, as a guide, 10^9 cells of a Gram-negative organism like *Escherichia coli* or *Pseudomonas pyocyanea* have a dry weight of *ca.* 0·32 mg.; contain *ca.* 56 μg. nitrogen and 8 μg. DNA; and have an extinction of *ca.* 1·6 with a 1 cm. path at 450 nm. (e.g. Schaechter, Maaløe & Kjeldgaard, 1958).

Dry weight

A sample of the culture is treated with formalin at a final concentration of 1 %, v/v, and centrifuged in a weighed tube to deposit the cells which are then washed once with 0·85 %, w/v, saline + 1 % formalin, and once with 0·05 % saline or with distilled water. The deposit is dried to constant weight in an oven at 105°. Before each weighing, the tube is placed in a desiccator over P_2O_5 until it has cooled to room temperature.

Chemical estimates

Estimates of bacterial nitrogen or protein are often used as indices of bacterial mass. Although this is satisfactory for many experiments, other estimates are sometimes more useful on occasions when a sizeable fraction of the cell nitrogen is in a structure whose formation varies independently of cytoplasmic growth, like the polyglutamic acid capsule of *Bacillus* species. Two alternatives are to measure bacterial phosphorus (Bennett & Williams, 1957) or bacterial DNA, which also estimates the number of genetically independent individuals and may be especially useful for organisms growing in clumps or chains.

Nitrogen estimations. The Kjeldahl method is widely used. The nitrogen in organic material is converted to ammonium sulphate by digestion with H_2SO_4 in the presence of a catalyst (see Steyermark, 1961; Jacobs, 1965). The ammonia can then be estimated:

(*a*) by volumetric titration following distillation, for example in the Markham still which measures 20 μg. N $\pm$ 1 % (Markham, 1942), or in a Conway micro-diffusion unit which measures 0·5 μg. N (Parker, 1961); or

(*b*) colorimetrically, after addition of Nessler's reagent, which measures 1 μg. N $\pm$ 4–5 %.

The second method is quicker and simpler, and is given here. It is obvious that the more sensitive the test, the more important it becomes to avoid contamination by extraneous nitrogen contributed by glassware, reagents, and the atmosphere, including tobacco smoke.

The colorimetric method requires the following reagents (Macfarlane & Grey, personal communication) of which 1, 3 and 4 are available from commercial sources:

(1) Nessler's reagent.

(2) Digestion mixture. Mix the following:

- (i) 1 ml. conc. H_2SO_4 (analytical or N-free grades) + 5 ml. glass-distilled water;
- (ii) 10 ml. 72 % perchloric acid (analytical grade) + 5·5 ml. glass-distilled water.

(3) Standard nitrogen solution, e.g. NH_4Cl.

(4) Gum ghatti solution (to prevent turbidity of Nessler's reagent). Cover 1 g. gum ghatti + 1 g. benzoic acid with 500 ml. glass-distilled water for 3–4 days at room temperature. Then dilute to 1 l. with glass-distilled water saturated with benzoic acid, and filter through cotton wool to remove undissolved solids.

The method is as follows:

(1) Measure the sample ($\leqslant$ 1 ml.) into a narrow Pyrex boiling tube or into a 10 ml. ground glass stoppered tube graduated at 5 ml. (ext. diameter 16 mm., length 100 mm.). Also put up 3 samples of the standard solution containing 5, 10 and 20 μg. N respectively, and set aside 2 tubes for blanks.

The tubes must be long enough for acid to condense in their upper part during heating in stage 3.

(2) Add 0·2 ml. digestion mixture to each tube.

(3) Heat the tubes at 200° on a digestion rack or in a heated block for 1 hr. in a fume cupboard.

(4) Remove the tubes from the heater and stand on the bench to cool to room temperature. As soon as possible, add *ca.* 1 ml. glass-distilled water and again cool to room temperature.

(5) Add 1·0 ml. gum ghatti solution and mix well.

(6) Add 1·5 ml. Nessler's reagent: if the stock solution is turbid, centrifuge and use the supernatant. Add sufficient glass-distilled water to bring the total volume to 5 ml.

(7) Measure the extinction at 425 nm., using a 1 cm. path. The intensity of colour due to N = extinction of the sample minus the mean extinction of the blanks.

The extinction is proportional to N concentration over the range, 5–40 μg. N.

Protein estimations. Of the many methods available, the colorimetric assay of Lowry *et al.* (1951) is widely used. The colour results from the biuret reaction of protein and Cu^{2+} in alkali, and from reduction of the Folin–Ciocalteau reagent by tyrosine and tryptophan.

The following reagents are needed:

(1) 2 %, w/v, Na_2CO_3 in 0·1 N-NaOH.

(2) 0·5 %, w/v, $CuSO_4.5H_2O$ in 1 %, w/v, aqueous Na or K tartrate.

(3) Alkaline copper solution: 50 ml. of (1) mixed with 1 ml. of (2) on the day of use.

(4) Diluted Folin reagent: Folin–Ciocalteau reagent obtainable commercially, diluted to make it 1 N in acid.

Place 3 ml. of (3) and up to 0·6 ml. of sample (15–300 μg. protein) in a 10 ml. tube. Mix. Stand at room temperature for at least 10 min. to dissolve the sample. Rapidly add 0·3 ml. of (4) and mix at once. After 30 min. or more, measure the extinction of 750 nm. or, if it is too great, at 500 mμ. For dilute samples containing less than 15 μg. protein/0·6 ml., mix 1·5 ml. sample with 1·5 ml. of double-strength (3) and proceed as before. Crystalline bovine albumin (Armour) can be used as standard.

If the sample does not dissolve in reagent (3), treat it initially with 0·1 ml. 1 N-NaOH followed, after 30 min. or more, by 1 ml. carbonate-copper solution (reagent 3).

Estimation of DNA and RNA (See Munro & Fleck, 1966). The preliminary steps are the same for both procedures. Add to the sample containing 5–500 μg. DNA or 10–30 μg. RNA sufficient 2·5 N-$HClO_4$ to give a final concentration of 0·25 N, chill for 30 min. and then centrifuge. Discard the supernatant. Dissolve the deposit including DNA and RNA by dispersing it in 0·5 ml. of 0·5 N-$HClO_4$ with a fine glass rod, add 3·5 ml. 0·5 N-$HClO_4$, and heat at 70° for 25 min. with occasional stirring. Centrifuge. Retain the supernatant and discard the deposit.

The estimation of DNA with diphenylamine depends on measurement of the blue colour formed with deoxy sugars. To prepare the reagent, dissolve 1·5 g. analytical grade diphenylamine in 100 ml. analytical grade glacial acetic acid and add 1·5 ml. conc. H_2SO_4. Store in the dark. If a blue colour develops during storage, redistilled acetic acid should be used. Add 0·1 ml. of 1·6 %, w/v, aqueous acetaldehyde to every 20 ml. stock solution on the day it is to be used. In the estimation, 2 ml. diphenylamine reagent is mixed with 1 ml. of the last supernatant and incubated at 30° for 16–20 hours. The extinction of the colour that develops is measured at 600 nm. Giles & Myers (1965) describe a modification that is both more sensitive and gives lower blanks.

A quick method is to omit acetaldehyde from the reagent, to heat at 100° for only 10 min., and to read the extinction after standing for 20 min. This method is only 20 % as sensitive as that using overnight incubation and is more liable to interference by other compounds.

In either method, include two blanks containing $HClO_4$ but no sample and a series of standards of known concentration. Deoxyadenosine is more convenient than DNA since its concentration can be accurately measured from its molar extinction (14,200 at pH 7 at 257 nm.). The standard tubes could then contain 0·02–0·2 μmole (5–50 μg.) deoxyadenosine and should yield a linear relation between concentration and extinction at 600 nm. in the assay.

The method measures only purine-bound deoxyribose, since pyrimidine-bound deoxyribose does not react under these conditions. The concentration of the unknown is obtained in terms of deoxyadenosine which, for double-stranded DNA, is approximately one half of the total nucleotides present. With single-stranded DNA, where purine and pyrimidines are not necessarily present in equal amounts, the total nucleotides can be calculated, given the base composition of the DNA in question.

The estimation of RNA with orcinol depends on the conversion of purine-bound ribose to furfural which reacts with orcinol to give a green colour: ferric chloride is present as a catalyst. 3 ml. of the supernatant obtained by treating the sample with $HClO_4$ is mixed well with 0·3 ml. of a freshly-prepared 10%, w/v, solution of orcinol in absolute ethanol and 3 ml. acidic $FeCl_3$ (330 mg. $FeCl_3.6H_2O$/l. conc. HCl). The tube is heated at 100° for 45 min., cooled and the extinction read at 660 nm. Also include two blanks not containing the sample and a series of standards of known concentration, e.g. 0·04–0·2 μmole adenosine (molar extinction, 15,000 at pH 7 at 259 nm.). Since RNA is largely single-stranded, its base composition must be known to estimate the amount present.

Many experiments are concerned, not with the total DNA or RNA, but with changes in their respective concentrations and it is then sufficient to express the results in terms of purine nucleotides.

Light-scattering

The amount of light scattered by a bacterial suspension may be proportional to its concentration expressed as mass or number, or to mean cell length, depending on the way in which the measurements are taken. The relationships observed in practice are best discussed after the optical theory which, though difficult in detail (see Oster, 1960; Powell, 1963; Koch & Ehrenfeld, 1968), is fairly simple in principle.

Consider a culture placed in a beam of parallel monochromatic light. The proportion of incident light failing to traverse the culture unchanged represents the sum of light absorbed and of light scattered by the organisms. Absorption of visible light is usually negligible, since most species are unpigmented and almost transparent (p. 142). Changes in transmission are therefore determined by scattering, due to reflection at the surface of the organisms and to diffraction within them. Consequently, there is no scattering if the refractive index of the organisms and their suspending medium is the same. The intensity of the light deviated at any particular angle to the incident rays may be plotted against the corresponding angle to show how scattering occurs. Such graphs reveal that, as this angle increases, the intensity does not fall smoothly but shows peaks and depressions caused by interference between rays arising from different points in the suspension. With large transparent particles, like bacteria illuminated by white light, the greatest intensity is immediately to each side of the emergent undeviated light at an angle of less than 90° to the axis (Fig. 1.3; Koga & Fujita, 1962; Powell, 1963).

The total amount of light scattered increases directly with the ratio, particle size:wavelength of incident light (cp. Taysum, 1956; Koch, 1961). It follows that scattering will be greater (*a*) at a given wavelength, the larger the organisms, and (*b*) the shorter the wavelength used with a given organism. The shortest wavelength commonly used is 450 nm. as measurements in the ultra-violet introduce technical complications. 650 nm. is preferable for broth cultures as the extinction of broth is then minimal.

In estimating bacterial concentration, either scattered light or undeviated light can be measured. The former is the more sensitive method at low cell concentrations. The receiver, whether it is the observer's eye or a photocell, will then be detecting the difference above zero intensity; whereas the amount of undeviated light differs from the blank reading by only a small amount that may be more difficult to measure. The reverse is true at high cell concentrations where measurements of undeviated light should be more sensitive. In practice, measurements of undeviated light invariably include some scattered light as all receivers accept light over an angle, a factor which necessarily leads to a decrease in sensitivity (see Powell, 1963, Table 1).

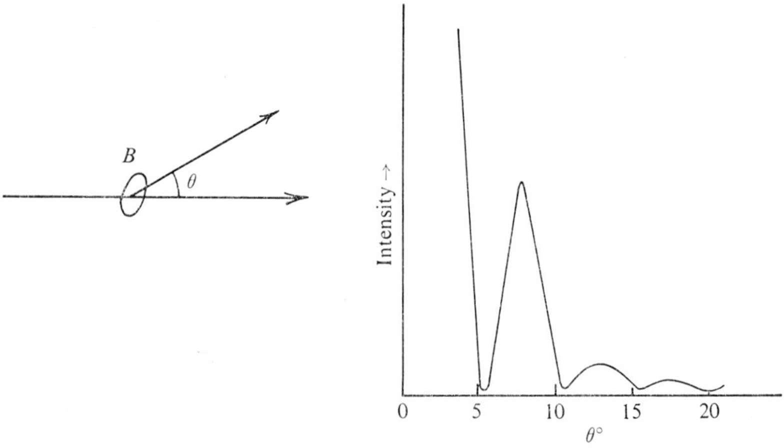

(*a*) (*b*)

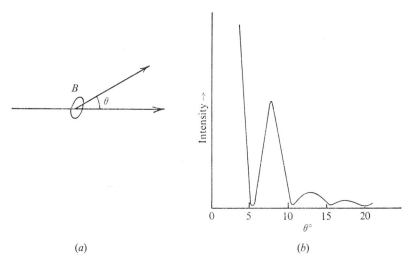

Fig. 1.3. Light-scattering by a bacterial suspension. (*a*) A ray encountering the bacterium, *B*, is deflected from its original path through the angle θ. (*b*) Angular distribution of scattered light. Ordinate: intensity at a given angle θ. Abscissa: values of the angle θ. (After Koga & Fujita, 1962.)

The numerous methods and types of apparatus are described by Oster (1960), and only a few of the most common are mentioned in the following sections.

Measurement of undeviated light. The incident light, I_0, and the transmitted light, I (that is, the light *not* scattered) are often related at low bacterial concentrations by the Lambert–Beer law:

$$I = I_0 10^{-\epsilon lc},$$

where ϵ is the extinction coefficient, l the depth of the suspension, and c the bacterial concentration.

$$\log (I/I_0) = -\epsilon lc \quad \text{or} \quad \log (I_0/I) = \epsilon lc.$$

Log (I_0/I), termed the *density* or *extinction* of the suspension, plotted against c, the bacterial concentration, gives a straight line whose slope is determined by ϵl. That is, the change in extinction produced by unit change in bacterial concentration depends on the depth of suspension (l) and on its

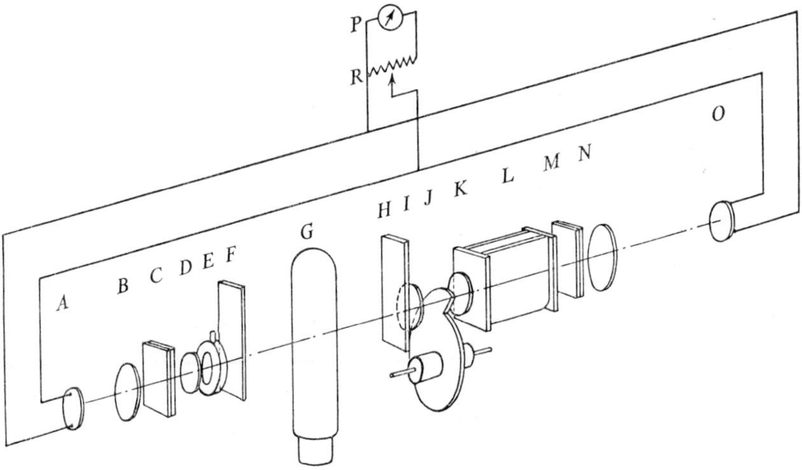

Fig. 1.4. The Spekker absorptiometer, showing the optical system and photocell circuit (by courtesy of Hilger and Watts Ltd). *A*: photocell. *B*: window. *C*: spectrum filter. *D*: lens. *E*: iris diaphragm. *F*: heat-absorbing filter. *G*: lamp. *H*: heat-absorbing filter. *I*: window. *J*: variable shutter controlled by a graduated drum (not shown) bearing density and percentage transmission scales. *K*: lens. *L*: cell for liquids. *M*: spectrum filter. *N*: lens. *O*: photocell. *P*: spot galvanometer. *R*: sensitivity control. The diagram does not show two shutters which prevent light reaching the photocells unless they are opened by depressing a horizontal bar on the front of the instrument.

The drum is set to 0 on the density scale; the iris, *E*, is closed; and the sensitivity control set to its minimum. The galvanometer should read 0 in the centre of the scale after it has been freed by releasing its clamp. The bacterial suspension is placed at *L*. After opening the shutters, the appropriate sensitivity for the sample is found by adjusting the sensitivity control until the light spot on the galvanometer passes from 0 to the end of the scale. The spot is then brought back precisely to 0 by opening the iris, *E*. The shutters are closed and the suspending liquid alone substituted at *L*. On opening the shutters, the light spot is deflected and is now brought back to 0 by moving the graduated drum which operates *J*. After closing the shutters, the optical density of the bacterial suspension is read from the appropriate scale on the drum.

intrinsic properties expressed by ϵ. The value of ϵ differs for different bacterial species (Brown, 1919–20) and even for the same species treated in different ways (Spaun, 1962). The practical point is that a new standard curve relating extinction to concentration has to be constructed for each species and each type of experiment.

The Lambert–Beer law does not hold at high bacterial concentrations, which are underestimated because of secondary scattering in the suspension which redirects light towards the receiver. The range of concentrations to which the law applies depends partly on the wavelength of the incident light (Gerrard, Parker & Porter, 1961).

A long-established subjective method is to match the bacterial suspension against standard 'opacity' tubes like those containing barium sulphate devised by Brown & Kirwan (1914–15) and Brown (1919–20) and now made by Burroughs Wellcome and Co., or the W.H.O. standard of small glass particles suspended in water (Maaløe, 1955). The comparison is made by holding the unknown suspension and the standard side by side against a white card ruled with a black line $\frac{1}{2}$ in. wide. The line is set at right angles to the tubes which are viewed by daylight while standing back to a window. With practice, the opacity of the unknown tube can be brought to within 10 % of that of the standard.

Photometers depending on visual measurements are not generally used but one instrument of this type which can be used successfully is the M.R.C. grey-wedge haemoglobinometer (Scullard & Meynell, 1966). This depends on bringing the two halves of a divided illuminated field to a uniform intensity as judged by eye. Light from the source reaches the eye by two paths, one through the suspension and the other through the suspending fluid alone, so compensating for fluctuations in the intensity of the source. The two halves of the field are equalized by interposing a wedge-shaped neutral density filter in one path.

The Spekker absorptiometer (Hilger and Watts) is often used (Fig. 1.4). There are two light paths, one passing through the test suspension. The intensity is measured by a variable diaphragm coupled to a scale from which the extinction is read.

Measurement of scattered light. Incident and scattered light are related by an equation resembling that derived from the Lambert–Beer law:

$$\log (I/I_0) = -klc.$$

Again, the equation holds for only a limited range of concentrations. When scattered light is measured, its intensity depends on the angle between the axis of the incident light and that of the receiver, owing to the differing degrees of scattering at different angles (Fig. 1.3).

The Pulfrich photometer fitted with a suitable chamber is occasionally used in microbiology (e.g. Powell & Stoward, 1962). Nephelometers are in common use where the incident light is at an angle to the photocell. A simple machine of this type is the Eel nephelometer (Evans Electroselenium Ltd, Harlow, Essex). The sample is contained in a round-bottomed

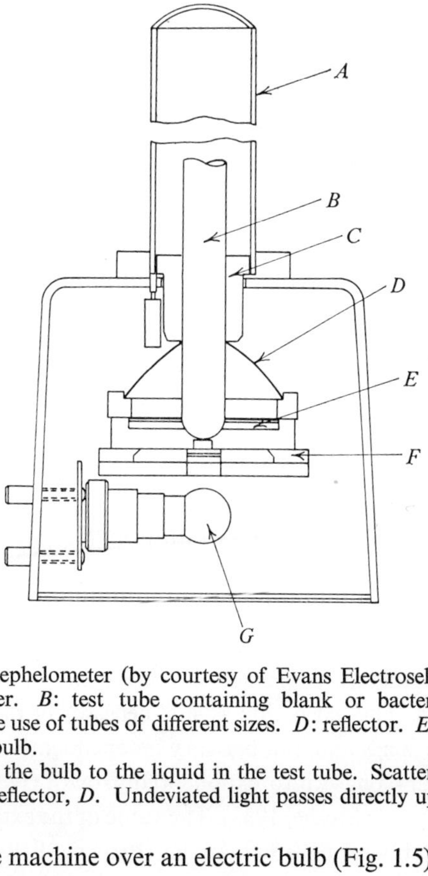

Fig. 1.5. The Eel nephelometer (by courtesy of Evans Electroselenium Ltd). *A*: removable metal cover. *B*: test tube containing blank or bacterial suspension. *C*: adaptor, allowing the use of tubes of different sizes. *D*: reflector. *E*: photocell. *F*: filter holder. *G*: electric bulb.

Light passes from the bulb to the liquid in the test tube. Scattered light reaches the photocell from the reflector, *D*. Undeviated light passes directly upwards.

tube placed in the machine over an electric bulb (Fig. 1.5). Light scattered at 45°–138° to the incident beam reaches a ring of photoelectric cells surrounding the base of the tube (Powell, 1963) and its intensity is measured on an accompanying galvanometer bearing an arithmetic scale ruled in arbitrary units whose zero reading is adjustable. The sensitivity is standardized by inserting a ground glass standard and by adjusting the galvanometer until the scale reading has a predetermined value. If the blank reading given by uninoculated medium cannot be brought to zero by the zero setting, the amount of scattering by the organisms is then taken as the scale reading for the culture minus that of the blank.

The Spekker absorptiometer can also be adapted for nephelometric readings.

The simultaneous use of two light paths, as in the grey-wedge photometer, one for the blank and the other for the bacterial suspension, eliminates errors due to changes in the intensity of the light source. Nevertheless, instruments with a single path like the Eel nephelometer, in which the blank and bacterial suspension are interposed alternately, are widely and successfully used.

In measuring either undeviated or scattered light, it is not always necessary to remove samples from the culture to special cuvettes—which incidentally avoids standardizing the cuvettes. Some instruments take test tubes and the culture can then be grown in a ⊥-shaped tube or in a flask with a horizontal side tube down which the medium is tipped. Cotton wool plugs should be avoided as fibres tend to enter the medium and disturb the readings, especially when the incident light enters the base of the tube. Another disturbing factor is swirling of the medium, which should be allowed to come to rest before readings are made.

Practical applications. The relation between bacterial concentration and the extinction, or the amount of scattered light, is governed partly by the state of the organisms. Three situations can be distinguished:

(1) In growing cultures, the extinction usually appears proportional to cell mass as measured by dry weight or bacterial nitrogen, regardless of the culture medium and of whether division is synchronous or asynchronous (Schaechter, Maaløe & Kjeldgaard, 1958; Burns, 1959). Measurements of transmitted light therefore provide a convenient and rapid means for determining mass. The extinction is not necessarily proportional to cell numbers because mass and number vary independently, as when cells pass from the stationary phase into exponential growth (Hershey & Bronfenbrenner, 1938; Alper & Sterne, 1933). The value of the extinction coefficient, ϵ, differs considerably for different species (Brown, 1919–20; Spaun, 1962).

The proportionality of extinction to mass is not observed with some species, for example staphylococci (Rogers, 1954); while theory suggests that for spherical bacteria the extinction should vary with the 4/3rd power of the mass (Koch, 1961).

The properties of scattered light appear to have been less studied. Its intensity may be proportional to cell numbers, not to cell mass, if measurements are made at 45°–90° to the incident light because the intensity does not increase as the cells enlarge between divisions (Powell & Stoward, 1962).

(2) Where the concentration of intracellular constituents changes because cells swell or shrink due to entry or loss of water. Thus, as Gram-negative (but not Gram-positive) organisms swell when the osmotic pressure of their medium is lowered, they scatter less light and the extinction falls (see Mitchell & Moyle, 1956; Mager *et al.* 1956). Such changes might obviously be misleading if samples from a culture were diluted in different media before readings were made (Bernheim, 1964).

(3) Where cells are constant in shape and composition but cease to be oriented at random in suspension. Changes in turbidity arising in this way are seen as swirling when a culture is tapped and viewed by eye. Cells also become more or less aligned when a suspension runs through a fine tube, and mean bacterial length can then be measured photometrically (Powell & Stoward, 1962).

TOTAL CELL COUNTS

The total cell count is usually determined by a calibrated counting chamber which is examined microscopically, or by a Coulter counter in which the organisms are counted electronically. Although counting chambers are less precise, they are both inexpensive and relatively simple to use and are unlikely to be supplanted by the Coulter counter except when sufficient counts are to be made on a single experimental system to justify proper calibration of the instrument. Other methods such as counts of stained films (p. 21) or the measurement of scattered light are sometimes used.

The total and viable counts on an exponentially growing culture are expected to be equal, since microscopy of growing cells shows that more than 98 % form microcolonies (see Powell, 1958). However, total counts obtained from chambers may be twice the viable count, a finding which originally suggested that up to 50 % of organisms might be dead but which now appears to arise from consistent overfilling of the counting chamber or to killing of some organisms in making the viable count (p. 25).

Counting chambers

A typical chamber has a central depression of known depth whose surface is ruled into squares of known area. A drop of bacterial suspension is placed on the ruled area and covered by a thick coverslip supported by the main part of the slide. The correct chamber depth is generally supposed to be achieved by working the coverslip to and fro on the slide until interference patterns—'Newton's rings'—appear between them. The organisms are then left to settle before counting. Motile or pathogenic organisms have to be killed by heat or by suspension in diluent containing formalin.

Although counting chambers are usually of the specified shape, Norris & Powell (1961) found that the depth of the fluid after filling was often 10–50 % greater, due to poor apposition of the coverslip to the slide, despite the appearance of interference patterns. These authors suggested that this aberration accounted for most of the discrepancy observed between viable and total counts, and recommended that the real depth of the chamber should be determined on each filling by interferometry.

Other sources of inaccuracy include loss of organisms by adhesion to the glass, and their agglutination in acid suspending liquids, such as saline that has been exposed to air. These difficulties are avoided by suspending the organisms in 0·85 %, w/v, saline containing a trace of anionic detergent (e.g. Teepol, I.C.I.) buffered to pH 7·5 with Na_2PO_4 and with formalin added to 0·5 %, v/v, to prevent the growth of contaminants (Norris & Powell, 1961). It is as well to examine the chamber microscopically before it is filled, as organisms may persist on the glass from previous counts unless the slide and coverslip are washed immediately after use to prevent drying of bacterial suspensions.

The Helber and the Petroff–Hausser counting chambers are in general use. The Helber chamber is 0·02 mm. deep and has a ruled area of large squares enclosing small squares whose sides measure 0·2 mm. and 0·05 mm. respectively. To avoid counting the same organism twice if it lies across the ruling between two small squares, all cells on the top and left-hand margin of each square are ignored when counting that square—although these cells are included in the counts for the appropriate neighbouring squares. A sample giving 5–15 organisms per small square (*ca.* 10^8 organisms/ml.) is convenient to count. A hand tally counter saves memorizing numbers as the counting is done.

The whole of a large square (0·04 sq. mm.), enclosing 16 small squares, of a Helber chamber is seen at once using a 16 mm. objective and a ×25 ocular. Unstained organisms can be counted by phase-contrast or by dark-field microscopy. The former is simpler, as cleanliness is not so essential, but some organisms may be invisible. Organisms can also be counted by transmitted light microscopy after staining, but, in our hands, this has never been as satisfactory as the other methods.

At least 600 organisms should be counted (p. 180). The chamber should be set up, say, four times, counting at least 150 organisms each time, rather than counting 600 organisms at once in the same chamber, since the estimated concentration depends on how closely the coverslip fits to the slide.

The count/ml. is calculated as follows. If 660 cells are counted in a total of 100 small squares of the Helber chamber, the average per small square is 6·6. The volume of a small square is 1/20,000 cu. mm. so that the observed total count $= 6{\cdot}6 \times 20{,}000$/cu. mm. $= 1{\cdot}32 \times 10^5$/cu. mm. or $1{\cdot}32 \times 10^8$/ml. That is, the count/ml. culture is equal to $2n \times 10^7$, where n is the mean number of organisms per small square.

When the count is very low, the mean number of organisms per square can be estimated from the proportion of squares containing no organisms, in the same way that the multiplicity of phage infection is estimated from the proportion of bacteria adsorbing 0 particles on p. 177 (see Tippett, 1932, for a general discussion of this form of count).

The Coulter counter. The bacterial suspension passes from one chamber to a second through a small orifice. As each bacterium, or any other particle, traverses the orifice, the resistance of the system changes momentarily and can be measured by electrodes placed in the two chambers. The magnitude of the voltage pulse thus produced is proportional to the volume of the particle. The threshold voltage recorded is adjustable so as to discriminate between the background noise and the particles of interest. Furthermore, by making repeated counts at progressively increasing thresholds, the distribution of particle volumes in the suspension is obtained (Harvey & Marr, 1966). The instrument therefore provides a very rapid means of counting large numbers of cells, so drastically diminishing sampling error in the count (and making technical error of greater importance) and for assessing their properties, provided it has been standardized for the organisms in question and that contamination by extraneous particles is avoided by filtering the culture medium. However, unless the aperture, threshold, and the suspending medium have been chosen correctly, the results are likely to be misleading and, for this reason, counting chambers are still to be preferred for the occasional experiment as well as for checking the degree of aggregation of the organisms by microscopy.

Counts of stained films

The Breed milk count is a well-known example. Using a fine-bore pipette or a microsyringe, a drop of known volume (e.g. 0·01 ml.) is spread over an area of 1 sq. cm. on a slide, fixed, stained, and counted. The total count per 0·01 ml. is the number of organisms per field divided by the area of the field in sq. cm. (determined by micrometry, p. 140).

The organisms are not randomly distributed on the slide, being more numerous at the centre than at the edges, and counts must therefore be made across the complete width of the drop to obtain an average (Hanks & James, 1940). Two other sources of error are the failure of some dead organisms to stain, and detachment of a fraction of organisms during fixation and staining.

Organisms in dilute suspension are countable after collection and staining on a membrane filter (Ecker & Lockhart, 1959*a*; Lumpkins & Arveson, 1968), although there are obvious difficulties in counting isolated organisms on a non-uniform background.

Precision of total counts

The precision of any total count is limited by two sources of error:

(1) *sampling error*, arising from the limited number of organisms counted; and

(2) *technical error*, like errors in the volume of the chamber or counting of the same organism twice.

The important difference between these causes of imprecision is that sampling error is unavoidable.

Sampling error arises from the fact that organisms are distributed at random in suspension. If a number of samples are taken from a culture, no one would be surprised to find that the first contained 25 organisms while the second contained 18 or 32, in the same way that six coins might give four heads when tossed on one occasion and two on the next. Exactly the same argument applies to the numbers of organisms in different squares of a counting chamber, each of which is analogous to a different sample of the culture. It is true that such differences become less as the average number of organisms per square increases, but some variation due to chance is always present (Fig. 6.2). If this is the only cause for differences between squares, the standard error (S.E.) of the mean number of organisms per square has its minimum value and equals the square root of the mean itself (p. 180). However, if technical errors occur, the S.E. is correspondingly increased.

Technical error may arise in counting the organisms, for example, because different parts of the chamber are not of the same depth or, judging from red cell counts in similar chambers, because of inaccuracy in observation which may have a strong personal bias that leads one observer to obtain consistently high counts whereas another's are consistently low (see Dacie & Lewis, 1963). Another source of error comes in setting up the chamber when the coverslip may be unpredictably distorted (Norris & Powell, 1961). The cumulative effect of technical error can make the S.E. of the observed mean 50 % greater than that predicted from random causes alone (Norris & Powell, 1961), although, in some instances, counts on individual squares of a given chamber show no more than random differences (see Eisenhart & Wilson, 1943).

VIABLE COUNTS

A viable count ostensibly measures the concentration of living organisms, but its significance necessarily depends on what is meant by 'living'. No satisfactory definition has ever been provided, with the result that a multitude of working definitions has been proposed in the past (see Wilson, 1922; Postgate, Crumpton & Hunter, 1961). Some depend on a characteristic, like the ability to form a colony visible to the naked eye, which would be universally accepted as a criterion of viability. Other definitions depend on less obvious characteristics, such as motility or fermentative ability (see Wilson, 1922) or resistance to staining (see Knaysi, 1935). None of these alternatives has been widely accepted, presumably because they rest on functions which, at present, are generally felt to be only distantly connected with our intuitive preconceptions of 'viability'. The usual working definition, accepted here, equates 'viability' with the power to form a macroscopic colony on nutrient agar or to produce visible turbidity in broth in a dilution count (p. 185).

The limitations of this definition should be appreciated. The qualification, 'macroscopic', is important, as instances of limited multiplication are known where cells divide a few times to produce a microcolony and then stop, as with irradiated yeast (p. 123) or bacteria treated with ferrous ions (Catlin, 1956). Furthermore, organisms which are dead according to this definition may still possess functions characteristic of viable cells, including the ability to support the growth of virulent phage and to form β-galactosidase upon exposure to an inducer. It is also the rule in bacteriology to find that the viable count is higher on some media than on others, especially when the organisms have been exposed to a bactericidal agent—whether it be radiation, heat, or a chemical (see Harris, 1963). Higher counts are usually obtained on 'rich' media, like digest agar, than on simple defined media, like glucose-salts medium. It follows that no viable count has any absolute significance, for its value often depends as much on the experimental conditions as on the state of the organisms.

The two major techniques are *colony counts*, in which nutrient agar is inoculated so as to obtain discrete colonies after incubation from whose numbers the viable count is calculated, and *dilution counts*, in which tubes of broth are inoculated with known dilutions of the culture, the viable count being estimated from the proportion of inoculated tubes becoming turbid at each dilution. Colony counts are preferable, as the viable count is estimated far more precisely than by the dilution method (p. 31). Whichever method is used, it yields only the count of 'viable particles' as distinct from the number of individual viable cells, since these are commonly in pairs or chains. In colony counts, the count is therefore often expressed in terms of 'colony-forming particles' rather than 'viable bacteria'. Such counts are generally determined after an arbitrary period of incubation, such as 18–24 hr. for *E. coli* growing on nutrient agar at 37°. However, in counting bacteria or spores surviving exposure to bactericidal agents like heat, streptomycin or ultra-violet irradiation, the colony count may increase for several days.

Colony counts

There are numerous methods that differ principally in the means used to obtain separate colonies for counting. In surface counts, the inoculum is spread on the surface of a well-dried nutrient agar plate, whereas in other methods the inoculum is mixed with molten agar held just above its setting point (45°–48°) which is immediately distributed and allowed to set at room temperature before incubation. Thus, the inoculated agar is poured into a Petri dish (pour plate method) or on to the surface of a nutrient agar plate (overlay method), or is spread out evenly on the wall of a test tube (roll tube method). The overlay method is possibly the easiest and quickest. The preliminary stages of all counts are the same up to inoculation of the medium.

Dilutions. Bacterial suspensions have usually to be diluted to obtain discrete colonies and, as the viable count may lie between 0 and, say, 2×10^9 organisms/ml., the range of dilutions required is often very large. A constant dilution factor is therefore usual and gives a series of dilutions whose concentrations form a geometric series: for example, a dilution factor of 10 and an initial concentration of 5×10^8/ml. gives 5×10^7, 5×10^6, ..., 5×10^{-1} organisms/ml. in 9 dilutions. The same dilution factor is preferable throughout an experiment as it reduces the chance of error when making a large number of counts. If the viable count is completely unknown, either a few samples are inoculated from each of a large number of closely spaced dilutions, knowing that at least one dilution must give a convenient number of colonies to count, or more samples can be taken from each of a smaller number of more widely spaced dilutions. In the second case, if one dilution gives colonies that are too crowded to count, the next dilution will be satisfactory, for, although it may give only a relatively small number of colonies per plate, this is compensated to some extent by having more plates at each dilution. When inocula are incorporated in agar, the volume of the inoculum can be varied to a greater extent than in surface counts. In either case, the precision of the count increases, the more colonies are counted, a reasonable number being 600 (p. 180).

Diluent. The composition and temperature of the diluent often lower the observed count, especially if the organisms are plated on selective media containing agents like bismuth sulphite (Crone, 1948) or sodium deoxycholate (Meynell, 1958), or if they have been exposed to a lethal agent like u.v. irradiation (Hollaender, 1943). The colony count may fall by half in 15 min. at room temperature if a broth culture is diluted in distilled water, tap water or saline (Wilson, 1922; Demain, 1958) or if organisms are transferred from one diluent to another (Gossling, 1959). Phage and animal viruses are well known to be rapidly killed by dilution and shaking in protein-free salts solutions unless special precautions are taken (Lark & Adams, 1953). The cause of such changes is often not known, but traces of Cu were found in one diluent and its effects successfully counteracted by adding 0·316 mM (0·004 %, w/v) EDTA (Postgate & Hunter, 1962). The count usually changes least in a diluent consisting of a balanced salts solution, buffered to prevent changes in pH during autoclaving and containing protein, like 10 %, v/v, broth or 0·01 %, w/v gelatin, which may protect the cells by removing toxic cations or by preventing surface denaturation during shaking (Adams, 1948). One formula is 'T2 buffer': NaCl, 4 g.: K_2SO_4, 5 g.; KH_2PO_4, 1·5 g.; Na_2HPO_4, 3 g.; $MgSO_4$, 0·12 g.; $CaCl_2$, 0·11 g.; and gelatin, 0·01 g./l. distilled water.

Enterobacteria growing at 37° die rapidly when transferred to cold diluent at 0°–4°, although stationary phase organisms are unaffected (Meynell, 1958). *Staphylococcus aureus* is resistant in both the exponential and the stationary phases of growth (Gorrill & McNeil, 1960).

Diluents are most conveniently set out in rimless tubes of $\frac{3}{4}$ in. diam. closed by loose-fitting metal caps that can be arranged with one hand, rather than by caps fitted with clips to hold them on the tube or by wool plugs. Capped tubes are conveniently sterilized and stored in wire racks taking up to 72 tubes each, but the actual dilutions are made more easily in a rack that allows the diluent to be seen in the bottom of each tube.

Pipetting. In making each dilution step, a sample of culture is transferred, usually by pipette, to a known volume of diluent. The actual volumes do not seem to affect the precision of the count, within limits, and a 1/10 dilution is as well made by adding 0·5 ml. to 4·5 ml. diluent as by adding 1 ml. to 9 ml. or 10 ml. to 90 ml. A 1/100 dilution is most simply made by adding 0·1 ml. with a 0·1 ml. single-mark pipette to 10 ml. diluent. A fresh pipette must always be used for each step when diluting vegetative bacteria. If this is not done, the count will be as much as 100 times too high because organisms retained on the wall of the pipette, presumably by electrostatic attraction, are not all washed out at each step. Organisms from one dilution may therefore be retained on the pipette wall for several dilution steps, only then to leave the pipette to produce a falsely high count (Ingram & Eddy, 1953). Spore suspensions behave in the contrary fashion, by becoming concentrated on dilution in acid-cleaned pipettes due to a spore-free film of water being left on the glass (Gerrard, Parker & Porter, 1961).

Pathogenic organisms should never be pipetted by mouth, and a rubber teat or a patent bulb pipette must be used. Moreover, with practice, pipetting by teat is as quick as by mouth. Teats cannot be used on pipettes with a small exit hole—'fast-running' pipettes must be specified. Single-mark container pipettes yield the stated volume when emptied completely, but some have an overlarge exit hole which allows infected suspension to slop out of the tip during handling. Delivery pipettes are often used as container pipettes by blowing out the residual liquid that should be left in the tip, but the consequent inaccuracy is probably small compared to other sources of error, particularly the random sampling error of the number of colonies (p. 182). For the same reason, grade B pipettes are sufficiently precise.

Calibrated loops made of heavy gauge welded platinum wire (Asheshov & Heagy, 1951) and dropping pipettes (p. 27) are sometimes used both for dilutions and plating.

Surface colony counts. For *spread plates* 0·1–0·4 ml. from each dilution is pipetted on to a nutrient agar plate and spread out evenly with a sterile spreader until it has been absorbed. Irregular counts follow if the plate is rubbed hard with the spreader after the suspension has been taken up by the agar. Spreaders are bent glass rods or Pasteur pipettes bent after the tip has been sealed. They are sterilized in tins by dry heat and discarded after use, or the same spreader can be sterilized repeatedly during the experiment by immersion either in a beaker of boiling distilled water containing ceramic fragments to prevent bumping or in ethanol which is burnt off the spreader in the pilot flame of a Bunsen burner. The last method is the most convenient, but is not safe with pathogens as the pool of ethanol inevitably becomes contaminated and drops may fall on the bench before the spreader is flamed.

In the *drop method* (Miles & Misra, 1938), samples from up to 10 dilutions are accommodated on a single plate of well-dried agar as discrete drops, usually of 0·02 ml., delivered with a calibrated dropping pipette. The drops are usually allowed to dry undisturbed but they can be dispersed by gently rocking the plate. The preparation of suitable pipettes follows below.

The advantages of this method are economy, particularly when the count is completely unknown, since a wide range of dilutions is accommodated on a single plate, and the ease with which differences between plates and also between inoculated and uninoculated agar are seen. For example, the minute colonies formed by abortive transductants of salmonella are far more easily detected than on spread plates.

A drawback is that organisms may divide before the drop is absorbed by the agar, so overestimating the count at the moment of sampling, and the method is therefore less reliable with rapidly dividing cultures than the pour plate or overlay methods.

No more than *ca.* 15 colonies per drop are countable, using a Gram-negative organism like *Escherichia coli*, without some colonies going undetected through crowding, although this difficulty is diminished, without affecting the viable count, if the colony size is lessened by incubation at 30° instead of 37°.

Preparation and use of dropping pipettes (Fildes & Smart, 1926). Pipettes are made from stainless steel tubing or from glass. If glass is used, a fine Pasteur pipette is drawn and pushed into a wire gauge until it is just held firmly. A nick is made in the glass with a glass file at its junction with the gauge, the pipette is withdrawn, and the distal part broken off. The end must be square and should be examined before use as it may chip during sterilization. A pipette delivering a drop of 0·02 ml. at a rate of 30 drops/min. has an external diameter of *ca.* 0·041 in. (Morse gauge, 57; Stubbs gauge, 57). Using a rubber teat to give a dropping rate of 60 drops/min., the gauges are 58 and 59 respectively.

The pipette is held vertically and drops allowed to fall from the tip on to the plate. The volume of each drop varies directly with the external diameter of the pipette, and inversely with the rate of dropping and the surface tension of the liquid. Fildes & Smart (1926) suggest correction factors for different liquids and also provide nomograms relating drop volume, dropping rate, and either the external diameter of the pipette or the size of wire gauge needed for its manufacture.

A form of surface count widely used in public health bacteriology entails the collection of suspended organisms on a *membrane filter* which is incubated in a Petri dish containing soft nutrient agar or an absorbent pad soaked in culture medium. The colony count may differ from that obtained from pour plates (p. 121).

Deep colony counts. The pour plate, overlay, and roll tube methods have already been mentioned. For *pour plates*, the inoculum is added to 20–30 ml. molten agar at 45°, mixed and at once poured into an empty Petri dish and allowed to set. Water-borne organisms may give a lower count by this method than on membrane filters owing to the sudden changes in temperature (Stapert, Sokolski & Northam, 1962). *Overlays* are made by inoculating a small volume of molten agar (e.g. 2·5 ml. of half-strength agar) which is poured on to a plate of full-strength nutrient agar. This method has the advantage, compared to pour plates, that the overlay sets almost at once and that the colonies are virtually in one plane which makes for easier counting. The surfaces of the plates must be horizontal or the overlay will be thicker at one side than the other which leads to an uneven distribution of colonies and, in plaque counts, to gross differences in plaque size and morphology. In the *roll tube method*, the inoculum is added to a small volume of molten full-strength agar in a test tube which is immediately turned horizontally and rotated until the agar has set in a thin film on the inner surface of the tube which is then incubated with its opening downwards. The method requires considerable practice and the colonies are difficult to count.

The three methods share the advantage that widely differing volumes can be inoculated from the same dilution, which ensures that an adequate number of colonies is obtained from some inoculated plates or tubes without having to use a small dilution factor. A larger number of colonies can be counted on each plate or tube than in surface counts as growth within the agar makes the colonies smaller.

Counts of microcolonies. The methods so far described yield a count only after a period of incubation, usually overnight at the least, that is sufficient for colonies to become visible to the naked eye. However, microscopical examination enables the count and the proportion of viable cells in a culture to be obtained after 2–6 hr. incubation. The inoculum is measured with a micro-pipette (Alper, 1957) and placed either directly on nutrient agar (Postgate, Crumpton & Hunter, 1961) or on unglazed cellophane resting on nutrient agar which is subsequently examined *in situ* (Powell, 1956*b*) or after removal, fixation, and staining (Alper, 1957). The total count is the total number of cells seen initially and the viable count the number of microcolonies formed after incubation.

Colony counters. Semi-automatic counters are a considerable help when counting large numbers of colonies or plaques. The simplest type consists of an impulse counter, a contact with the agar plate, and a manual probe. The contact and probe should be of platinum or some metal that does not corrode when sterilized by flaming. When the probe touches the colony, the circuit closes and the counter advances by 1. Suitable counters have a fairly rapid response; a simple means for resetting the numbers to zero; and quiet, but not absolutely silent, operation so that one can tell if the counter trips correctly when the probe touches the colony.

A disadvantage of this circuit is that the colonies or plaques are marked by the probe and become difficult to count a second time. This is avoided by counting on the back of the plate, using as probe a pen with a spring-loaded nib. The nib is pressed on the glass and marks it opposite the colony counted, and, at the same time, displaces the nib holder upwards to close the circuit. When the pressure is released, the nib returns to its initial position and cuts the circuit.

Colonies are distinguished more easily when viewed by indirect transmitted light against a dark background, as in the illuminator described by Moore & Taylor (1950).

Viable counts by the dilution method

The culture is serially diluted as for a colony count and a number of tubes of medium then inoculated from a few of the dilutions. After incubation, all the tubes inoculated with concentrated suspension will probably be turbid, while, at low concentrations, all will be clear since their inocula did not contain any viable organisms. At intermediate concentrations, some tubes will be turbid and others will be clear. The proportion of tubes becoming turbid at each dilution is necessarily related to the viable count of the undiluted culture. In practice, the relationship is usually given by the first term of the Poisson series and, provided suitable checks show this to be so, the viable count is rapidly estimated from the results by the use of tables (see pp. 192, 231).

Precision of viable counts

The precision of any viable count, like that of a total count (p. 22), is limited by sampling error, again unavoidable, and by technical error. Their relative importance has often been discussed (see Wilson, 1922; Eisenhart & Wilson, 1943), but, as the magnitude of technical error depends largely on the individual, whose performance almost certainly varies from day to day, the conclusions are unlikely to have much general significance. Moreover, it will be seen below that the inescapable presence of sampling error lessens the effect of technical error on the overall precision of the count.

In a colony count, it is evident that when the average inoculum per plate contains only a small number of organisms, the actual number in individual inocula must differ by chance because the organisms are randomly distributed in suspension. The numbers of colonies formed on each plate must therefore always differ to some extent, even when each is inoculated in exactly the same way. The imprecision introduced by such chance fluctuations in bacterial numbers is partly overcome by counting a large number of colonies and by then calculating the viable count from the mean number of colonies per plate (= total number of colonies counted on all plates/total number of plates). However, if another set of plates is inoculated from the same dilution, the mean per plate would probably differ from that given by the first set, just as the number of colonies on a single plate usually differs from that on another. What is therefore required is the precision with which a count of a given total number of colonies estimates the true mean per plate.

If Cm is the total number of colonies counted on C plates and m is the mean number of colonies per plate (p. 182), the standard error of Cm is known to be $\sqrt{Cm}$ if technical error is absent. The value of Cm will lie in the range $(Cm+1{\cdot}96\sqrt{Cm})-(Cm-1{\cdot}96\sqrt{Cm})$ in 95 % of counts (p. 202). Consequently, the 95 % confidence limits for the observed mean per plate, m, are $(Cm+1{\cdot}96\sqrt{Cm})/C$ and $(Cm-1{\cdot}96\sqrt{Cm})/C$.

The precision of the count can be expressed by S.E. $(Cm)/Cm\ (=\sqrt{Cm}/Cm)$ and is plotted against Cm in Fig. 6.2. This shows that the value of $\sqrt{Cm}/Cm$ decreases (i.e. the precision increases) as Cm increases; and also that although the precision at first increases rapidly as Cm increases from zero, it subsequently increases relatively slowly when Cm exceeds 600–1000. The precision gained by counting more colonies is therefore less and less, the larger the total becomes. The precision can also be expressed by calculating the standard error as a percentage of the total. Thus its value for a total of 600 colonies is $(\sqrt{600}/600) \times 100 = 4{\cdot}08\ \%$, and, for a total of 1000 colonies is $(\sqrt{1000}/1000) \times 100 = 3{\cdot}16\ \%$.

A count of 1000 colonies is hardly laborious, but is nevertheless of far greater precision than an extensive dilution count. Suppose each inoculum taken from a culture had a viable count of 1000. In a plate count, the S.E. would be $\sqrt{1000}$, or 31·6 as seen above. However, in an extensive dilution count using a dilution factor of 2, 10 tubes of medium at each dilution, and at least 10 dilutions, the S.E. of the log count would be 0·095 (Table 6.5). That is, the S.E. of the count would be of the order of 150, or about 15 % of the mean. (Note that because the confidence limits are symmetrical about the log count, they are asymmetrical about the count itself: p. 193.)

The reason for the greater precision of a colony count lies essentially in the greater number of observations from which it is calculated. A count of 1000 colonies is derived from the response of innumerable sites on the agar surface, on 1000 of which colonies form (see p. 183). In a dilution count, each site is a tube of broth which may or may not become turbid after inoculation. As relatively few tubes are used, a count by the dilution method is, in effect, derived from the presence or absence of bacterial growth in a far smaller number of sites than is feasible in a colony count. Since the sampling error decreases as the number of observations increases, a colony count necessarily carries a smaller sampling error than a dilution count.

The same considerations apply to titrations of phages or animal viruses, which can be done either by plaque or local lesion counts analogous to colony counts (p. 182), or by dilution methods (pp. 185, 205).

The numbers of colonies formed on different plates inoculated in the same way often differ no more than is accounted for by chance. However, greater differences tend to occur when mixed bacterial cultures are plated (see Eisenhart & Wilson, 1943), or when a pure culture is plated on a low concentration of an antibiotic or on a selective medium like deoxycholate agar (p. 63).

Technical error must clearly account for the above-random variation that is usually observed between counts on the same culture made from separate sets of dilutions. The error introduced by technical causes may be systematic, in that the count is consistently too high or too low, or random, if the error produces either effect.

Errors in making dilutions probably have a systematic bias as the diluent is often set out with the same pipette, and the dilutions are done by the same person pipetting in the same way. If the dilution factor was 1 % in error in each dilution step, being, for example, 9·9 instead of 10, the real degree of dilution in n steps would be $9{\cdot}9^{-n}$. The total error therefore depends also on the number of steps. With 6 steps, $9{\cdot}9^{-6} = 9{\cdot}41 \times 10^{-5}$ compared to an intended dilution of 10^{-6}. The percentage error is

$$\frac{10^6 - 9{\cdot}41 \times 10^5}{10^6} \times 10^2 = 5{\cdot}9\,\%.$$

Random errors, for example in the volumes plated from a given dilution, do not increase the total error as much as might be supposed owing to the presence of sampling error. The total error equals

$$\sqrt{\{(\text{random technical error})^2 + (\text{sampling error})^2\}}.$$

The sampling error in a count of 1000 colonies produces a standard error that is 3·16 % of the mean, as seen in the preceding section. Thus, if there was a random technical error of the same size, the S.E. would be $\sqrt{(3{\cdot}16^2 + 3{\cdot}16^2)} = 4{\cdot}47\,\%$ of the mean, and not $3{\cdot}16\,\% + 3{\cdot}16\,\% = 6{\cdot}32\,\%$, as might be thought.

Table 1.1. *Logarithms to base* 2 (computed by Finney, Hazlewood & Smith, 1955)

The Table gives the logarithm of every integer from 100 to 999. The logarithms of other numbers are obtained by using the values for $\log_2$ for powers of 10 given at the foot of the Table.

Example 1. Required, $\log_2$ of 133,000.

$$\begin{aligned} 133{,}000 &= 133 \times 10^3 \\ \log_2 133{,}000 &= \log_2 133 + \log_2 10^3 \\ &= 7{\cdot}055 + 9{\cdot}966 \\ &= 17{\cdot}021. \end{aligned}$$

Example 2. Required, $\log_2$ of 87·1.

$$\begin{aligned} 87{\cdot}1 &= 871/10 \\ \log_2 87{\cdot}1 &= \log_2 871 - \log_2 10 \\ &= 9{\cdot}767 - 3{\cdot}222 \\ &= 6{\cdot}545. \end{aligned}$$

On the rare occasions when it is worth expressing the count to more than three significant figures, the $\log_2$ is obtained from $\log_{10}$.

Example 3. Required, $\log_2$ of 16,740.

$$\begin{aligned} \log_{10} 16{,}740 &= 4{\cdot}2237 \\ \log_2 x &= 3{\cdot}32193 \log_{10} x \\ \log_2 16{,}740 &= 3{\cdot}32193 \times 4{\cdot}2237 \\ &= 14{\cdot}03. \end{aligned}$$

Table 1.1 (*cont.*)

	0	1	2	3	4	5	6	7	8	9
100	6·644	6·658	6·672	6·687	6·700	6·714	6·728	6·741	6·755	6·768
110	·781	·794	·807	·820	·833	·845	·858	·870	·883	·895
120	·907	·919	·931	·943	·954	·966	·977	·989	7·000	7·011
130	7·022	7·033	7·044	7·055	7·066	7·077	7·087	7·098	·109	·119
140	·129	·140	·150	·160	·170	·180	·190	·200	·209	·219
150	7·229	7·238	7·248	7·257	7·267	7·276	7·285	7·295	7·304	7·313
160	·322	·331	·340	·349	·358	·366	·375	·384	·392	·401
170	·409	·418	·426	·435	·443	·451	·459	·468	·476	·484
180	·492	·500	·508	·516	·524	·531	·539	·547	·555	·562
190	·570	·577	·585	·592	·600	·607	·615	·622	·629	·637
200	7·644	7·651	7·658	7·665	7·672	7·679	7·687	7·693	7·700	7·707
210	·714	·721	·728	·735	·741	·748	·755	·762	·768	·775
220	·781	·788	·794	·801	·807	·814	·820	·827	·833	·839
230	·845	·852	·858	·864	·870	·877	·883	·889	·895	·901
240	·907	·913	·919	·925	·931	·937	·943	·948	·954	·960
250	7·966	7·972	7·977	7·983	7·989	7·994	8·000	8·006	8·011	8·017
260	8·022	8·028	8·033	8·039	8·044	8·050	·055	·061	·066	·071
270	·077	·082	·087	·093	·098	·103	·109	·114	·119	·124
280	·129	·134	·140	·145	·150	·155	·160	·165	·170	·175
290	·180	·185	·190	·195	·200	·205	·209	·214	·219	·224
300	8·229	8·234	8·238	8·243	8·248	8·253	8·257	8·262	8·267	8·271
310	·276	·281	·285	·290	·295	·299	·304	·308	·313	·317
320	·322	·326	·331	·335	·340	·344	·349	·353	·358	·362
330	·366	·371	·375	·379	·384	·388	·392	·397	·401	·405
340	·409	·414	·418	·422	·426	·430	·435	·439	·443	·447
350	8·451	8·455	8·459	8·464	8·468	8·472	8·476	8·480	8·484	8·488
360	·492	·496	·500	·504	·508	·512	·516	·520	·524	·527
370	·531	·535	·539	·543	·547	·551	·555	·558	·562	·566
380	·570	·574	·577	·581	·585	·589	·592	·596	·600	·604
390	·607	·611	·615	·618	·622	·626	·629	·633	·637	·640
400	8·644	8·647	8·651	8·655	8·658	8·662	8·665	8·669	8·672	8·676
410	·679	·683	·687	·690	·693	·697	·700	·704	·707	·711
420	·714	·718	·721	·725	·728	·731	·735	·738	·741	·745
430	·748	·752	·755	·758	·762	·765	·768	·771	·775	·778
440	·781	·785	·788	·791	·794	·798	·801	·804	·807	·811
450	8·814	8·817	8·820	8·823	8·827	8·830	8·833	8·836	8·839	8·842
460	·845	·849	·852	·855	·858	·861	·864	·867	·870	·873
470	·877	·880	·883	·886	·889	·892	·895	·898	·901	·904
480	·907	·910	·913	·916	·919	·922	·925	·928	·931	·934
490	·937	·940	·943	·945	·948	·951	·954	·957	·960	·963
500	8·966	8·969	8·972	8·974	8·977	8·980	8·983	8·986	8·989	8·992
510	·994	·997	9·000	9·003	9·006	9·008	9·011	9·014	9·017	9·020
520	9·022	9·025	·028	·031	·033	·036	·039	·042	·044	·047
530	9·050	9·053	9·055	9·058	9·061	9·063	9·066	9·069	9·071	9·074
540	·077	·079	·082	·085	·087	·090	·093	·085	·098	·101
550	9·103	9·106	9·109	9·111	9·114	9·116	9·119	9·122	9·124	9·127
560	·129	·132	·134	·137	·140	·142	·145	·147	·150	·152

Table 1.1 (*cont.*)

	0	1	2	3	4	5	6	7	8	9
570	·155	·157	·160	·162	·165	·167	·170	·172	·175	·177
580	·180	·182	·185	·187	·190	·192	·195	·197	·200	·202
590	·205	·207	·210	·212	·214	·217	·219	·222	·224	·226
600	9·229	9·231	9·234	9·236	9·238	9·241	9·243	9·246	9·248	9·250
610	·253	·255	·257	·260	·262	·264	·267	·269	·271	·274
620	·276	·278	·281	·283	·285	·288	·290	·292	·295	·297
630	·299	·301	·304	·306	·308	·311	·313	·315	·317	·320
640	·322	·324	·326	·329	·331	·333	·335	·338	·340	·342
650	9·344	9·347	9·349	9·351	9·353	9·355	9·358	9·360	9·362	9·364
660	·366	·369	·371	·373	·375	·377	·379	·382	·384	·386
670	·388	·390	·392	·394	·397	·399	·401	·403	·405	·407
680	·409	·412	·414	·416	·418	·420	·422	·424	·426	·428
690	·430	·433	·435	·437	·439	·441	·443	·445	·447	·449
700	9·451	9·453	9·455	9·457	9·459	9·461	9·464	9·466	9·468	9·470
710	·472	·474	·476	·478	·480	·482	·484	·486	·488	·490
720	·492	·494	·496	·498	·500	·502	·504	·506	·508	·510
730	·512	·514	·516	·518	·520	·522	·524	·526	·527	·529
740	·531	·533	·535	·537	·539	·541	·543	·545	·547	·549
750	9·551	9·553	9·555	9·557	9·558	9·560	9·562	9·564	9·566	9·568
760	·570	·572	·574	·576	·577	·579	·581	·583	·585	·587
770	·589	·591	·592	·594	·596	·598	·600	·602	·604	·605
780	·607	·609	·611	·613	·615	·617	·618	·620	·622	·624
790	·626	·628	·629	·631	·633	·635	·637	·638	·640	·642
800	9·644	9·646	9·647	9·649	9·651	9·653	9·655	9·656	9·658	9·660
810	·662	·664	·665	·667	·669	·671	·672	·674	·676	·678
820	·679	·681	·683	·685	·687	·688	·690	·692	·693	·695
830	·697	·699	·700	·702	·704	·706	·707	·709	·711	·713
840	·714	·716	·718	·719	·721	·723	·725	·726	·728	·730
850	9·731	9·733	9·735	9·736	9·738	9·740	9·741	9·743	9·745	9·747
860	·748	·750	·752	·753	·755	·757	·758	·760	·762	·763
870	·765	·767	·768	·770	·771	·773	·775	·776	·778	·780
880	·781	·783	·785	·786	·788	·790	·791	·793	·794	·796
890	·798	·799	·801	·803	·804	·806	·807	·809	·811	·812
900	9·814	9·815	9·817	9·819	9·820	9·822	9·823	9·825	9·827	9·828
910	·830	·831	·833	·834	·836	·838	·839	·841	·842	·844
920	·845	·847	·849	·850	·852	·853	·855	·856	·858	·860
930	·861	·863	·864	·866	·867	·869	·870	·872	·873	·875
940	·877	·878	·880	·881	·883	·884	·886	·887	·889	·890
950	9·892	9·893	9·895	9·896	9·898	9·899	9·901	9·902	9·904	9·905
960	·907	·908	·910	·911	·913	·914	·916	·917	·919	·920
970	·922	·923	·925	·926	·928	·929	·931	·932	·934	·935
980	·937	·938	·940	·941	·943	·944	·945	·947	·948	·950
990	·951	·953	·954	·956	·957	·959	·960	·961	·963	·964
	10	10^2	10^3	10^4	10^5	10^6	10^7	10^8	10^9	10^{10}
	3·322	6·644	9·966	13·288	16·610	19·932	23·253	26·575	29·897	33·219

2

BACTERIOLOGICAL CULTURE MEDIA

Although numerous culture media have been devised—over 2000 were classified by Levine & Schoenlein in 1930—their essential ingredients are relatively few in number. They range from solutions of inorganic salts, suitable for autotrophs, to the complex media made from digested meat enriched with blood or serum that are traditionally used for the isolation of highly exacting pathogens like streptococci. The main distinction is between defined media, like glucose-salts medium (p. 37), whose major constituents are known, and those of undefined composition, like the peptones made from partly hydrolysed protein (p. 44).

Which type of medium is used depends largely on the purpose of the experiment. Herbert (1961*a*) has argued strongly for the general use of defined media, but it must be recognized that they have many practical drawbacks. The organisms usually differ phenotypically from those grown in an undefined medium like broth, for example in composition and division rate; their growth is more easily inhibited by over-aeration or toxic cations or by an imbalance between some of the constituents, notably amino acids; some bacteria need a large number of growth factors and, indeed, for some no defined medium is yet known (see Hanks, 1966). All these disadvantages could probably be overcome in time, but, meanwhile, undefined media like digest broth are still in general use although they introduce an uncontrolled factor into the experiment.

The quantitative analyses published for many undefined media represent a step towards standardization. The figures have to be treated with caution, for independent analysis shows that batches of the same medium differ considerably in composition. Furthermore, many essential constituents are usually ignored, despite evidence that even those required in trace amounts, like cations or vitamins, may be present in unduly low concentration (p. 47). One may then find that bacterial growth is normal although processes like sporulation or phage adsorption are inhibited (Weinberg, 1955; Hershey, Kalmanson & Bronfenbrenner, 1944).

DEFINED MEDIA

The essential constituents for a non-exacting heterotroph, like *Escherichia coli*, are inorganic salts and sources of carbon and nitrogen, typically glucose and $(NH_4)_2SO_4$. Buffer and a chelating agent, such as citrate, are often added. Exacting strains will grow on these media if the appropriate growth factors are added.

The degree of care required in making and dispensing defined media depends largely on the type of experiment. Genetic experiments using fermentation markers or requirements for amino acids or purines can be done with unwashed agar, analytical grade reagents, containers with rubber liners, and water distilled without special precautions, for example in a Manesty steam still. Although such distilled water is usually satisfactory it may be contaminated with oil, anti-rust compounds (Gifford, 1960), and inhibitory fatty acids (Price & Gare, 1959). Greater care is needed in quantitative work or where vitamins and other growth factors active in minute concentrations are concerned. Washed agar (p. 53), recrystallized reagents, water triple-distilled in glass, and all-glass containers may then be necessary. The reagents should be dispensed with spatulas freed from organic matter by flaming to red heat. Deionized water is sometimes inhibitory and, if the resin is contaminated by micro-organisms, may contain significant amounts of vitamins: these can be removed with activated charcoal. Cation deficiencies may appear if deionized water is used for culture media usually prepared from water taken from a steam still. In washing glassware the difficulty is to remove dirt and organic matter without leaving antibacterial agents on the glass. Chromic acid and household cleansers, particularly detergents, are strongly adsorbed and troublesome to wash off, and often cause sporadic inhibition of growth, for example in a few tubes of a batch or at one side of a plate. The special detergents sold for laboratory work are said to be free of this disadvantage. One method for routine work is to scrub wet plates with the solid hexametaphosphate sold for water-softening, followed by rinsing in dilute HCl and two changes of distilled water. Pipettes taken from lysol or Chloros are washed in tap water, soaked in 4 %, w/v, Na_2CO_3, rinsed in tap water, soaked in 1 pt. conc. HCl + 9 pt. tap water, washed repeatedly in tap water in an automatic washer, and, finally, rinsed in distilled water before drying.

Phosphate-buffered salts medium

Many media of this type have been described, such as M9 medium (Adams, 1959) and those given here:

(1) After Tatum & Lederberg (1947).

Carbon source	2 g.	NH_4Cl	5 g.
K_2HPO_4	3 g.	NH_4NO_3	1 g.
KH_2PO_4	1 g.	Na_2SO_4	2 g.
$MgSO_4.7H_2O$	0·1 g.	Distilled water	1 l.
		(Agar	15 g.)

(2) Davis & Mingioli (1950).

Carbon source	2 g.	$(NH_4)_2SO_4$	1 g.
K_2HPO_4	7 g.	Na citrate.$3H_2O$	0·5 g.
KH_2PO_4	3 g.	Distilled water	1 l.
$MgSO_4.7H_2O$	0·1 g.	(Agar	15 g.)

The carbon source, salts, and agar are dissolved and autoclaved separately at 121° for 15 min. Phosphate precipitates with certain grades of agar, even when mixed after autoclaving, and may form small crystals along loop tracks and within the medium that are easily mistaken for microcolonies. This is particularly prone to occur with M9 medium. The pH of distilled water should always be checked before dissolving agar for, if it has become acid by absorbing CO_2 from the air, the agar often hydrolyses on autoclaving and becomes too soft. If the water should be acid, it can be boiled to drive off CO_2 or else a little tris can be added. The carbon source is usually made up at 100 times the final concentration (e.g. 20 %, w/v, glucose), while the salts are prepared at four times their final concentration. To prepare the medium, 1 vol. of salts solution, supplemented with the carbon source and any growth factors like amino acids, is added to 3 vols. of either (1) distilled water for liquid medium, or (2) molten 4/3rd strength agar (2 %, w/v, in distilled water) previously cooled to 50°. The concentrations of supplements are discussed in the next section.

Note:

(1) The agar, salts solution, and carbon source must not be autoclaved together, or the medium often becomes inhibitory (p. 55).

(2) Citrate is added to prevent sequestration of metal ions as phosphate and appears essential for some strains. Alternative chelating agents are EDTA (1–0·001 g./l.) which can be autoclaved in the medium or sodium hexametaphosphate (4 g./l.) which is sterilized by filtration. The complexes formed by these agents provide a reserve from which free cations arise by dissociation to be used by the organisms. The 'growth factors' produced

by autoclaving glucose in phosphate buffer sometimes prove to be chelating agents (Mayer & Traxler, 1962; Hanks, 1966). Excess chelation will inhibit growth (see Lankford, Kustoff & Sergeant, 1957).

(3) Citrate is used as carbon source by some species like *Salmonella typhimurium*, and the concentration present in Davis & Mingioli medium is sufficient to allow visible growth in liquid or on solid medium when the principal carbon source cannot be used. If this is a disadvantage, there are three possibilities: to buffer with tris in place of phosphate (v.i.); to try omitting citrate; or to substitute EDTA or hexametaphosphate.

(4) Many essential ions are not included in these media, although bacteria contain substantial amounts of Ca, Fe, Zn, Mn and Cu (Rouf, 1964). The phosphate-ammonia mixture should be supplemented with Mg and Fe at least (Young, Begg & Rentz, 1944), rather than relying on their chance presence in the other constituents or the glassware (e.g. Pappenheimer & Johnson, 1936; Spitznagel & Sharp, 1959). Mixtures of trace elements are sometimes added, for example: MgO, 10 g.; $CaCO_3$, 2 g.; $FeCl_3.6H_2O$, 5·4 g.; $ZnSO_4.7H_2O$, 1·44 g.; $MnSO_4.4H_2O$, 1·11 g.; $CuSO_4.5H_2O$, 0·25 g.; $CoSO_4.7H_2O$, 0·28 g.; H_3BO_3, 0·062 g.; and $Na_2MoO_4.2H_2O$, 0·49 g. dissolved in 950 ml. distilled water + 50 ml. conc. HCl to give a stock solution of 200 times working strength (Powell & Errington, 1963*a*).

For methods of removing trace metals from defined media and from water, see Donald, Passey & Swaby (1952) and Holt, Lux & Valberg (1963).

(5) If cells grown overnight with aeration on glucose as carbon source are subcultured to fresh medium, there is usually an appreciable lag before multiplication starts. This is eliminated by having less glucose in the overnight culture (up to 0·5–1·5 g./l. with *Escherichia coli*), for, in such cultures, multiplication stops when the glucose is exhausted and often begins immediately more glucose is added. Lag is also shortened by including organic acids like glutamic or aspartic (Morrison & Hinshelwood, 1949).

Stationary phase cells grown in broth have a lag of several hours when subcultured to glucose-salts medium.

(6) With cultures in liquid salts medium, adequate aeration is essential if multiplication is to occur at anything near its maximum rate. On the other hand, excessive aeration inhibits growth (p. 72).

(7) The carbon source has a considerable effect on the growth rate, which is usually greater with glucose than with other compounds like lactose or succinate (Schaechter, Maaløe & Kjeldgaard, 1958).

(8) The deliberate addition of traces of broth to solid medium is often

useful, for example 0·01 %, w/v, Bacto Nutrient Broth. It enables non-exacting mutants to emerge after exposure of an exacting strain to a mutagen (the phenomenon of phenotypic lag: Demerec & Cahn, 1953). And, in isolating recombinants, it allows the selected clones to form full-size colonies after 18 hr. incubation, although the colonies of the parental strains remain pinpoint. The number of recombinant colonies may also be increased (Riley & Pardee, 1962). Certain strains, notably *E. coli* K12 58.161 and *Salm. typhi*, may show a low efficiency of plating on glucose-salts medium unless a trace of broth is present.

Other complex supplements are available which have a more defined composition. Acid-hydrolysed casein lacks tryptophan (p. 45); peptone treated with hydrogen peroxide lacks tryptophan, methionine, cystine, and tyrosine (Lyman *et al.* 1946); and peptone and casein hydrolysates are free of riboflavin, folic acid, biotin, pantothenic acid, and nicotinic acid after two extractions with 2 %, w/v, activated charcoal, for example at pH 3 (Roberts & Snell, 1946).

Tris-buffered salts medium (Hershey, 1955)

Carbon source	2 g.	$MgCl_2$	0·095 g.
Tris (0·1 M)	12·1 g.	KH_2PO_4	0·087 g.
NaCl	5·4 g.	Na_2SO_4	0·023 g.
KCl	3·0 g.	$FeCl_3$	0·00016 g.
NH_4Cl	1·1 g.	Distilled water	1 l.
$CaCl_2$	0·011 g.		

The pH should be adjusted to 7·4 with HCl and the medium sterilized by autoclaving at 121° for 15 min.

It is simplest to prepare two salts solutions as follows:

Salts solution A (ca. 4 times working strength)

Tris	48·4 g.	NH_4Cl	4·4 g.
NaCl	21·6 g.	Distilled water	960 ml.
KCl	12·0 g.		

Adjust the pH to 7·4 with HCl.

Salts solution B (100 times working strength)

$CaCl_2$	1·1 g.	$FeCl_3$	0·016 g.
$MgCl_2$	9·5 g.	Conc. HCl	50 ml.
KH_2PO_4	8·7 g.	Distilled water	1 l.
Na_2SO_4	2·3 g.		

The acid is added to prevent precipitation.

960 ml. of solution *A* and 40 ml. of solution *B* are mixed to give 1 l. of complete salts solution at 4 times the working strength. This is sterilized by

autoclaving at 121° for 15 min. One volume of this solution is added to 3 vol. of either distilled water or molten 4/3 normal strength agar. Once again, the carbon source, salts, and agar should not be autoclaved together.

Supplementation of defined media with growth factors

Identification of nutritional requirements. A preliminary identification is often possible by supplementing with mixtures of different classes of growth factors (Lederberg, 1950):

(1) 0·1 %, w/v, acid-hydrolysed casein + tryptophan as a source of amino acids, though vitamins are also likely to be present (p. 45);

(2) a mixture of water-soluble vitamins, each at 1 μg./ml. medium;

(3) 0·3 %, w/v, yeast extract, which supplies a variety of factors, including purines, pyrimidines, amino acids, peptides and vitamins;

(4) 0·1 %, w/v, alkali-hydrolysed yeast nucleic acid, as a source of mononucleotides and nucleosides which satisfy the growth requirements of many purine- and pyrimidine-requiring strains. The starting material is usually a preparation such as Yeast Nucleic Acid (B.D.H.) which dissolves in water made just alkaline with NaOH or NH_4OH. The mononucleotides or nucleosides of the purine and pyrimidine bases are easily obtained by heating 2 %, w/v, sodium nucleate solution in 0·1 N-NaOH at 100° for 2·5 hr. in a sealed tube. The free bases are not obtained since ribosidic linkages are relatively stable in alkali. Free adenine and guanine can be prepared by mild acid hydrolysis, but pyrimidine mononucleotides are hydrolysed only under more severe conditions.

Supplements are tested either in liquid medium, which may, however, become turbid due to the emergence of non-exacting mutants, or on solid medium. Here the supplements are either spotted as solids on an overlay (p. 28) heavily inoculated with the test organism (auxanographic method: Pontecorvo, 1949); added as drops of the stock 1 % solution to the inoculated surface of unsupplemented defined medium; or added to wells cut with a flamed cork borer in a plate whose surface has already been inoculated. If wells are used, their bases should be sealed with a drop of molten agar to prevent leakage between the medium and the Petri dish and each should contain a few drops of the concentrated 1 %, w/v, stock solutions of amino acids and purines described below. For clear-cut results, the surface of the medium must be evenly inoculated, for example by flooding with the test culture and removing excess suspension by pipette. A positive response appears as a halo of growth round the growth factor, from which it may be separated by a zone of inhibition, and is seen most distinctly against a dark background. Occasionally, growth occurs only on a line between two factors, indicating that the organism requires both.

Supplementation to meet a known requirement. If preliminary tests suggest a particular type of requirement, like a vitamin or an amino acid, the strain is tested against pure compounds included in the concentrations given in the next section, or against mixtures of pure compounds grouped so as to include most of the common multiple and alternative requirements that arise from the nature of biosynthetic pathways, e.g. a multiple requirement for aromatic amino acids may be satisfied by shikimic acid (see Ames & Martin, 1964).† A third test can finally be run against individual members of each group, either singly or in permutation. One method for permuting growth factors is to arrange them in a chess-board so that each appears alone and in different combinations (Holliday, 1956).

The concentrations of single compounds required in the medium depend on the strain and the yield of organisms, but the figures given below provide a guide. Stock solutions of 100–1000 times working strength are convenient. A nomogram relating the concentrations of the stock solutions, degree of dilution and the required concentration is given on p. 42.

Amino acids. 10 μg./ml. medium—up to 20 times this concentration is used when an excess is needed in assays (see Kavanagh, 1963, p. 622). DL-Compounds are usually adequate: D-isomers may be inhibitory (Grula, 1960). The literature suggests that the preparations now sold can be accepted as pure.

Stock solutions contain 0·1–1 g./100 ml. Dissolve alanine, asparagine, arginine hydrochloride, glutamine, histidine hydrochloride, isoleucine, leucine, lysine hydrochloride, methionine, threonine, valine, phenylalanine, serine, proline, hydroxyproline, and glycine in water. Other acids are less soluble but dissolve in acid or alkali. Cystine and tryptophan are usually dissolved in 0·2–0·4 N-HCl; and aspartic acid, tyrosine and glutamic acid in 0·2–0·4 N-NaOH.

Purines, pyrimidines, nucleotides and nucleosides. 10 μg./ml. medium. Stock solutions contain 0·1–1 g./100 ml. Dissolve adenine, guanine,

† *Some multiple requirements.* Isoleucine + valine; phenylalanine + tyrosine (+ tryptophan + *p*-aminobenzoic acid); adenine + thiamine; methionine + lysine.

Some alternative requirements. Serine/glycine; cystine/methionine; lysine/threonine; proline/glutamic acid/α-keto glutaric; α-amino butyric acid/isoleucine; purines or pyrimidines and their nucleosides and nucleotides; cystine/cysteine.

Lederberg (1950) suggests testing with the following groups of amino acids: (1) lysine, arginine, methionine, cystine; (2) leucine, isoleucine, valine; (3) phenylalanine, tyrosine, tryptophan; (4) histidine, threonine, glutamic acid, proline, aspartic acid; (5) alanine, glycine, serine, hydroxyproline.

thymine, and cytosine in 0·1 N-HCl; and xanthine and uracil in 0·1 N-NH_4OH.

Vitamins. 1 μg. or less/ml. medium. Tentative figures are 1 μg./ml. of thiamine, pantothenate, riboflavin, nicotinic acid, choline, and pyridoxamine; 0·05 μg./ml. of folic acid and *p*-amino benzoic acid; and 0·005 μg./ml. of biotin. For stock solutions, dissolve thiamine hydrochloride, calcium *d*-pantothenate, nicotinic acid, pyridoxamine hydrochloride, *p*-amino benzoic acid, and choline chloride in water; riboflavin in 0·02 N acetic acid; and folic acid in 0·01 N-NaOH, subsequently diluted to 0·001 N for stock. These solutions are stable for at least a month at 4°.

Fatty acids. For many acids, the concentration stimulating growth is extremely small (e.g. 10^{-6} M) and is only slightly less than that causing

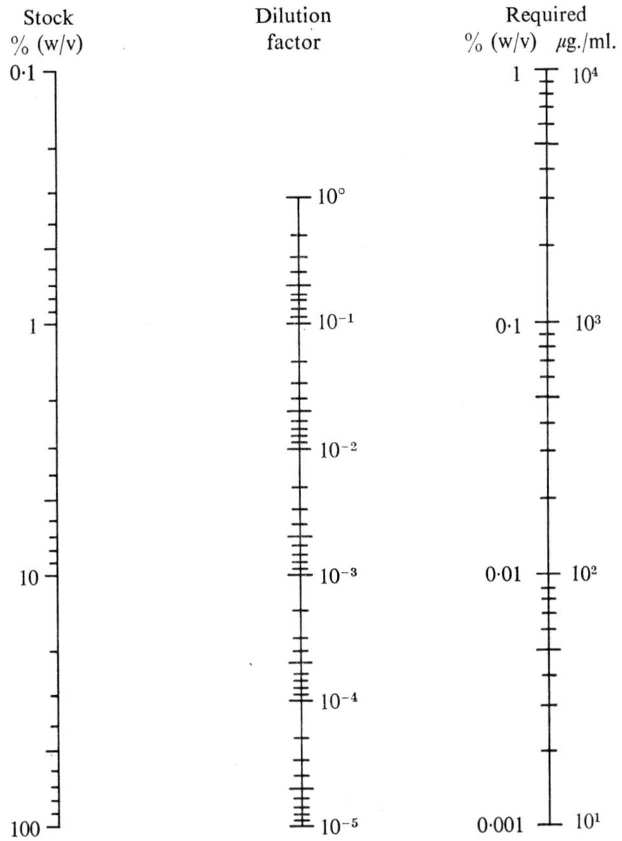

Fig. 2.1. Nomogram giving dilution factors for stock solutions.

inhibition. Their effects are therefore most easily tested by adding them at, say, 10^{-4}M to medium containing an absorbent like serum or charcoal (p. 56). The sodium salts may be water-soluble; otherwise, the compounds are dispersed in 95 %, v/v, ethanol and diluted in buffer.

Experience suggests that stock solutions are sterile if the reagents are dissolved in sterile diluent in sterile bottles and are heated for 3 min. in a bath of boiling water. Heat-labile compounds like glutamine and thiamine should be sterilized by filtration. Riboflavine and ascorbic acid are rapidly destroyed in solution by light and by air respectively. Many sulphur-containing compounds, like sulphite, sulphide, or cysteine, are unstable (Postgate, 1963).

Interpretation

The determination of unknown growth requirements often encounters difficulties, particularly with species whose wild type has multiple requirements. An organism like *Bacillus anthracis* may grow on acid-hydrolysed casein or on an equivalent mixture of amino acids, yet no clear cut requirement for any single amino acid may be demonstrable when each is omitted in turn from the mixture. Auxotrophic mutants of non-exacting species like *E. coli* generally present fewer problems although, even here, perhaps only 30 % of a batch of uncharacterized mutants will respond to one or other of the supplements mentioned above. False negative results may be obtained due to inhibition by some supplements, like adenine and certain amino acids, notable valine and cystine (see Rowley, 1953). A non-exacting strain is therefore needed as a control. Growth requirements may depend on the other growth factors present, the carbon source, and the temperature of incubation (see Lichstein, 1960). Growth may occur in the presence of a given compound if a large number of organisms is inoculated, as in plate tests, but not with a small inoculum, a result which suggests that the organisms co-operate to overcome an inadequacy of the medium (p. 55). Certain amino acids, such as cystine, histidine or glycine, may be required solely for chelation of cations and can be replaced by chelating agents like EDTA (Johnson, 1954; Demain, 1958; Anagnostopoulos & Spizizen, 1961). If a supplement is effective only in high concentration, it is probably acting non-specifically, for example as an absorbent, or is contributing a growth factor which it contains in small amounts. Serum is a notorious example, for it is known to absorb fatty acids (and is replaceable by charcoal: p. 55), to act as a buffer (Holt, 1962), and to contribute at least three components for the growth of mycoplasma (Smith,

1960), in addition to essential fatty acids like oleic or stearic for other organisms (Cohen, Snyder & Mueller, 1941; Boné & Parent, 1963). In the last instance, it may be replaceable by bile, a useful source of water-soluble lipids (Hutner, 1942). A relatively pure supplement like glutamic acid, when used at 5 mg./ml. medium, was found to supply methionine (McDonald, 1963). Yeast extract may provide, not organic compounds, but metals like Zn and Mo, a possibility detected by incinerating the extract and testing the ash (Grant & Pramer, 1962). (In dry-ashing, losses by volatilization are minimized by heating at the lowest possible temperature. The sample is heated gently in a silica or platinum dish until organic matter is destroyed and is then left overnight in a muffle furnace at not more than 420°. See Report, 1960*a*. The ash is then dissolved in dilute HCl and the solution added to the medium in a concentration equivalent to that of the original organic supplement.)

UNDEFINED MEDIA BASED ON PROTEIN HYDROLYSATES

These contain a mixture of the breakdown products of hydrolysed proteins like casein or mammalian muscle. The principal methods of hydrolysis are:

(*a*) acid hydrolysis, usually by HCl, which is confined to the preparation of complete hydrolysates of partially purified proteins like casein or gelatin; and

(*b*) enzymic hydrolysis by pepsin, pancreatin, trypsin, or papain, added as either partially purified or crude preparations, like minced pig stomach or pancreatic extract (Cole & Onslow, 1916). Enzymic digestion yields partial hydrolysates, referred to as 'tryptone' or 'peptone', which generally give better growth than complete hydrolysates or mixtures of amino acids. The method of preparation is milder, so that labile growth factors are presumably preserved; while many organisms grow better on media containing small peptides, possibly because these can be assimilated when the corresponding amino acids cannot (see Leach & Snell, 1960). The degree of hydrolysis of commercial preparations is indicated in the published analyses by the distribution of nitrogen among various arbitrary classes of protein breakdown products, such as primary and secondary proteoses, and peptones, defined by their solubilities in water, $(NH_4)_2SO_4$, etc. The 'proteose peptones' sold are less hydrolysed than 'peptone'. The total N and amino-N contents are also determined, for example by a Kjeldahl method and by formol titration respectively.

Various types of digests often differ markedly, especially when used in liquid form. Exponential growth often does not occur at the maximum rate, and the growth curve may consist of a series of linear portions of decreasing slope, indicating successive exhaustion of various nutrients. This can usually be overcome by adding glucose and by increased aeration, though foaming is a serious difficulty. On solid medium, the differences are less important. Thus, in a viable count, the number of colonies is often the same, although the colonial diameters may differ. Many commercial media are prepared so as to give the correct pH when dissolved in water. Some are not, however, and it is important to check this in each case as the solutions are often very acid.

Casein digests

Casein, obtained from milk caseinogen by acid precipitation, yields hydrolysates contaminated with growth factors. Both acid and enzymic hydrolysates are used as sources of amino acids or peptides. 'Vitamin-free' brands are sold, but, nevertheless, some authors purify them before use.

Acid-hydrolysed casein. Casein is heated with strong HCl until completely hydrolysed to its constituent amino acids: for example 200 g. casein may be autoclaved with 280 ml. 6 N-HCl for 45 min. at 120° or refluxed with N-HCl for 18 hr. There is complete destruction of tryptophan, a severe loss of cysteine, and minor losses of serine and threonine. The hydrolysate is initially deep black, and subsequent steps in preparation usually include decolorization and removal of some of the chloride contributed by the HCl. One method involves distillation of residual HCl, precipitation of Cl as $PbCl_2$ which also decolorizes the solution, and removal of Pb^{3+} with BaS and H_2SO_4 (Mueller & Johnson, 1941). The method of purification, especially the use of charcoal, may considerably alter the nutritional properties of the hydrolysate (see Barton-Wright, 1952). Thus, Ford, Perry & Briggs (1958) removed residual vitamins from 50 g. 'vitamin-free' hydrolysate, previously dissolved in 700 ml. water and adjusted to pH 3·8 with acetic acid, by adding 10 g. activated charcoal, stirring for 1 hr., and filtering (e.g. under suction through a Whatman no. 50 paper). The filtrate was brought to pH 6·5 with 10 N-KOH, treated with a further 5 g. charcoal, and again filtered. The volume of the filtrate was finally made up to 1 l. with water.

Enzymic hydrolysates of casein. These are partial hydrolysates containing varying proportions of small peptides and amino acids, including tryptophan. The usual pH for hydrolysis by papain is between 5 and 7, and by pancreatic extract, *ca.* 8·0.

The 'vitamin-free' enzymic digest described by Roberts & Snell (1946) is made with pancreatin (e.g. B.P., U.S.P.). Gradually add 120 g. 'vitamin-free' casein to 2 l. 0·8 %, w/v, $NaHCO_3$ until the casein is completely dissolved. Then add 600 mg. pancreatin suspended in 15–20 ml. water and enough sulphur-free toluene to form a thin surface layer and to act as a disinfectant. Incubate for 48 hr. at 37°. Toluene is next removed by exposure to flowing steam in an autoclave for 20–30 min. Cool the digest to room temperature, adjust the pH to 6·0 by adding *ca.* 7 ml. glacial acetic acid, and filter under suction. Add 60 g. activated charcoal, stir for 30 min., and again filter using Filter-Cel if necessary. Add *ca.* 75 ml. glacial acetic acid to the filtrate to bring the pH to 3·8 and mix in 24 g. activated charcoal. After 30 min. stirring, filter off the charcoal. Pass 50–75 ml. water through the residue obtained in each filtration and pool with the bulk of the digest which is finally diluted to 2·4 l. This corresponds to 50 mg. casein/ml. digest, and the digest is discarded unless at least 40 mg. solids/ml. are present. For a crude digest, the method is interrupted after toluene is removed.

Muscle digests

The proteolytic enzymes most commonly used are trypsin, derived from mammalian pancreas, and papain, from the fruit of the papaya.

Tryptic digestion is frequently done by Hartley's (1922) method, or by the modification of Pope & Smith (1932), using a pancreatic extract prepared according to Cole & Onslow (1916). Although this method is extremely effective, it is troublesome as the extract is made from fresh pig's pancreas and the digestion takes longer than with papain. Other preparations of trypsin, including the purified enzyme, do not yield a satisfactory medium. The pharmacopoeial preparation, Liq. Trypsinae Co., B.P., contains glycerol, which may be inconvenient.

Papain is considerably easier to work with owing to its greater heat-stability and indifference to pH. The digestion can therefore be rapidly performed at temperatures up to 60° and at pH between 6 and 8, whereas trypsin is inactivated fairly rapidly even at 50° and the pH has to be kept at 8·0. A number of methods for papaic digestion have been published but that described below has proved satisfactory using the powdered papain sold commercially and does not require large volumes of liquid to be held at a high temperature.

In all methods it is usual, after digestion, to boil at *ca.* pH 5·0 to flocculate undigested protein, which is filtered off, and then to adjust the pH to

8·0–8·5 and to boil for 30 min. (or to autoclave at 121° for 15 min.) to remove phosphates that otherwise precipitate during subsequent autoclaving at pH 7·4. However, other compounds besides protein and phosphate are removed, including haematin (Herbert, 1949) and significant amounts of cations. Heating at high pH has been used to lower the iron concentration of casein hydrolysate (Mueller & Miller, 1941), and the Mn and Mg deficiencies ($< 10^{-6}$ M) found in some media presumably arose in this way (Webb, 1948; Weinberg, 1955). Thymine is also frequently lacking. Many nutrient agars have NaCl added to 0·5 %, w/v, although it is usually only needed to prevent haemolysis when blood is added in preparing blood agar (Cruikshank, 1965). Omitting NaCl entirely is the simplest way to stop *Proteus* swarming: if it fails, 4 %, w/v, NaCl can be tried (Kopper, 1962).

All muscle digests probably contain fermentable carbohydrate initially, just as do those prepared from soybean meal, although this is removed from peptones sold for fermentation tests. The traditional means for removing sugars is to incubate the filtered broth with live yeast, for example 60 g. (2 oz.) baker's yeast with 60 l. broth at pH 7·0 for 1 hr. at 30° (Pope & Smith, 1932).

Papaic and Hartley's digest broth do not usually need buffering, provided glucose or other sugars are not added. The titration curve is approximately linear between pH 5 and 8, and the buffering capacity equals that of 0·05 M Sørenson buffer at pH 6·8.

Papaic digest broth. Mix 1 lb. (2·2 kg.) of minced meat with 1 l. tap water + 4 ml. 10 N (40 %, w/v) NaOH. Suspend 2 g. powdered papain (e.g. B.D.H.) in 50 ml. tap water, either by putting the powder in a mortar and mixing in drops of water with a pestle, or by dispersing the powder with an electric homogenizer. Add the suspended papain to the mince and stir well. Heat to 60°. Transfer the mince to 37° for 4 hr., stirring occasionally. Then adjust the pH to 5·0 by adding HCl or 10 ml. glacial acetic acid, boil for 5 min., and filter hot under gravity through a coarse hardened paper (e.g. 'Hyduro 904' made by J. Barcham Green Ltd, Maidstone, Kent). Make up the filtrate to its original volume with distilled water. Adjust the pH to 8·5 with 10 N-NaOH, boil for 30 min. or autoclave at 121° for 10 min. to precipitate phosphates, and filter hot through the same grade of paper. Bring the pH to 7·5, distribute, and autoclave at 121° for 5 min. The broth is unnecessarily concentrated for many purposes and can be diluted 1/2 to 1/3 with distilled water before sterilization.

The routine can be interrupted at two points. The mixture of mince, alkali, and papain can be left at 4° overnight and heated to 60° the following morning. The filtrate at pH 5·0 can safely be left at 4° for days if it is saturated with $CHCl_3$.

Digest agar is prepared by adding agar to the filtered broth at pH 5·0, bringing the pH to 8·5 with NaOH, and autoclaving at 121° for 15 min. to melt the agar and to precipitate phosphates. On removal from the autoclave, the medium is stirred well to disperse the agar and finally clarified by filtration through paper pulp (p. 53). The filtrate is adjusted to pH 7·4, distributed, and sterilized by autoclaving at 121° for 15 min.

Preparation of nutrient agar plates. Bottles of stock nutrient agar are first melted by immersion in boiling water, steaming, or autoclaving. Before the molten agar can be poured into plates, it should be cooled to 50°–60° to lessen the amount of condensation formed during setting. Bubbles are removed by flaming the agar surface with a Bunsen just after pouring. If plates are needed on the same day, they usually require drying at 37° for 1–2 hr. with the lids removed and the agar inverted; but, if they are to be stored, losses from contamination and drying are avoided by first leaving the closed plates at 37° overnight and then storing them in polythene bags. Plates to be used for counts by the overlay method must be poured on a horizontal surface. Perhaps the commonest fault in pouring plates of all kinds is to use too little medium. It is almost always better to use too much rather than too little: 35 ml. in a 9 cm. plate is usually ample.

Formulae based on commercial products

Peptone water

Peptone	10 g.
NaCl	5 g.
Distilled water	1 l.

Dissolve the peptone and salt, adjust the pH to 7·4, and sterilize by autoclaving. For *peptone agar*, double the amount of peptone and add 1·5 %, w/v, powdered agar. Phosphates do not usually precipitate to an inconvenient extent in either liquid or solid medium if distilled water is used, so avoiding filtration. Peptone water alone usually gives poorer growth than papaic or tryptic digest broth; for example, the viable count of an unaerated overnight culture of *Salmonella typhimurium* is *ca.* 2×10^8/ml. compared to $1{\cdot}5 \times 10^9$/ml. in broth.

Meat extract broth

Peptone	10 g.
Meat extract	10 g.
NaCl	5 g.
Distilled water	1 l.

Meat extract usually contains sufficient calcium to produce a heavy precipitate of phosphate when autoclaved with peptone at pH 7·4. The medium, which is initially acid, therefore needs to be autoclaved at 121° for 15 min. at pH 8·5 to bring down phosphates which are filtered off before the medium is finally sterilized by autoclaving. Solid medium is made by adding agar, as for digest agar.

Tryptone-yeast extract medium. One useful formula is *L* broth, as Ca^{2+} can be added to 0·005 M to allow adsorption of phage P1 without precipitating as phosphate.

Tryptone	10 g.
Yeast extract	5 g.
NaCl	5 g.
Glucose	1 g.
Distilled water	1 l.

The pH should be 7·0. Glucose is added from a sterile 20 %, w/v, solution to the autoclaved medium. For *L* agar, the medium is solidified by adding 12 g. agar/l., and glucose added to the molten medium just before pouring plates.

Undefined supplements

Meat infusion. Meat extract. All infusions are aqueous extracts of minced tissue. The conditions of extraction are a matter of choice. The usual practice is to suspend fat-free meat in tap water (e.g. 500 g./l.) and to leave the mixture at 4° overnight or at 60° for 3 hr. with occasional stirring. Residual meat is removed by filtration through a coarse filter. The filtrate contains heat-coagulable protein which is removed by boiling for 15 min., followed by passage through a fine filter. The infusion is usually combined with an enzymic digest like peptone before use.

Wright's infusion was made by dissolving peptone in the suspension of meat in water which, after extraction, was brought to 100° by steaming for 30 min. before meat particles were filtered off (Wright, 1933; see p. 58).

Meat extract is usually a thick paste obtained by concentrating the infusion *in vacuo*. A typical brand is 'Lab Lemco' sold by Oxo Ltd (for analysis, see Sykes, 1956).

Yeast extract. This provides vitamins and other growth factors, including purines and pyrimidines, for media based on meat or casein hydrolysates. It may incidentally act as a source of cations (p. 44).

Commercial preparations are made from bakers' or washed brewers' yeast, which is either hydrolysed with HCl or proteolytic enzymes, or extracted with water after lysing the cells at 55° (see Sykes, 1956). The liquid extracts are concentrated by drying to a powder or by evaporation to a thick paste (e.g. 'Marmite', obtainable before NaCl is added, and 'Yeastrel').

Published analyses of autolysates show a high concentration of vitamins and amino acids, including tryptophan (Sykes, 1956). The exact composition seems unimportant when used at a concentration of 0·3 %, w/v, for the extract probably provides a gross excess of growth factors like vitamins which are required by bacteria only in trace amounts.

One recipe for yeast extract is: Crumble 250 g. pressed bakers' yeast into 1 l. distilled water, boiling over a naked flame. Stir continuously until frothing ceases (about 5 min.). Filter through paper pulp in a Buchner funnel. Store in Winchester bottles in the dark, after saturating with $CHCl_3$ and shaking vigorously. The extract is diluted 1/10 in the medium before sterilization.

Boiling as a means of extracting growth factors is not as incongruous as it might seem, for it is possible that factors like the coenzyme I required by *Haemophilus influenzae* are otherwise inactivated by enzymes simultaneously released from the yeast (Farrer, 1946).

Liver preparations. This is included in media for *Brucella*, but the rationale is unknown. It consists of an aqueous infusion (see Huddleston, Hasley & Torrey, 1927) or a digest, such as one of the brands of Hepar Proteolysatum B.P. sold for the treatment of pernicious anaemia. Other preparations are sold by the usual firms.

Brain and brain infusion. Pieces of brain were frequently included in the early media for anaerobes but have been superseded by muscle fragments or reducing agents (p. 81). An aqueous infusion of calf brain still appears in media like Brain Heart Infusion (Difco), recommended for organisms like streptococci and gonococci. A method of preparation is given by Pelonze & Viteri (1926). It is uncertain if the function of the brain is nutritive or to provide serum and lipids that counteract the toxicity of fatty acids present in infusion and broth.

Red cell extracts. These provide haematin for exacting strains like *Pasteurella pestis* (Herbert, 1949), and in addition, coenzyme I for *Haemophilus influenzae* which requires both factors. It enables these species to grow on digest broth which lacks haematin because it is incidentally removed with phosphate following digestion (p. 47). Well-known examples are Levinthal's medium and Fildes' peptic digest (Fildes, 1920). Fildes notes that ox and sheep red cells are relatively resistant to trypsin but not to pepsin. Both methods of preparation evidently prevent inactivation of coenzyme I by the agent present in sheep serum and red cells (Waterworth, 1955).

CONTROL OF pH IN CULTURE MEDIA

All cultures tend to change in pH during growth. If peptone is the principal nutrient, the pH rises due to alkali formed in deamination of amino acids; if fermentable sugars are present, the pH falls. Changes in pH can be limited to some extent by including buffers in the medium. An alternative to a buffer is sodium formate or succinate which is metabolized by many species with release of Na^+ that helps to neutralize acid formed from sugar. Chalk was often included in older formulae for the same reason. If these are inadequate, more effective control is achieved by adding acid or alkali to the culture from an automatic titrator (e.g. Edebo *et al.* 1962).

Most buffers are effective over a range of 2 pH units, $\simeq$ pK $\pm$ 1, where pK is the pH at which the buffer salt is 50% dissociated. Thus, the two buffers most commonly used in culture media, phosphate (pK_1 = 6·8) and tris (tris(hydroxymethyl) aminomethane: pK = 8·2), are useful for pH 5·8–7·8 and 7·2–9·2, respectively. Several new buffer salts with pK near 7·0 are described by Mallette (1967). Citric acid-phosphate acts over a far greater range since there are 3 and 2 dissociating groups, respectively (pK 3·1, 4·8, 5·4; and 1·9, 6·7). The equations relating pH of a buffer solution and the amount of acid or alkali added to it are discussed on p. 75, while Table 2.1 gives the composition of phosphate and other buffers.

Some buffers inhibit bacterial growth and, as the undissociated and dissociated forms of a salt often differ in this respect, the inhibitory effect may depend on pH in so far as this determines the degree of dissociation. Both citrate and phosphate trap cations and may thereby inhibit processes like phage adsorption or enzyme function. Inorganic phosphate buffers, like Sørensens, precipitate as calcium phosphate when added to meat digest broth or to nutrient agar, particularly on heating, but not in *L* broth or *L* agar (p. 49). No precipitation occurs in broth with sodium glycerophosphate (pK_1 = 6·8), which forms a soluble calcium salt, or with tris. Bicarbonate (pK = 6·3, 10·3) is frequently used for buffering tissue cultures and the amount that has to be added to the medium and the appropriate concentration of atmospheric CO_2 can be read off Fig. 3.3 on p. 73.

SOLIDIFYING AGENTS

Agar-agar, introduced by Frau Hesse in 1881 (see Hitchens & Leikind, 1939), gelatin, and silica gel are the only substances now generally used, although others tested in the past include carragheen, sodium pectate, starch gel, and methyl cellulose. Any of these can be used in Petri dishes or, if large volumes of medium are being used, in large bottles (e.g. Roux bottles) or covered enamel trays whose lids have been lined with absorbent paper to take up condensation. The depth of medium should not be greater than, say, 1·5 cm. for the yield of cells does not increase proportionately. The bacterial growth is harvested in buffer by means of a bent glass spreader or, if the agar is in bottles, by adding glass balls of 0·5–1 cm. diam. and allowing them to roll to-and-fro over the surface of the medium.

Large volumes of nutrient agar are expensive and sometimes inconvenient to handle. Cells can also be grown to extremely high density in liquid cultures with adequate aeration and pH control, either in glass vessels (Edebo *et al.* 1962) or in some form of dialysis apparatus where the organisms are separated by a semi-permeable membrane from a much larger volume of sterile medium forming part of an open (Sterne, 1958) or a closed culture system (see Gerhardt & Gallup, 1963).

Agar-agar

Agar, a polysaccharide isolated from seaweed, is sold as a yellowish-white powder (see Crowle, 1961). It has the useful properties of forming gels with water that melt at *ca.* 100°, solidify at 40°–45°, and resist digestion by all but a few microbial species. Heat-labile compounds or living organisms survive addition to molten agar at 45°, provided it is cooled at once. Its gelling power is not affected by several cycles of melting and solidification at the usual pH of 7·0–7·5, so that, in emergency, agar can be reclaimed and used again after purification (Roe, 1942). However, agar is easily hydrolysed so that it will not gel after autoclaving in acid solution, for example in stored distilled water that has absorbed CO_2 from the air.

The usual concentration is 1–2 %, w/v, according to the brand, and is easily dissolved by autoclaving at 121° for 15 min. Batches of the same brand may differ in gelling power, and some affect colonial morphology, notably in making genotypically smooth enterobacteria form rough-looking colonies (Brodie & Shepherd, 1950; Fulthorpe, 1951).

The solutions do not need filtration routinely, except when dissolved in digest broth or meat extract medium when phosphates may precipitate with

Ca from the agar (p. 37). Solutions are then usually filtered through a layer of paper pulp, $\frac{1}{4}$–$\frac{1}{2}$ in. thick, built up on a layer of gauze in a Buchner funnel. The funnel, pulp, and receiver are heated by steaming or by running through boiling distilled water before taking the hot molten agar from the autoclave to be filtered immediately under suction. Clarified grades are sold for special purposes, like gel precipitation tests.

Some of the normal constituents may be troublesome. Amounts of fatty acids toxic to many bacterial species are usually present but can be inactivated (p. 56) or, less conveniently, extracted by refluxing with methanol (Ley & Mueller, 1946). Ions may interfere with haemolysin production by *Staphylococcus aureus* (see McIlwain, 1938), immunophoresis (see Crowle, 1961) and with bacterial growth (see Corwin *et al.* 1966). Sulphonated polysaccharides inactivate many animal viruses and prevent plaque formation in agar-containing medium; these are removed by protein precipitates, like acid-precipitated serum, or diethylaminoethyl (DEAE) dextrans (Schulze & Schlesinger, 1963). Vitamins and other growth factors may be sufficiently abundant to upset genetic experiments, although ordinary grades of agar are usually satisfactory for amino acid- or purine-requiring strains. If necessary, however, growth factors are removable by washing.

Washed agar. One method is to wash a known weight of agar repeatedly (e.g. 10 times in 2 days) with distilled water at room temperature and then to filter. Most of the water is easily decanted as the agar remains granular and sinks to the bottom of the vessel. It is next covered with 95 %, v/v, ethanol for 12 hr. at room temperature, filtered, resuspended in 95 % ethanol for 4 hr. at room temperature, and again filtered. Finally, the agar is added to boiling 95 % ethanol, brought to the boil, and filtered. The washed agar is spread out and dried at room temperature. Aqueous extraction may be sufficient for some purposes as in preparing Noble agar.

Gelatin

Gelatin, a protein prepared from skin and bones, is now used only for special purposes, as its gels melt at *ca.* 25°, which is less than the usual incubation temperature of 37°, and are liquefied by proteolytic organisms (see Gershenfeld & Tice, 1941). As with agar, colonies that usually appear smooth may appear rough on some batches, due, in this case, to an increased content of fatty acid (Rook & Bruckman, 1953).

Semi-solid agar. Gelatin is currently used in the isolation of motile mutants from non-motile cultures (Edwards & Ewing, 1962) but may also select mutant sex factors (p. 279):

Bacto-peptone	10 g.
Bacto-gelatin	80 g.
Bacto-agar	4 g.
NaCl	5 g.
Distilled water	1 l.

The mixture of solid ingredients can be stored in bottles. Immediately before use, the correct volume of distilled water is added and the medium autoclaved at 121° for 15 min. On removal from the autoclave, the medium is mixed well by shaking the bottle vigorously, and allowed to cool to 60° before it is poured. If it is poured at a higher temperature, excessive condensation appears on the lid of the plate and on the surface of the medium which cannot be removed by drying at 37°, as with nutrient agar. The source of the ingredients is important and the above formula has been satisfactory.

The plate is inoculated at one side. At 37° the gelatin is molten and the medium soft enough for motile, but not non-motile, organisms to swarm from the primary site of inoculation. At 20°, when the gelatin is solid, even motile organisms are restrained so that the plate can be preserved simply by transfer from 37° to room temperature. This medium is more convenient and more selective for motile organisms than soft agar alone. When selecting mutants carrying new flagellar antigens, the antiserum for the parental flagellar type is included in the medium at 10 times its agglutinating titre in tube tests.

The medium can be replaced by sterile strips of Whatman No. 1 filter paper laid at right angles across a ditch cut in a plate of ordinary nutrient agar. After inoculation at one end, only motile organisms cross the ditch in the paper. If necessary, the paper can be soaked in antiserum of the titre used in slide agglutination tests. The method although useful for *Salmonella* is unsuitable for some motile genera, including *Pseudomonas* (Jameson, 1961).

Silica gel

This is silicon dioxide, SiO_2, prepared initially as a sterile sol from a solution of sodium silicate which is gelled before use by adding electrolytes with the medium.

The sol is prepared by running an aqueous solution of sodium silicate at *ca.* 10 %, w/v, through a cation-exchange column to give a clear, almost colourless, sol containing *ca.* 3 %, w/v, SiO_2. Various resins have been used and can be regenerated: Soucol (Taylor, 1950), Zeo-Karb 215 (Selkon & Mitchison, 1957) and Amberlite IR 120 (H) (Sørensen, 1962). The sol has a pH of *ca.* 3·0, is stable for 1–3 weeks, and can be sterilized by autoclaving at 121° for 15 min. The gelled solid medium is prepared at room temperature by mixing a sterile electrolyte solution, often a salts solution like those given on p. 37, with sufficient sol to give a final concentration of *ca.* 1·5 %, w/v, SiO_2. The medium is left to gel at room temperature for 1–2 hr.

INHIBITORY BATCHES OF CULTURE MEDIUM

Growth sometimes does not occur in a medium like nutrient broth that is usually satisfactory. A common finding is that a small inoculum dies out although inoculation of a large number of organisms is followed by their multiplication (e.g. O'Meara & Macsween, 1937; Proom *et al.* 1950; Kindler, Mager & Grossowicz, 1956; Rowatt, 1957*a, b*). This indicates that the efficiency of plating is low or that the inoculated organisms co-operate to render the medium suitable for their growth, for example by absorbing inhibitory amounts of Cu or fatty acid (von Hofsten, 1962; Davis, 1948) or by supplying growth factors like glutamine or haematin that are lacking in the medium (Fildes & Gladstone, 1939; Herbert, 1949).

There is no single cause for failure of growth, but each of the four inhibitory factors described below affects several species and has been repeatedly discussed. Less expected causes reported in the literature include contamination of media by formaldehyde derived from filter paper and, possibly, by an anti-rust compound in the steam supply to the autoclave (Gifford, 1960).

Glucose heated in medium

Inhibition as well as stimulation of growth has often been noted when glucose is autoclaved with salts medium or broth (see Guirard, 1958, p. 266; Bowers & Williams, 1962).

Inhibition is not due to 'caramelization', for media remain inhibitory if the brown material is removed with charcoal. The critical factors appear to be heating at an alkaline pH in the presence of phosphate. Thus, a solution of glucose, phosphate, and peptone becomes inhibitory after autoclaving at 122° for 15 min. at pH 7·2, but not at pH 5·4 (Lewis, 1930). The cause of inhibition is not clear. Glucose may act specifically in a reaction involving phosphate which results either in conversion of medium nitrogen to a non-utilizable form (Lewis, 1930), or in formation of an inhibitor (Finkelstein & Lankford, 1957). Reducing sugars, including glucose but not non-reducing sugars like sucrose, react on heating with amino acids through a Maillard-type reaction. This may produce a deficiency of the acids concerned: examples are tryptophan, L-lysine, DL-methionine (Patton & Hill, 1948) and cysteine (Lankford, Ravel & Ramsey, 1957; Stokes & Bayne, 1958). In this case, normal bacterial growth occurs if the medium is appropriately supplemented.

Fatty acids

Many inhibitory effects are attributable to the fatty acids present in almost all bacteriological materials, including peptone, casein, agar and cotton wool (see Pollock, 1948; Nieman, 1954). The usual result is failure of growth but, occasionally, specific functions like sporulation or capsule formation are inhibited (Hardwick, Guirard & Foster, 1951; Meynell & Meynell, 1964). Nutrient agar inhibits the growth of species like *Mycobacterium tuberculosis*, *Haemophilus influenzae* and *Neisseria gonorrhoeae* unless fatty acids are removed either by refluxing the agar with a fat solvent like methanol (Ley & Mueller, 1946) or by adding an agent which binds fatty acid. For example, 0·5 %, w/v, activated charcoal (the fine grades sold for chromatography are very convenient); 0·15 %, w/v, soluble starch†; anion-exchange resins (Sutherland & Wilkinson, 1961); 10 %, v/v, serum; or serum albumin fraction V (other proteins and serum fractions are relatively ineffective: Davis & Dubos, 1947).

Tubes plugged with cotton wool and sterilized by dry heat at 160°–180° for 1½ hr. show a greasy film on their inner surface which contains sufficient acid to inhibit bacterial growth in broth subsequently added to the tubes (Wright, 1934; Pollock, 1948). Similarly, filtration through cotton wool renders culture media inhibitory (Boyd & Casman, 1951). Even glass-distilled water may be contaminated (Price & Gare, 1959).

Mere extraction of the culture medium is not always sufficient to allow growth, and an absorbent, like serum or charcoal, has to be present during incubation. The reason may be either that inhibitory amounts of acid are formed during growth or that traces of free acid are required for growth and are maintained at a suitable concentration by dissociation from their

† Lintner's soluble starch is starch slightly depolymerized by acid. It dissolves in hot water and can be autoclaved in a 3 %, w/v, solution for 15 min. at 121° without gelling; the sterile solution is then added to sterile molten agar before pouring. It flocculates when autoclaved in nutrient agar in a concentration of 0·15 %, w/v, and should therefore first be added to boiling liquid medium which is filtered if the starch and medium are later to be autoclaved together.

complex with the absorbent (Davis & Dubos, 1947). It is known that these acids may stimulate or inhibit growth, according to their concentration (see Nieman, 1954). Their growth-promoting effects are most simply detected by adding known acids to medium containing purified absorbent to eliminate their toxic, but not their nutritional, effects.

Inhibition by fatty acid cannot be deduced solely because an agent like charcoal permits growth, for many absorbents are probably contaminated by these acids and other growth factors unless they are specially purified; for example normal rabbit serum is required by *Leptospira pomona* because it supplies essential fatty acids and is ineffective after extraction with fat solvents (Johnson & Gary, 1963). Charcoal, as supplied, is heavily contaminated with lipid and ions, notably Fe: organic compounds are destroyed by heating in a silica vessel in nitrogen; while ions are removed by boiling with concentrated HCl followed by repeated washing on a filter with boiling distilled water until the filtrate is neutral.

Inadequate reduction of the medium

Various observations have been explained by this hypothesis:

(1) the failure of small inocula to grow in broth unless it has been heated just before inoculation, and, conversely, rapid reversion of heated broth to an inhibitory state on storage (Wright, 1933; Dubos, 1930; O'Meara & Macsween, 1937);

(2) the failure of nominally aerobic species to grow on nutrient agar in air unless 0·25 %, w/v, sodium sulphite is added (see Drennan & Teague, 1917);

(3) growth in the depth but not at the surface of a stab culture of an aerobic species (Schütze & Hassanein, 1929);

(4) the toxicity of some peptones unless they are heated during preparation with an aqueous suspension of meat, which contains strong reducing systems (Wright, 1933).

Some of these findings are undoubtedly genuine instances of 'inadequate reduction'. The species studied by Drennan & Teague (1917) and by Schütze & Hassanein (1929) included *Pasteurella pestis* which is now known to require haematin for aerobic but not for anaerobic growth on digest agar (Herbert, 1949). As haematin is removed from digest agar by the method of preparation (p. 47), the organisms inevitably fail to grow from small inocula in air, although large inocula co-operate successfully (p. 183). Even a non-exacting aerobe like *Escherichia coli* may fail to grow in salts medium unless a reducing agent is added (p. 72).

Wright considered that inhibition was due to a constituent of peptone that could be reversibly oxidized and reduced, only the oxidized form being toxic. However, the toxicity of some peptones on heating was then found to be due to Cu^{2+} (O'Meara & Macsween, 1937), probably derived from metal containers, which combined with meat fragments and was filtered out of the medium when this was prepared by Wright's method (p. 49). Later work showed that the toxicity of Cu^{2+} was here linked to the decomposition of cystine leading to the formation of colloidal sulphur or CuS (Woiwod, 1954). Colloidal sulphur was also implicated in the toxicity of peptone by Schuhardt *et al.* (1952) and by Rowatt (1957*b*) and may also be formed in the presence of Fe^{3+} (Konowalchuk *et al.* 1954) or of Cu^{2+} and pyridoxal (Traxler & Lankford, 1957).

The Cu content of current commercial peptones is generally satisfactory, though this considered in isolation may be misleading as the inhibitory effects of Cu are also determined by the pH of the medium and the closure of the culture vessel (see Sykes, 1956). Thus, anaerobic cultures of *Escherichia coli* in glucose-salts medium are more readily inhibited than those incubated aerobically (Gorini, 1961).

Peroxide

Some inhibitory effects have been attributed to peroxide, since they are overcome by adding to the medium agents like catalase, MnO_2 (1 g./l.), haematin, serum, activated carbon, or reducing agents (Table 3.3). Peroxide has been thus detected in culture media exposed to ultraviolet light (Wyss *et al.* 1948) or sunlight (Burnet, 1925), especially when eosin or other fluorescent dyes are present; during the preparation of papaic digest broth and other media, particularly after autoclaving (Proom *et al.* 1950; Rowatt, 1957*a*); and when citrate is autoclaved with Mn^{2+} in synthetic medium (Barry *et al.* 1956). The amount of peroxide varies, for, on some occasions, only catalase-negative strains are inhibited, while, on others even catalase-positive species fail to grow. In the experiments of Proom *et al.* (1950), peroxide was formed after exposure to air, since only surface, not deep, colonies were inhibited.

pH INDICATOR MEDIA

The importance of these media lies in the wide use of fermentative ability as a marker in genetics and in taxonomy.† Although fermentation can be

† The colloquial terms, 'fermentation' and 'sugar', are used, although they are often incorrectly applied to oxidative metabolism and to alcohols, like mannitol.

detected by a specific indicator, like ONPG used for strains forming β-galactosidase, the usual method of detecting fermentation is to include an orthodox pH indicator in the medium, and therefore depends on a net fall in pH due to more acid than alkali being formed as the culture grows. Not infrequently, however, the pH rises although the sugar is fermented because of the amount of alkali produced by deamination of amino acids (e.g. Merrill, 1930; Wedum, 1936; Knight & Proom, 1950). The same effect is seen on plates of a simple indicator medium when the colour of a fermenting colony fails to develop completely because it is placed near a large number of non-fermentors. This difficulty is avoided in the special solid media described later.

Such errors are unimportant with pure cultures when the test is used solely for identification, provided the results are consistent, but are misleading when the object is to determine which sugars are attacked. The organisms should then be tested for their ability to grow on the sugar as sole carbon source; the concentration of sugar can be assayed before and after growth of the culture; or a test not dependent on cell multiplication can be tried (Clarke & Cowan, 1952). The amount of acid formed depends partly on the manner in which the sugar is metabolized, for relatively little is produced by the purely aerobic mechanism found in some genera like *Pseudomonas* (Hugh & Leifson, 1953). Other species, like *Escherichia coli*, metabolize sugars anaerobically to produce organic acids, but, if oxygen is available, these are in turn broken down to water and CO_2 which escapes from the medium. The acid produced is therefore less under aerobic than under anaerobic conditions, for example at the surface compared to the depths of a thick agar slope.

Pure cultures

These are often tested in liquid medium, though solid medium may be more sensitive.

Liquid indicator media. The usual liquid medium is 1 %, w/v, peptone water, pH 7·4, containing 1 %, w/v, sugar and an indicator. The complete medium is sterilized by tyndallization. Autoclaving for 10 min. at 115° is often used without misleading results, though heating is known to alter many sugars, especially in alkaline solution. A particular hazard is the conversion of one sugar to another, for example the interconversion of mannose, glucose and fructose. The fact that autoclaved medium appears satisfactory suggests that fermentation tests provide only an insensitive guide to the state of the sugar.

Indicators include neutral red (0·0005 %, w/v), phenol red (0·005 %, w/v), bromocresol purple (0·002 %, w/v) and chlorophenol red (0·001 %, w/v). Andrade's indicator is sometimes used in peptone water.

Preparation of Andrade's indicator. Dissolve 0·5 g. acid fuchsin in 100 ml. distilled water, add 17 ml. 1 N-NaOH, and leave overnight at room temperature. The solution should be straw yellow next day: if it is brownish, add a little 1 N-NaOH and stand again. This is diluted 1/100 in medium adjusted to pH 6·8 to allow for the alkalinity of the indicator.

Andrade's indicator and neutral red fade in stored medium. Indicators may also be decolorized in anaerobic cultures, as many are affected by a fall in E_h. The pH is then determined by shaking the culture in air or by adding fresh indicator.

Some species, such as *Streptococcus pyogenes* and *Neisseria meningitidis*, also require serum, as in Hiss' serum water: 1 vol. sterile sheep or ox serum and 3 vol. of either distilled water or 0·1 %, w/v, peptone water sterilized by tyndallization. The serum must previously be heated to 60° for at least 1 hr. to inactivate naturally occurring amylase and maltase which otherwise give false positive results with starch and maltose (Goldsworthy, Still & Dumarescq, 1938; Hendry, 1938).

With liquid media, the results may depend on the closure, due to retention of CO_2. Thus, tightly closed screw-capped bottles or medium sealed with paraffin may give positive results when loosely closed bottles or unsealed medium do not (Marcus & Greaves, 1950; Hayward, 1957).

Gas production is often determined in the same test by placing a small tube, sealed at its upper end, within the medium (Durham's tube). The air is expelled during autoclaving or by tipping the culture bottle. Gas production is indicated by the formation of a bubble of gas within the tube after incubation. False positives occur if tubes, stored for some days at 4° after autoclaving, are incubated at 37°, as sufficient air may be absorbed during storage to form a detectable bubble in the tube when it is warmed (Archambault & McCrady, 1942). Sporadic false negative results have been noted if the medium and the sugar solution are sterilized separately and then mixed, as the sugar does not always diffuse into the Durham's tube (Black & Dillon, 1947).

Hydrolysis of ONPG. Lactose fermentation can also be detected by hydrolysis of *o*-nitrophenyl-β-D-galactopyranoside (ONPG). This is colourless but on hydrolysis by β-galactosidase, yields a yellow product. Apart from its importance in quantitative studies of β-galactosidase synthesis, the reaction is used to detect the enzyme in those numerous naturally-occurring species of enterobacteria that appear non-lactose-fermenting solely because they lack an active permease system (Lowe, 1960; Lapage & Jayaraman, 1964). The usual technique is to mix 3 vol. of 1 %, w/v, peptone water, pH 7·5, with 1 vol. sterile ONPG solution (0·6 %, w/v, ONPG in 0·01 M-Na_2HPO_4, pH 7·5, sterilized by filtration). The mixture keeps for a month at 4°. In the test, the medium is inoculated and incubated overnight at 37°: a positive result is shown by a yellow colour which develops in as little as 3 hr. if a large inoculum is used. The cells do not require to be lysed as in a quantitative assay. Permease-negative strains can also be detected by a pH change either in 0·1 % peptone water containing 5 %, w/v, lactose or on bile salt indicator plates. Either of these methods might obviously be applicable to other sugars.

Solid indicator media. Solid medium can also be used for pure cultures provided the nutrient base is free from fermentable sugars: an example is peptone agar (p. 48) containing an indicator to which autoclaved sugar solution is added before the plates are poured. The indicators used for peptone water are suitable but mixed indicators showing a rise as well as a fall in pH are feasible, for instance neutral red and crystal violet (0·001 %, w/v, of each: Grunbaum & Hare, 1902); acid fuchsin and methylene blue (0·05 % and 0·005 %, w/v, respectively at pH 6·6: Jeter & Wynne, 1949); or water blue (0·006 %, w/v), phenol red (0·01 %, w/v) and α-naphthol-phthalein (0·01 %, w/v) at pH 7·2 (Juhlin & Ericson, 1961).

Perhaps the most sensitive method for detecting fermentation is to use solid medium put up in tubes with a slant and deep butt, and containing an indicator, the least amount of peptone compatible with growth in order to minimize alkali formation (e.g. 0·2 %, w/v), and a relatively high concentration of sugar (1–2 %, w/v). The surface of the slant is inoculated with a loop and the butt is stabbed. The advantage of solid compared to liquid medium is that mixing does not occur, so that acidity may be detectable in one part of the slope when the remainder is alkaline (Hugh & Leifson, 1953). Gas formation is indicated by fragmentation of agar in the butt.

Mixed cultures

Mixtures are separated and tested at the same time by using plates of indicator medium. The simplest formula is peptone agar containing sugar and indicator, but has the disadvantage that the pH of colonies may be altered by diffusion of acid or alkali from neighbouring areas. The following media are strongly recommended, as the altered dye is confined to each colony which is thus clearly distinguished from its neighbours.

Sectors and mixed colonies are easily seen. A trace of buffer (0·01 M) often improves definition.

Eosin–methylene blue (EMB) medium

Peptone	10 g.	Eosin Y	0·4 g.
NaCl	5 g.	Methylene blue	0·065 g.
K_2HPO_4	2 g.	Agar	15 g.
Sugar	10 g.	Distilled water	1 l.

The initial pH should be 7·3.

The eosin, methylene blue, nutrient base, and sugar solution are best prepared and sterilized separately, and mixed just before pouring plates.

The stock dye solutions (100 times working strength) may not need to be sterilized.

Eosin Y (Gurr)	4 g.
Distilled water	100 ml.
Methylene blue	0·65 g.
Distilled water	100 ml.

Note:

(1) The dye concentrations have been altered since the original description as the older brands were not pure. The ratio of eosin: methylene blue now recommended is 5·8:1 (0·04% and 0·0065%, w/v, respectively as above: see Levine, 1943).

(2) The medium is initially a dense blackish-blue colour. Non-fermenting colonies are colourless while those of fermentors are black with a metallic sheen. These changes occur because the dyes are present, not as a mixture, but in combination in the medium. Their compound precipitates at low pH, although the individual dyes are soluble; and at high pH, it is colourless, whereas methylene blue is blue (Wynne, Rode & Hayward, 1942).

(3) Gram-positive organisms are at least partially inhibited. Some strains of Gram-negative genera, including *Salmonella* and *Shigella*, are also inhibited, but grow if reduced methyl green is substituted for methylene blue though Gram-positive organisms are still inhibited (Fredericq, 1947).

(4) Bacterial lysis or 'leakage' causes blackening of colonies, and EMB medium has been used for isolating F+ salmonellas (Zinder, 1960) and non-lysogenic colonies from irradiated lysogenic cultures (Zinder, 1958).

(5) The dyes can also be used in the following defined medium which permits selection of clones by their nutritional characteristics and scoring of their fermentation markers in one step (Lederberg, 1949). Sodium succinate is included as a carbon source for organisms unable to attack the sugar: their colonies remain colourless.

Sugar	10 g.	$(NH_4)_2SO_4$	5 g.
Sodium succinate	5 g.	Eosin Y	0·4 g.
NaCl	1 g.	Methylene blue	0·065 g.
$MgSO_4$	1 g.	Agar	15 g.
K_2HPO_4	2 g.	Distilled water	1 l.

Solutions of the carbon source, salts, dyes and agar should be sterilized separately and mixed before plates are poured (cf. salts medium, p. 37).

Bile salt indicator media

Sodium deoxycholate	2·5 g.	Sugar	10 g.
Neutral red	0·025 g.	Nutrient agar	1 l.

The following bile salt, indicator and sugar solutions are made up separately and mixed with molten nutrient agar just before pouring plates. The initial pH is 7·2–7·5.

Sodium deoxycholate	10 g.
Distilled water	100 ml.

Sterilize by heating at 60° for 60 min. Keep at room temperature. The solution gels at 4° but clears on warming to 60°.

Neutral red	1 g.
Distilled water	100 ml.

Sterilize by autoclaving at 121° for 15 min. The medium is then prepared by mixing:

Molten nutrient agar, 7·2	925 ml.
10% deoxycholate solution	25 ml.
1% neutral red solution	2·5 ml.
20% sugar solution	50 ml.

Note:

(1) The medium is initially transparent and reddish brown. Acid-producing colonies are bright red and are also opaque due to precipitation of deoxycholate when the pH falls below *ca.* 6·8. The colour is restricted to the colony but a red precipitate may form in the surrounding medium; this is limited by including 0·01 M tris. Non-fermentors form colourless transparent colonies.

(2) One suitable nutrient agar is Oxoid Blood Agar Base No. 2. With some agars, crystals form in the agar after incubation, accompanied by a dense deposit of finer crystals where the agar is touched by a loop.

(3) This medium differs from Leifson's medium used for isolation of salmonellas in not containing added citrate (Leifson, 1935). Nevertheless, some strains of *Escherichia coli* as well as Gram-positive genera are inhibited by deoxycholate. Dehydrocholate is said not to inhibit *E. coli* and

Staphylococcus aureus and can even be used with *Streptococcus pneumoniae* if the concentrations of bile salt and neutral red are reduced (Morse & Alire, 1958). These authors also use dehydrocholate and neutral red in glucose-salts medium.

(4) If the deoxycholate concentration exceeds 0·6 %, w/v, 'cryptic' strains that normally appear non-fermenting owing to lack of the appropriate permease nevertheless ferment the sugar, presumably due to slight cell damage caused by the bile salt (Coetzee, 1962).

(5) Streptomycin can be included for selection of streptomycin-resistant clones but, if added from a solution of streptomycin sulphate in water, its concentration should not exceed 200 μg./ml. medium. Otherwise, the acidity of the solution precipitates the deoxycholate.

(6) Salmonellas that have been subjected to cold shock (Meynell, 1958) or exposed to penicillin (Hartman, personal communication) have an extremely low efficiency of plating on deoxycholate agar compared to nutrient agar.

Calcium carbonate indicator medium

Peptone	20 g.	Precipitated chalk	10 g.
Bromothymol blue	0·05 g.	Agar	15 g.
Sugar	10 g.	Distilled water	1 l.

The initial pH should be 7·2.

Make up and sterilize separately: (1) peptone agar containing bromothymol blue; (2) 20 %, w/v, sugar solution; (3) $CaCO_3$ as a 10 %, w/v, autoclaved suspension in water to be diluted 1/10 in the molten agar just before pouring. The chalk tends to settle out unless a stabilizing agent is added. Keston & Rosenberg (1967) prepare the nutrient base at twice working strength in a 0·6 %, w/v, solution of high viscosity carboxymethylcellulose, mix it with an equal volume of a 0·4 %, w/v, suspension of $CaCO_3$ in water, and autoclave.

Note:

(1) The medium is initially opaque and greenish blue. Acid-producing colonies turn yellow and may be surrounded by a clear zone in which the chalk has been dissolved. If the chalk is omitted, acid produced by fermenting colonies spreads through the medium and, consequently, non-fermenting colonies may appear yellow (Wade, Smiley & Boruff, 1946).

(2) Many organisms, such as streptococci and staphylococci, will grow that are inhibited by EMB or deoxycholate medium (Korman & Berman, 1958).

Tetrazolium agar, for detection of non-fermenting colonies (Lederberg, 1948)

Tetrazolium solution	10 ml.
20 %, w/v, sugar solution	50 ml.
Nutrient agar	1 l.

The tetrazolium and sugar solutions are sterilized individually and added to the molten agar just before plates are poured. The stock tetrazolium solution (Zamenhof, 1961) contains:

2:3:5-triphenyltetrazolium chloride	0·5 g.
Distilled water	100 ml.

Sterilize by filtration.

The nutrient base considerably affects the results of the test. A satisfactory formula is:

Nutrient broth	25 ml.
2 %, w/v, agar in distilled water (p. 37)	75 ml.

Note:

(1) The medium is initially colourless and transparent. Non-fermenting colonies are opaque red following reduction of tetrazolium at alkaline pH to the insoluble product, formazan. Sectors and mixed colonies are distinguishable. Acid-producing colonies are either colourless or have rose-coloured centres, as no reduction occurs at acid pH. The medium is not completely satisfactory, for acid-producing colonies gradually turn red on prolonged incubation, and the present formula is most efficient with *ca.* 700 colonies per plate. Nevertheless, it is considerably more efficient for the identification of non-fermentors than previous media.

(2) This concentration of tetrazolium will inhibit *Streptococcus pyogenes* and *Bacillus subtilis* but not enterobacteria or *Staphylococcus aureus* (May *et al.* 1960).

Endo medium. This medium originally contained lactose and enabled lactose-fermenting and non-lactose-fermenting clones to be distinguished by the colours of their colonies. It differs from the media described above in that the change in colour depends on the production of acetaldehyde, not acid, by fermenting organisms. The medium is initially pink and consists essentially of nutrient agar, a sugar, and basic fuchsin decolorized with sulphite (Schiff's reagent). Non-fermentors form colourless colonies, while those of fermenters are red due to the complex formed by Schiff's reagent and acetaldehyde derived from breakdown of the sugar.

Table 2.1. *Buffer mixtures*

Mixture	*A*	*B*
1. Citric acid–phosphate	$Na_2HPO_4.2H_2O$ (0·2 M: 35·6 g./l.)	Citric acid. H_2O (0·1 M: 21 g./l.)
2. Sodium acetate–acetic acid	Na acetate. $3H_2O$ (0·2 M: 27·22 g./l.)	Acetic acid (0·2 M: 11·5 ml. glacial +988·5 ml. water)
3. Phosphate	$Na_2HPO_4.2H_2O$ (0·2 M: 35·6 g./l.)	$NaH_2PO_4.2H_2O$ (0·2 M: 31·2 g./l.)
4. Tris-HCl	HCl (0·2 M: 18 ml. 36% HCl+982 ml. water)	Tris (0·2 M: 24·2 g./l.)

Mixture ...	1	2	3	4
	ml. of *A*			
pH				
2·2	0·40	—	—	—
4	1·24	—	—	—
6	2·18	—	—	—
8	3·17	—	—	—
3·0	4·11	—	—	—
2	4·94	—	—	—
4	5·70	—	—	—
6	6·44	3·7	—	—
8	7·10	6·0	—	—
4·0	7·71	9·0	—	—
2	8·28	13·2	—	—
4	8·82	19·5	—	—
6	9·35	24·5	—	—
8	9·86	30·0	—	—
5·0	10·30	35·2	—	—
2	10·72	39·5	—	—
4	11·15	41·2	—	—
6	11·60	45·2	—	—
8	12·09	—	4·00	—
6·0	12·63	—	6·15	—
2	13·22	—	9·25	—
4	13·85	—	13·25	—
6	14·55	—	18·75	—
8	15·45	—	24·50	—
7·0	16·47	—	30·50	—
2	17·39	—	36·00	22·10
4	18·17	—	40·50	20·70
6	18·73	—	43·50	19·20
8	19·15	—	45·75	16·25
8·0	19·45	—	47·35	13·40
2	—	—	—	10·95
4	—	—	—	8·25
6	—	—	—	6·10
8	—	—	—	4·05
9·0	—	—	—	2·50
ml. of *B*	20–*A*	50–*A*	50–*A*	25
Make up to:	—	100 ml.	100 ml.	100 ml.

3

OXYGEN, CARBON DIOXIDE AND ANAEROBIOSIS

OXYGEN

The oxygen demand of aerobes growing in liquid culture is met initially by O_2 dissolved in the medium, which, as it is used, is replaced by oxygen from the atmosphere. It is often found, however, that owing to the extremely low solubility of O_2 (0·2 mM/l. broth at 25°), the demand of a dense rapidly growing culture cannot be satisfied without taking steps to increase the rate at which dissolved O_2 is replenished (see Richards, 1961; Lockhart & Squires, 1963; Finn, 1967). The usual method is aeration but it may sometimes not increase the critical factor—the concentration of dissolved O_2—because, for example, air introduced by bubbling escapes too rapidly from the medium. The efficiency of oxygenation is measurable by the sulphite and other methods (p. 71) and allows different culture systems to be compared.

The O_2 demand/ml. culture equals the demand/cell × cell numbers/ml. The maximum demand per culture usually comes towards the end of exponential growth. However, the demand per *cell* is greatest in the lag phase, due solely to the increased size of individual cells, as the demand per unit *weight* of organisms remains constant throughout the growth cycle (Hershey & Bronfenbrenner, 1938; Grieg & Hoogerheide, 1941). Values observed for *Escherichia coli* are 2–11 cu. mm. $O_2 \times 10^{-7}$/cell/hr. compared to a constant 5 cu. mm. $O_2 \times 10^{6}$/g. bacterial nitrogen/hr. (see also Martin, 1932; Clifton, 1937). The rate of bacterial respiration is virtually independent of the concentration of dissolved O_2 until this reaches extremely small values, e.g. 5×10^{-3} mM/l., so that the medium does not necessarily have to be kept saturated with O_2 (Finn, 1967).

The rate at which oxygen passes from the atmosphere into solution in the medium depends on several factors:

(1) the degree of oxygen saturation of the medium, expressed by $C^* - C_L$, where C^* is the oxygen concentration at which atmosphere and medium are in equilibrium and C_L is the actual concentration in the medium. In aeration, C^* is the O_2 concentration of air;

(2) the area, a, of the interface between atmosphere and medium; and

(3) the ease with which oxygen crosses this interface, denoted by the factor K_L. The larger K_L, the greater the ease of passage.

The rate of O_2 absorption by the medium therefore equals $K_L a(C^* - C_L)$. The product, $K_L a$, measures the capacity of the culture system to absorb O_2 when run in a given way and is usually termed the O_2 absorption rate constant (mM/l./min./unit difference in O_2 concentration). Alternatively, the O_2 absorption rate ($K_L aC^*$; mM/l./min.) is measured, as in the sulphite method where C_L, the concentration of dissolved O_2, is zero.

Increasing the efficiency of oxygenation

Oxygen absorption is increased by increasing C^*, a, or K_L.

Increase in C^*, the partial pressure of O_2 in the atmosphere. This method is rarely used.

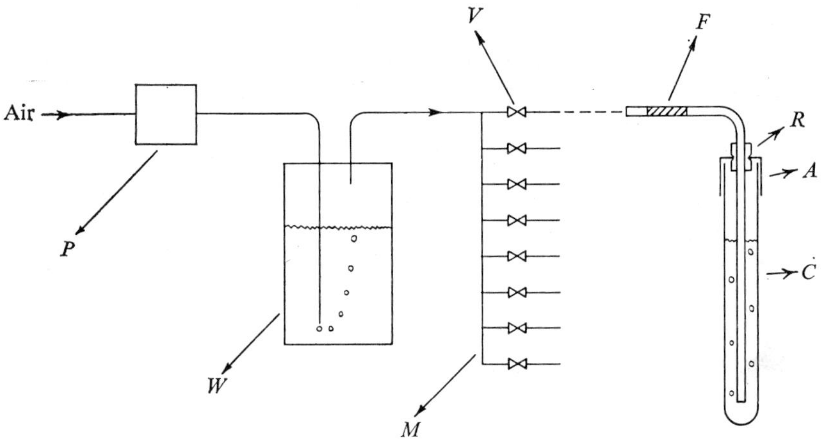

Fig. 3.1. A simple air line for the aeration of laboratory cultures. *P*: pump, e.g. as sold for aquaria. *W*: wash bottle, to prevent evaporation of the culture. *M*: manifold bearing valves, *V*, each of which can incorporate a flowmeter. *F*: air filter of non-absorbent cotton wool. *R*: rubber tubing surrounding the bubbling tube passing through the aluminium cap, *A*. *C*: culture medium.

In assembling the culture tube, a hole is punched in the aluminium cap, a piece of rubber tubing inserted, and the bubbling tube pushed through the lumen so as to make a close fit. After inserting an air filter of non-absorbent cotton wool in the sidearm, the tube, cap and glass tubing are sterilized in the autoclave. The size of the bubbles can be decreased by narrowing the tip of the glass tube or by fusing it to a sintered glass block.

Increase in a, the interfacial area. The usual means is bubbling or shaking. Bubbling is easily achieved by an aquarium pump (Fig. 3.1). The incoming air is sterilized by passage through non-absorbent cotton wool held between two constrictions in glass tubing. If the apparatus is in a hot room, a wash bottle is needed to saturate the air supply and prevent evaporation of the medium. The rate of flow to individual cultures is controlled by needle valves and can be measured with a flowmeter. Small bubbles are more effective than large, since the interfacial area is greater, and are often produced by introducing air through a coarse sintered glass disc or cylinder (pore size *ca.* 50 μ). However, if the disc forms the base of the culture vessel, it may trap up to 90 % of the organisms within 30 min. of starting aeration (Ginsberg & Jagger, 1962). A horizontal baffle placed just above the air inlet prevents bubbles escaping too quickly from the medium and from clustering in the centre of the vessel. A stirrer has the same effect and, used with vertical baffles projecting inwards from the sides of the vessel, gives extremely high aeration rates (Table 3.1).

Table 3.1. *Oxygen absorption rates in laboratory cultures*

Vessel	Volume of medium		Air flow (l./min.)	K_LaC^* (mM-O_2/l./min.)	Reference
18 × 150 mm. test tube	10 ml.	Stationary	—	0·03	1
Erlenmeyer flask 500 ml.	20 ml.	Stationary	—	0·32	1
Erlenmeyer flask 500 ml.	20 ml.	Eccentric shaker (250 r.p.m.)	—	1·1	1
Indented Erlenmeyer flasks 500 ml.	20 ml.	Eccentric shaker (250 r.p.m.)	—	2–9·5	1
1000 ml.	50 ml.	Reciprocal shaker (80–100 strokes/min.)	—	0·78–1·5	2
1000 ml.	200 ml.	Reciprocal shaker (80–100 strokes/min.)	—	0·22–0·78	2
		Stirred:			
Baffled tank 3·5 l.	1460 ml.	750 r.p.m.	5·8	3·6	3
		1100 r.p.m.	6·1	6·33	3

1, Smith & Johnson (1954); 2, Corman *et al.* (1957); 3, Maxon & Johnson (1953).

Aeration by shaking usually improves, the larger the vessel, the smaller the volume of medium, and the more vigorous the movement (see McDaniel, Bailey & Zimmerli, 1965*a*, *b*). A considerable increase is produced by vertical baffles placed within the flask or formed by indenting the glass (Table 3.1; Fig. 3.2).

Increase in K_L, the ease with which O_2 passes from atmosphere to medium. Foaming often hinders the aeration of shaken cultures, especially with protein-containing medium like digest broth, and at high cell concentrations (Starks & Koffler, 1949). Apart from hindering the passage of atmospheric O_2 into the culture, there is the hazard of contaminated medium

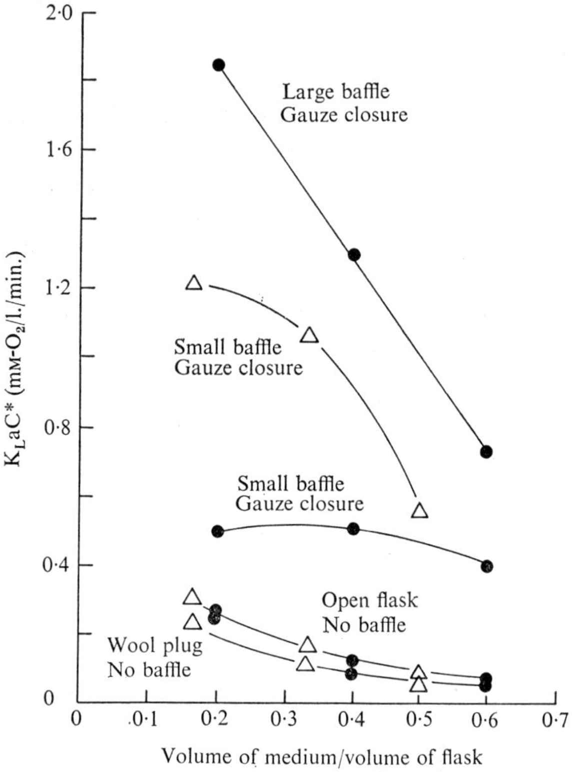

Fig. 3.2. Aeration in shaken conical flasks (Gaden, 1962). Ordinate: O_2 absorption rate. Abscissa: proportion of total flask volume occupied by medium. The volumes of the flasks were 300 ml. (△) and 500 ml. (●). The O_2 absorption rate increases, the less the flask is filled, and the larger the baffle.

rising out of the culture vessel. Anti-foaming agents are fairly successful and act in concentrations of a few parts per million; examples are polypropylene glycol, PPG2000 (Shell Chemical Co.) and Silicone Antifoam RD (Midland Silicones Ltd; see Herbert *et al.* 1965). A crude method is to dab sterile silicone grease on the inside of the culture vessel just above the medium. Accumulation of foam is also prevented in special culture vessels by centrifugal force (Perret, 1957) or by returning foam to the base of the medium with inflowing air (Hedén, 1957).

Aeration may be limited by the closure of the vessel, particularly by cotton wool plugs if they become wet (Corman *et al.* 1957; Schultz, 1964), and it is preferable to close flasks by layers of cotton between gauze tied to the neck (Gaden, 1962) or, with saprophytes, by a raised clip-on metal cap.

Measurement of O_2 absorption rates by the sulphite method

The principle is that sulphite is oxidized to sulphate in the presence of a Cu catalyst. The volume of O_2 absorbed by a known volume of sulphite solution is thus measurable by titrating the amount of sulphite before and after a known period of aeration (Cooper, Fernstrom & Miller, 1944). The value of $K_L aC^*$ (mM-O_2/l./min.) can then be calculated. The reaction rate is largely independent of the sulphite concentration.

Tap water is placed in the vessel and Na_2SO_3 crystals added to *ca.* 1 N-SO_3^{2-} ion with $CuSO_4$ to at least 0·001 M-Cu^{2+}. When these have dissolved, an initial sample is taken at once and a second sample after a known period of aeration. Using a pipette flushed with nitrogen, each sample is run into fresh standard I_2 solution, and residual I_2 titrated either by standard thiosulphate with a starch indicator or by a colorimetric method (Ecker & Lockhart, 1959*b*). The O_2 absorbed is calculated from the volume of thiosulphate required for the second sample minus that required for the initial sample. One mole O_2 absorbed is equivalent to two equivalents of iodine or to four equivalents of thiosulphate according to the following equations:

$$O_2 + 2Na_2SO_3 = 2Na_2SO_4,$$
$$Na_2SO_3 + I_2 + H_2O = Na_2SO_4 + 2HI,$$
$$I_2 + 2Na_2S_2O_3 = Na_2S_4O_6 + 2NaI.$$

The method is at present widely used, but its results have to be interpreted with care as it measures O_2 absorption by an aqueous solution of sulphite, not by a culture medium containing organisms and their metabolic products. It is perhaps for this reason that the estimates of $K_L aC^*$ given by the sulphite method may be twice those given by other methods (Pirt & Callow, 1958).

The sulphite method clearly cannot be used to determine the O_2 concentration in a culture at a given moment although this can now be done by potentiometric techniques (see Lockhart & Squires, 1963; Moss & Saeed, 1967). Reliable oxygen electrodes have only recently been devised (see Herbert, Phipps & Tempest, 1965).

Overaeration

Multiplication of aerobes is easily inhibited by aerating too vigorously, especially in salts medium and at cell concentrations of less than 10^7/ml. (Winslow, Walker & Sutermeister, 1932; Lwoff & Monod, 1947). The remedy is to aerate cautiously at all times; not to aerate broth cultures until they are definitely turbid; and not to aerate cultures in salts medium until 1–1·5 hr. after inoculation when the cells should have begun to grow exponentially.

The causes of inhibition are unknown, but two possibilities are CO_2 deficiency, since the CO_2 concentration of air is less than that optimal for bacterial multiplication (Kempner & Schlayer, 1942), and an E_h too high for bacterial growth. Either is more likely to occur in salts medium than in broth which contains both CO_2-sparing compounds and reducing systems that tend to keep the E_h at a favourable level (Lepper & Martin, 1930). Thus, prolonged lag in salts medium is abolished by 0·01 %, w/v, Na_2S (Lwoff & Monod, 1947; p. 78).

CARBON DIOXIDE

Carbon dioxide is an essential growth factor for many species as their multiplication is prevented if CO_2 is removed by bubbling liquid cultures with CO_2-free air or by exposing plates to compounds that absorb CO_2 (see Valley, 1928; van Niel *et al.* 1942). The CO_2 requirement of freshly isolated pathogens may be still more strict, for they often fail to grow even in air. The best-known examples are *Neisseria* and *Brucella*, but other species like pneumococci may be equally dependent on CO_2 added to the atmosphere (Auger, 1939). The requirement of *Escherichia coli* for CO_2 in synthetic medium disappears if certain organic acids, such as aspartic, glutamic or α-ketoglutaric, are present (Lwoff & Monod, 1947; Ajl & Werkman, 1949).

The optimal CO_2 concentration is usually said to be about 5 %, v/v, but it should be appreciated that organisms may not assimilate CO_2 directly from the atmosphere but as bicarbonate ion (HCO_3^-) formed by solution of CO_2 in the medium. Theory shows that the HCO_3^- concentration depends not only on CO_2 concentration but also on several other factors, including the pH and temperature of the medium. When CO_2 and HCO_3^- are in equilibrium, they are related approximately by the Henderson–Hasselbalch equation:

$$\log [HCO_3^-] = pH - pK' + [\log CO_2], \qquad (3.1)$$

where $[HCO_3^-]$ is the molar concentration of bicarbonate in the medium; pH is that of the medium; $pK' = -\log([H^+][HCO_3^-]/[CO_2])$; and $[CO_2]$ is the molar concentration of CO_2 in the medium and equals $5{\cdot}87 \times 10^{-7} P\alpha CO_2$.

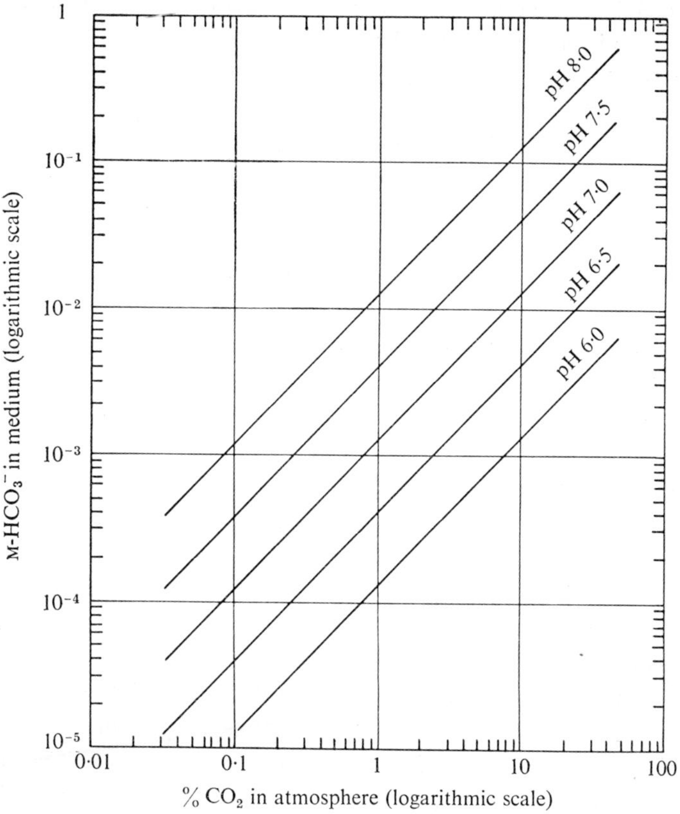

Fig. 3.3. Relation between bicarbonate, CO_2, and pH at 37°. Ordinate: bicarbonate concentration in medium. Abscissa: % CO_2 in atmosphere. The lines for different pH are calculated from the Henderson–Hasselbalch equation (p. 72) assuming $\alpha = 0{\cdot}56$ ml./ml. and the $pK' = 6{\cdot}32$, as for distilled water. The calculations are only approximate, as the values of the parameters depend on the pH and on the composition of the medium. If the pK′ is 6·1, as for plasma, the bicarbonate concentrations shown here should be multiplied by antilog $(6{\cdot}32 - 6{\cdot}10) = 1{\cdot}66$.

P is the atmospheric pressure; CO_2 is the % CO_2, v/v, in the atmosphere at pressure P; and α is the solubility in ml./ml. of CO_2 in the medium at the temperature concerned (Umbreit, Burris & Stauffer, 1964, chapter 2). The values of pK′ and α depend somewhat on pH, temperature, and the presence of other solutes, but their values for water at 37° are 6·32 and 0·56 ml./ml. respectively.

Some implications of equation (3.1) are most easily seen if log [HCO_3^-] at a given pH and temperature is plotted against log % CO_2. This gives a straight line, and if similar lines are drawn for other pH at the same temperature, a set of parallel lines is obtained (Fig. 3.3). This shows that a given concentration of [HCO_3^-] is produced by an infinite number of combinations of pH and CO_2; the higher the pH, the less CO_2 is required (cf. Meynell & Meynell, 1964). The value of [HCO_3^-] is *ca.* 0·02 M for 5 %, v/v, atmospheric CO_2 equilibrated at 37° with medium at pH 7·5.

Each straight line in Fig. 3.3 is the relation that would be observed if the medium was of infinite buffering capacity, since its pH remains the same whatever the CO_2 concentration. In practice, buffering capacity is limited and all media become somewhat more acid on exposure to CO_2, due to formation of H_2CO_3. Some effects ascribed to CO_2 may therefore be due to a fall in pH unless this point is considered. When the pH is controlled, CO_2 concentrations exceeding 50 %, v/v, at pH 6·6, corresponding to 0·05–0·10 M-HCO_3^-, inhibit many bacterial species (Coyne, 1933).

At lower concentrations, bicarbonate provides a useful buffer whose capacity is greatest at pH of 6·3 and 10·3 (pK_1 and pK_2 respectively). Cultures must be incubated with an atmosphere containing added CO_2, and the concentrations required for pH 6–8 can be taken from Fig. 3.3. For pH $\simeq$ 9·5, see Delory & King (1945).

Various concentrations of CO_2 can be produced (1) by using an anaerobic jar evacuated to a pressure corresponding to the appropriate amount of residual air and by then letting in pure CO_2 until the jar pressure returns to atmospheric (see Fig. 4.10). The simplest CO_2 reservoir is a football bladder fitted with a glass tap that is filled from a cylinder fitted with a reducing valve; (2) by putting the cultures in a closed jar with a burning candle which gives *ca.* 2·5 %, v/v, CO_2 and 17 %, v/v, O_2 (Morton, 1945); (3) by CO_2 buffers based on diethanolamine which produce concentrations between atmospheric and 2–5 %, v/v (Umbreit, Burris & Stauffer, 1964).

ANAEROBIOSIS

The traditional broad division of bacterial species into aerobes and anaerobes is based on their apparent tolerance of O_2 during growth; but, since the distinction was originally drawn by Pasteur (1861), it has become clear that it reflects the effect of O_2, not on the organisms directly, but on the degree of oxidation of their culture media. The difference initially observed between aerobic and anaerobic species arose because it so happened that aerobes could grow in most of the conventional media exposed to air, whereas anaerobes failed to grow until the medium became reduced following exclusion of air. Oxygen in itself has no direct bearing on the suitability of a medium, for anaerobes are able to grow even in aerated medium, provided reducing agents are added (Kliger & Guggenheim, 1938). In such cultures, the oxidation-reduction systems presumably take an appreciable time to come to equilibrium so that the medium can be, at one and the same time, almost saturated with O_2 and yet remain sufficiently reduced to allow the growth of anaerobes.

Oxidation-reduction potential

An oxidation is defined as a reaction in which electrons are lost, for example $Fe^{2+} \rightarrow Fe^{3+} + e$; conversely, electrons are gained in reduction. An oxidizing agent therefore accepts electrons from a compound being oxidized.

Any compound has a certain tendency to be oxidized and can be thought of as possessing a corresponding pressure of electrons available for donation. If the solution of such a compound is arranged to form a half-cell in a circuit with a different half-cell of known potential, a potential difference is set up whose magnitude is related to the oxidizing (or reducing) power of the compound which can therefore be expressed in volts. The accepted standard half-cell is the normal hydrogen electrode (potential taken as zero), but, for practical reasons, the hydrogen electrode is usually replaced by another electrode such as the calomel or the Ag/AgCl electrode to which its potential bears a known relation.

An oxidation-reduction (O-R) system is described by constants and equations broadly resembling those relating pH and degree of dissociation of a buffer salt. Thus, the O-R potential, E_h, of a half-cell at a given temperature and pH is given by

$$E_h = E'_0 + \frac{k}{n} \log_e \frac{[\text{Ox}]}{[\text{Red}]}. \qquad (3.2)$$

k is a constant whose value depends in part on the temperature; n is the number of electrons involved in the reaction (e.g. n is 1 for $Fe^{2+} \rightarrow Fe^{3+}$, but is 2 for many O-R indicators); and E'_0 equals the E_h at which the compound is 50 % reduced. That is, when [Ox] = [Red], just as pK equals the pH at which [BA] = [HA] in the usual equation for a buffer salt, pH = pK + log ([BA]/[HA]). The plot of E_h against % reduction is sigmoid and so resembles the plot of pH against % dissociation of a buffer salt, though its shape is not constant and depends on the value of n (Fig. 3.4). For a given value of n, however, the shape of the curve is always the same and its position on the abscissa of Fig. 3.4 is determined by the value of E_0', the E_h at the point of inflection of the curve, just as pK corresponds to this point on the pH scale for a buffer. The degree of stability of an O-R system is un-

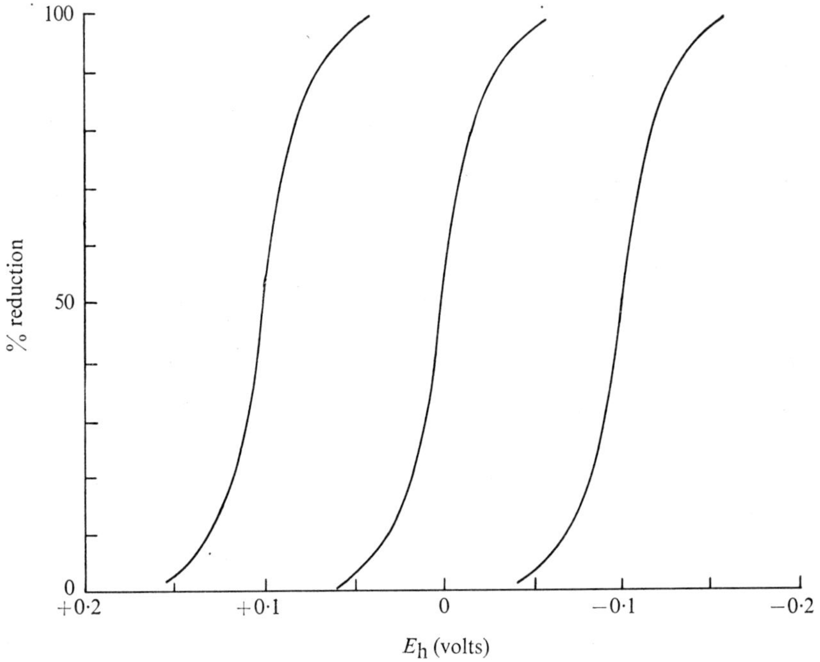

Fig. 3.4. Relation between degree of reduction and oxidation-reduction potential (E_h) for three hypothetical compounds of $E_0' = +0{\cdot}1$, 0·0, and $-0{\cdot}1$ V. respectively. All the curves have the same shape (equation (3.2)), but their positions on the abscissa differ according to E_0' which equals the E_h at which each compound is 50% reduced.

related to E_0', just as the capacity of a buffer is unrelated to its pK, and is usually referred to as the degree of 'poising'. An important difference between O-R and acid-base systems is that the first are 'sluggish' and take some time to come to equilibrium, a point to be remembered when measuring E_h.

Measurement of E_h. Either indicators or a potentiometric method can be used, as with pH.

Indicators are characterized by their E_0' values, the E_h at which each is 50% reduced, as in Fig. 3.5 where E_0' is shown for different pH since its value is partly determined by the hydrogen ion concentration. The E_h of a system at known pH can therefore be determined by measuring the degree to which it reduces standard indicator dyes, most of which are coloured when oxidized and colourless when reduced. The percentage reduction is

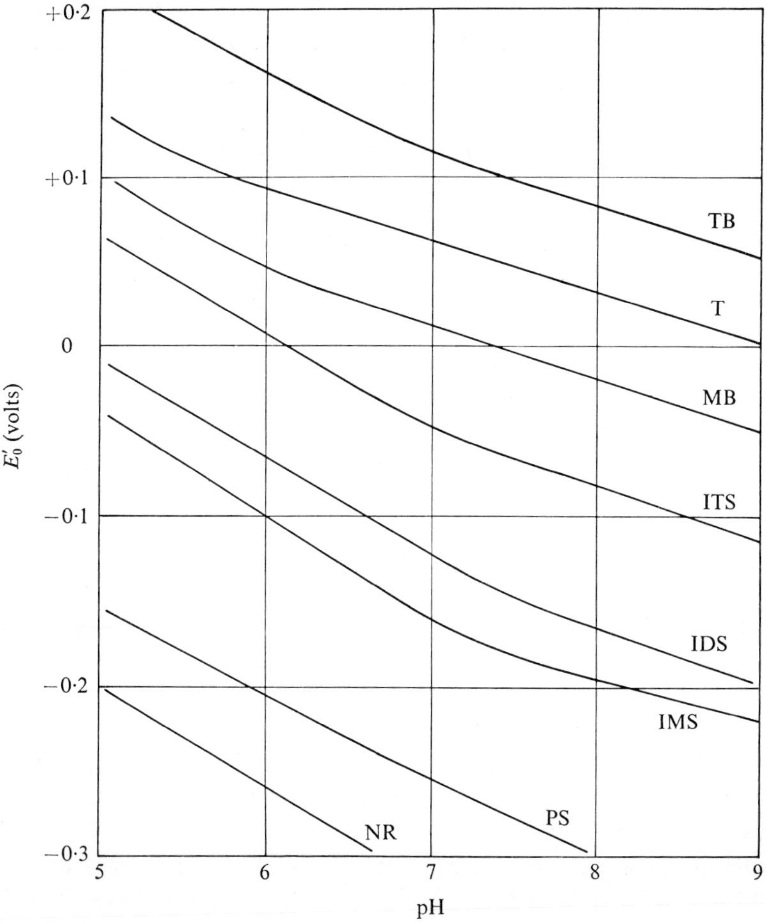

Fig. 3.5. Values of E_0' between pH 5 and 9 for some oxidation-reduction indicators. Ordinate: E_0'. Abscissa: pH. TB: toluylene blue. T: thionine. MB: methylene blue. ITS: indigo-tetrasulphonate. IDS: indigo-disulphonate. IMS: indigo-monosulphonate. PS: pheno-safranine. NR: neutral red.

used to obtain the E_h by applying Table 3.2, which shows the corresponding values of $(E_h - E_0')$ computed for the usual indicators that gain two electrons on reduction (i.e. for $n = 2$ in equation (3.2)).

The behaviour of O-R indicators differs from that of the usual pH indicators in several respects which complicate the measurements. As many systems are only weakly poised, their E_h may be altered by the indicator itself; this should therefore be used in the lowest concentration possible. Reduction may take a considerable time to occur in a sluggish system. Furthermore, the indicator may inhibit the organisms: methylene blue and resazurin are the only O-R indicators generally used in culture media.

Table 3.2. *Relation of percentage oxidation of an O-R indicator to oxidation-reduction potential at* 37°

% indicator oxidized	$E_h - E'_0$	% indicator oxidized	$E_h - E'_0$
99	+0·061	45	−0·003
98	+0·052	40	−0·005
95	+0·039	35	−0·008
90	+0·029	30	−0·011
85	+0·023	25	−0·015
80	+0·019	20	−0·019
75	+0·015	15	−0·023
70	+0·011	10	−0·029
65	+0·008	5	−0·039
60	+0·005	2	−0·052
55	+0·003	1	−0·061
50	zero		

Potentiometric measurements employ an inert electrode, often platinum, immersed in the culture and coupled to a standard half-cell of known potential. For details of apparatus and technique, see Hewitt (1950) and Spiegler & Wyllie (1956).

E_h of culture media. The bacteriological significance of O-R potential lies in the dependence of growth on successive enzymic reactions, each of characteristic E'_0, which will not occur unless the medium has a suitable E_h. This is immediately evident with obligate anaerobes which fail to grow in media at $E_h >$ *ca.* −0·1 V., whereas the E_h of sterile aerated broth is +0·1 to +0·2 V. (Table 3.3, p. 81) although it falls slowly to *ca.* −0·1 V. on removal of air (Knight, 1930*a*). Obligate aerobes are equally dependent on the E_h although this is not so apparent because culture media like digest broth are usually suitable without special adjustment. Even so, the inhibition of growth observed during overaeration may be due to a rise in E_h to an unfavourable value (p. 72). This is particularly likely to occur in salts medium which is only weakly poised (Ward, 1938). Both aerobes and anaerobes can to some extent bring the E_h of their medium to a value suitable for their growth, but their ability to do so depends on their concentration.

Thus, a given number of clostridial cells may fail to grow when dispersed in a tube of broth but succeed when concentrated by centrifugation.

The E_h of a culture medium is the net result of the reaction of its constituents with the oxidizing or reducing agents to which it is exposed. The usual method of lowering E_h by removing O_2 acts indirectly by decreasing oxidation of the medium; but E_h can equally well be lowered by adding a reducing agent to the medium, even if this be aerated (Kliger & Guggenheim, 1938). The E_h falls in growing cultures of either aerobes or anaerobes and, in fact, may fall more rapidly on oxygenating the culture of an aerobe, owing to the increase in bacterial metabolism that follows an improved O_2 supply (Hewitt, 1950). The E_h, unlike pH, is not easily maintained at a predetermined value. Digest broth and serum stabilize it at values higher than those of infusion broth or serum-free medium (Hewitt, 1950, figs. 28 and 29). It can also be controlled by passing traces of O_2 into a medium whose E_h would otherwise fall (Knight, 1930*b*), and a self-regulating system of this kind is described by Dobson & Bullen (1964). Alternatively, H_2 or O_2 can be generated in the medium by electrolysis (Hanke & Katz, 1943) although this has the drawback that toxic amounts of HOCl may be formed if much Cl^- is present (Sadoff, Halvorson & Finn, 1956).

Anaerobic media

The practical problems are the lowering of E_h and its subsequent maintenance at a low value. Details are given of two methods that produce a low E_h immediately without rendering the medium unsuitable for many experiments by including soft agar, etc. A few of the many other methods are mentioned briefly (see Hall, 1929; Willis, 1960).

Anaerobic jars. The cultures are incubated in sealed jars from which atmospheric O_2 is removed by conversion to water in the presence of a catalyst and added H_2 (see Laidlaw, 1915). If immediate anaerobiosis is needed, liquid media should be heated at 100° for 5–30 min. to remove dissolved air and cooled to room temperature before being put in the jar.

Two types of jar are used:

(*a*) That sold by Baird and Tatlock which uses a patent ('cold') palladium catalyst active at room temperature (Heller, 1954). This catalyst needs renewal from time to time but has the advantage of avoiding heating and the danger of explosion.

(*b*) The McIntosh and Fildes jar, which uses a heated palladium catalyst (Fildes & McIntosh, 1921).

A control tube containing an O-R indicator should be used with either jar to show that anaerobiosis is produced. One formula contains equal volumes of 0·006 N-NaOH (0·024 %, w/v); 0·015 %, w/v, aqueous methylene blue; and 6 %, w/v, glucose solution. A crystal of thymol is also added as antiseptic. For use, a tube containing 5 ml. of the blue solution is heated in boiling water until it becomes colourless, and, after cooling to room temperature, is placed in the jar. The solution remains colourless in an anaerobic atmosphere (Fildes & McIntosh, 1921). This indicator has to be renewed fairly frequently because it becomes impossible to reduce. A longer-lasting formula has glucose, 4 g.; Na_2HAsO_4, 2 g.; methylene blue, 0·003 g. dissolved in 100 ml. distilled water (Ulrich & Larsen, 1948).

The most convenient source of hydrogen is a cylinder fitted with a pressure-reducing valve from which the gas is transferred to the anaerobic jar either through a reservoir or by a football bladder. A bladder fitted with a glass stopcock is convenient and, after being emptied by compression, is filled from the reducing valve on the cylinder. The fittings on the cylinder and the jar may differ in size. The simplest way to make the necessary connections is to attach a glass or plastic conical connector to the bladder and to fit rubber tubing to the H_2 cylinder and to the valves of the jar. The connector can then be pushed home into the various pieces of tubing, although they may differ in bore.

Jars fitted with the *cold catalyst* are made from metal with an external indicator fitted to a tube let into the side. The lid is also of metal and carries on its outer surface two gas valves and a central boss. The inner surface carries the catalyst. The lid is sealed to the jar by a rubber gasket and secured by a clamp catching the edges of the jar and carrying a screw which bears on the boss.

The cultures are placed in the jar, the lid secured, and one valve connected to a mercury manometer while air is evacuated through the other until the pressure within the jar has fallen to *ca.* 40 cm. Hg. A convenient vacuum line is shown in Fig. 4.10. The latter valve is then shut, the H_2 source connected, and the valve reopened to let the H_2 into the jar. After 15–20 min., anaerobiosis is achieved. This point can be checked by interposing a wash bottle between the H_2 source and the jar. After the catalyst has acted for *ca.* 15 min., the H_2 supply is turned off for some minutes and is then turned on again. If no bubbling occurs in the wash bottle, no H_2 is entering the jar, and so all O_2 must have been removed.

The construction of *the McIntosh and Fildes jar* is much the same except that the lid also carries on its outer surface two electric terminals connected to a heating element placed on the inner surface which is surrounded by the catalyst of palladinized asbestos. The boss on the lid is spring-loaded in order to release the pressure that develops within the jar if the mixture of O_2 and H_2 should explode. The spring should therefore not be fully compressed by the screw holding the lid in place.

After the partially evacuated jar is filled with H_2, the current is switched on and the catalyst allowed to act for 15–20 min. The heating element requires a 12 V. supply. This is obtained from 240 V. mains by a resistance or transformer, or by putting three 65 W. carbon filament lamps in parallel in the mains supply to the jar. The lamps incidentally show when the current is on.

The catalyst is surrounded by a gauze envelope to prevent spread of an explosion from the element into the body of the jar on the principle of the Davy lamp. However, the gauze is often damaged if the heavy lid is placed with its inner surface downwards on the bench, so that the jar should be kept in a strong container or in a separate room while the current is on.

Addition of a reducing agent to the medium. Details of some of non-inhibitory reducing agents are given in Table 3.3.

Table 3.3. *Reducing agents suitable for culture media*[1]

Medium	Reducing agent (w/v)	pH	E_h (V.)	Reference
Broth[2]+0·1 % (w/v) agar	0·5 % glucose	7·6	−0·1	Reed & Orr (1943)
Broth[2]+0·1 % (w/v) agar	0·1 % Na thioglycollate	7·6	−0·2	Reed & Orr (1943)
Broth[2]+0·1 % (w/v) agar	0·1 % ascorbic acid	7·6	−0·2	Reed & Orr (1943)
Broth[2]+0·1 % (w/v) agar	0·1 % cysteine	7·6	−0·2	Reed & Orr (1943)
Supplemented salts medium	0·02–0·04 % sodium formaldehyde sulphoxylate	6·8–7·1	$< -0·36$	Hutner (1950)

[1] Many other reducing agents have been used, for example, 5 μM/ml. Na_2S (Grossmann & Postgate, 1953) and 0·002 M sodium mercapto-acetate (Briggs *et al.* 1955).

[2] The initial E_h of digest broth is +0·1 to +0·3 V. (Knight, 1930*a*; Hewitt, 1950).

The E_h of liquid medium rapidly falls to the expected value if it is heated to 100° for some minutes to drive off dissolved O_2. The low E_h so produced is maintained by putting up the medium in deep layers or in screw-capped bottles; or by sealing it with petroleum jelly or 'vaspar', a mixture of equal parts of jelly and paraffin oil. Oil alone is ineffective due to convection (see Hall, 1929). Oxygen can also be excluded from the medium by gassing the culture vessel with nitrogen prepared by passing commercial 'oxygen-free' nitrogen through a combustion tube holding Cu or Fe filings heated to 600°–700° (see also Umbreit, Burris & Stauffer, 1964).

No reducing agent is known that permits surface growth of anaerobes on solid medium incubated in air. This can only be achieved by combining a reducing agent with a special Petri dish (Brewer, 1942) or by covering inoculated agar containing a reducing agent with a glass slide or glazed cellophane to exclude air.

A well-known liquid medium based on a reducing agent contains:

Sodium thioglycollate	1 g.
Agar	1 g.
(Glucose	5 g.)
Resazurin	0·001 g.
Nutrient broth	1 l.

This formula differs from that given by Brewer (1940) in replacing methylene blue by resazurin as O-R indicator because it is less inhibitory (Pittman, 1946). Its oxidized form is blue; the intermediate product resorufin, which is formed in an irreversible reaction, is pink; and the reduced form, dihydroresorufin, is colourless (see Twigg, 1945). Reduced medium is therefore colourless and turns pink, but never blue, on oxidation. Glucose is added as nutrient. The medium is distributed in depth in tubes or screw-capped bottles and, after sterilization, remains reduced for at least a month in tubes stored at room temperature. Oxidized medium can be reduced by heating to 100° in a water bath (Skerman, 1953).

The low E_h maintained by this medium depends entirely on the thioglycollate, as the rate of oxidation by air is not decreased by including glucose or agar (Skerman, 1953). Nevertheless, if thioglycollate is omitted, the medium supports the growth of at least some anaerobes, presumably because the agar limits mixing of the broth, whose E_h in the depths of the tube therefore falls progressively because of its own reducing systems. Once bacterial growth is under way, further reduction occurs, and anaerobes spread through the medium (Hitchens, 1921). A similar mechanism enables obligate anaerobes to grow in the interstices of a mass of powdered chalk or glass added to ordinary broth incubated in air (see Hall, 1928).

The range of conditions present in thioglycollate soft agar medium has made it useful for the isolation of certain aerobic pathogens, including meningococci and streptococci, which may not grow at once on blood agar incubated in air (Foley & Schaub, 1944), and in sterility testing where thioglycollate inactivates mercurial disinfectants (p. 114).

In other anaerobic media, the E_h is lowered by including minced meat, which contains reducing agents (Robertson's meat medium: see Lepper & Martin, 1930), or iron filings or tacks, which incidentally produce a brown precipitate of FeO in the medium (Hayward & Miles, 1943). For numerous reducing agents and special culture vessels, see Hall (1929) and Willis (1960).

4

STERILIZATION

Although often regarded as nothing more than a technical necessity, sterilization continues to yield fascinating problems at every level. To take two examples: it was hardly self-evident that modern autoclaves would have greater difficulty in sterilizing a solitary package than a full load or that the parameters defining heat inactivation of a single-stranded DNA phage should be almost unaffected by moisture. These are contemporary questions, but others of no less interest have been discussed since the start of the century, often with great insight. In the laboratory, sterilization is largely concerned with the preparation of experimental material, like glassware or culture media, and with rendering it harmless after use. The usual means include the destruction of organisms by heat or chemical disinfectants, and, in preparative work, their removal by filtration. These are only a few of the methods that have been described, but some of the most efficient, such as gamma radiation or the bactericidal gases, are too dangerous for everyday use and need specialized equipment (see Symposium, 1961). The methods used with different materials are summarized in Table 4.6, p. 125.

The kinetics of disinfection are considered before the individual methods, for, although these presumably kill organisms by different means, the course of disinfection is often remarkably similar whatever agent is used. A fuller account of the various survival curves is given in chapter 6.

KINETICS OF STERILIZATION

Survival curves

If a population of organisms is exposed to a lethal agent, the proportion surviving at any time can be plotted against the exposure to give a *survival curve* (e.g. Figs. 4.1, 6.13). This is often exponential, as a straight line is obtained when the logarithm of the proportion of survivors is plotted against the exposure expressed in arithmetic units of time (Figs. 4.1 *c*, 6.7 *a*). Exponential curves are so common that certain authors have believed that they reflect a fundamental process common to all disinfectants (see Rahn, 1945). However, there are undoubtedly many cases in which the curve is not exponential (see Humphrey & Nickerson, 1961), since the same method of plotting gives a curve with a shoulder, a steadily decreasing slope, or even a sigmoid shape.

The biological significance of these curves is not yet certain, although the hypothetical approach to this problem is illustrated by two well-known models which postulate that killing is due either to stochastic or to deterministic events in the cells (see also p. 225). The simplest form of the first hypothesis assumes that each organism is killed by a single lethal event occurring at random. Examples of this model occur in the target theories of radiobiology (p. 196) and in the chemical reaction model of disinfection discussed later. A more sophisticated model of a stochastic process is provided by a 'birth-death' process where the outcome of the encounter between organism and disinfectant is assumed to be governed, not by a single random event but by a succession of such events (p. 227; Frederickson, 1966). In all these forms of stochastic model, however, an exponential survival curve implies that the culture is uniformly susceptible to the disinfectant (p. 196). Non-exponential curves are accounted for by introducing extra assumptions while retaining the central postulate that each lethal event occurs at random. Thus, a curve that is concave upwards is attributed to differences in susceptibility of individual organisms in the culture (p. 196). One that is convex is held either to indicate the necessity for several events in each organism before death occurs (Rahn, 1945; p. 199) or to arise from technical causes like clumping of organisms, a lag before the lethal agent reaches the sensitive part of the cell, or to heat-activation of spores (Shull & Ernst, 1962).

This type of hypothesis might reasonably apply to a bactericidal agent like radiation where the lethal event could be a single change produced in an essential molecule, such as the deoxyribonucleic acid of the bacterial genome. But many disinfecting agents presumably kill the cell following the interaction of large numbers of molecules and the deterministic alternative is therefore often considered, more on formal grounds than because of its plausibility. The deterministic model postulates that each organism is killed by the joint action of a very large number of 'events', each occurring deterministically, so that x events would inevitably occur in each organism exposed to a given concentration of a disinfectant for a given time. This view is therefore completely opposed to that of the first hypothesis in which the number of events can never be predicted, since each is assumed to occur at random. If the organisms were all equally susceptible, the survival curve would be perpendicular to the time axis, the survival being

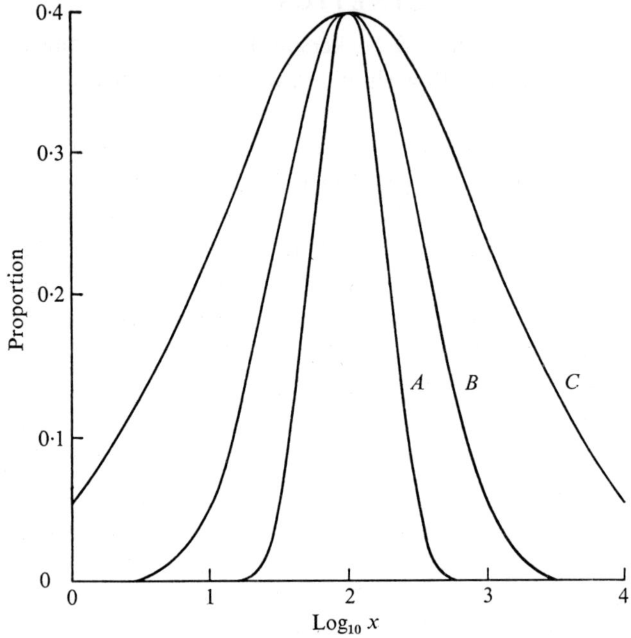

Fig. 4.1(*a*)

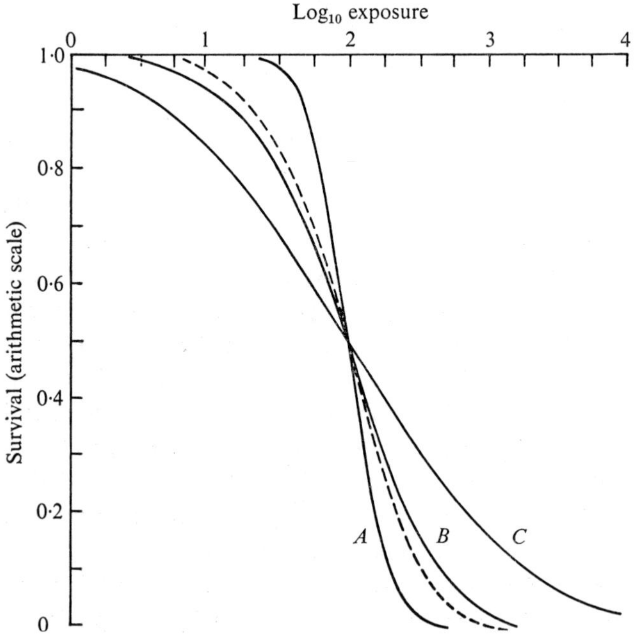

Fig. 4.1(*b*)

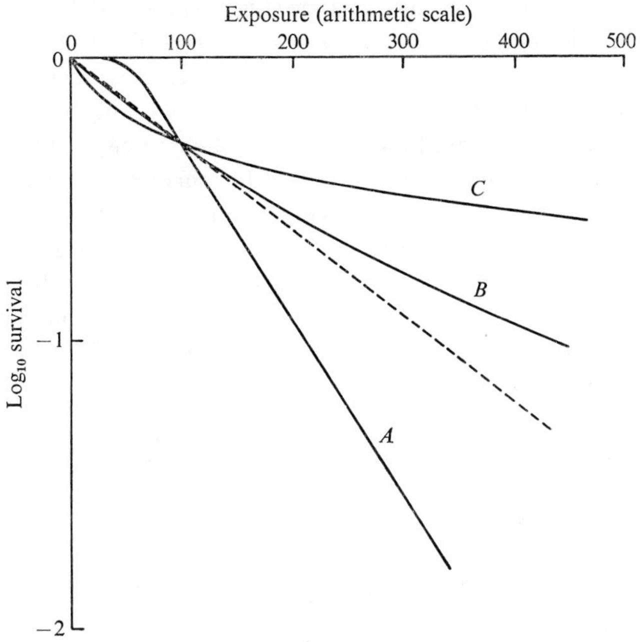

Fig. 4.1(*c*)

Fig. 4.1. Hypothetical survival curves derived from the deterministic hypothesis for disinfection.

(*a*) Three log-normal distributions of susceptibilities. Ordinate: proportion of culture having a given degree of susceptibility. Abscissa: susceptibility, i.e. the number, x, of lethal events required to kill an organism (arbitrary scale). The standard deviations of the curves are: *A*, 0·25; *B*, 0·5; and *C*, 1·00.

(*b*) and (*c*) Survival curves derived from the three distributions shown in Fig. 4.1*a*. The dashed line is an exponential curve having the same mean survival time. (*b*) Ordinate: survival. Abscissa: log exposure. (*c*) Ordinate: log survival. Abscissa: exposure.

100 % while x, the number of events per cell, is less than the lethal figure, and subsequently falling to 0 %. No survival curves of this sort are observed in practice, as all have some degree of slope which this hypothesis attributes to differences between the susceptibilities of individual organisms. The shape of the survival curve is therefore held to be determined solely by the distribution of individual susceptibilities, each of which is taken to be inversely proportional to the time a given organism survives the disinfectant. The distribution of survival times often appears to be log-normal, since probit survival plotted against the logarithm of time of exposure often gives a straight line (Fig. 6.13*a*). Three log-normal distributions, *A*, *B*, and *C*, are shown in Fig. 4.1*a* with the corresponding integrated survival curves

plotted either with survival against log time (Fig. 4.1 *b*) or with log survival against arithmetic time (Fig. 4.1 *c*). The latter curves also show the exponential curve having the same mean survival time. It is clear that the three curves derived from log-normal distributions could easily be confused in practice with curves derived on other assumptions: thus, in Fig. 4.1 *c*, curve *A* resembles a multi-hit curve (Fig. 6.8); curve *C* resembles two exponential curves meeting at an angle (Fig. 6.7 *b*); and curve *B* resembles an exponential curve (see Irwin, 1942). It is therefore extremely doubtful if the fundamental mechanism of sterilization can be ascertained from the shapes of survival curves. Moreover, survival curves may differ in shape in experiments on the same system apparently performed under identical conditions (Withell, 1942*a*, *b*; Eddy, 1953).

Killing considered as a chemical reaction

These models derive from the earliest studies of killing by heat or chemicals (see Chick, 1930) and, although insufficient in certain respects, they still provide a frame of reference in which to consider the kinetics and molecular basis of sterilization (Clark, 1937; Pollard, 1953, 1964; Johnson, Eyring & Polissar, 1954; Woese, 1960; Hiatt, 1964).

In the simplest model of chemical disinfection, killing is assumed to result from the reaction of x molecules of disinfectant, D, with y molecules of a substrate, O in the organism:

$$xD + yO \rightarrow D_x O_y.$$

The reaction rate is determined by the xth and yth powers of the molar concentrations of D and O, respectively, and the reaction is said to be of the order $x+y$. In sterilization, the molar concentration of D vastly exceeds that of O and may therefore be assumed to remain constant. The reaction rate for a given D is then, approximately,

$$\frac{d[O]}{dt} = -k\,[O]^y.$$

When $y = 1$, a constant proportion of viable organisms die in each interval of time and the survival curve is exponential (Fig. 6.7 *a*). Such a reaction is termed 'pseudo first order' for, although it results from union of the two reactants, D and O, and is of order $x+1$, its kinetics resemble those of a true first order reaction, $d[O]/dt = -k[O]$, in which only one reactant is involved.

Everyday experience tells one that the more concentrated a disinfectant, the more rapidly it usually acts. In practice, disinfectant concentration, C, and the time, t^*, corresponding to a given survival are found to obey the following relationship:

$$C^n t^* = a \quad \text{or} \quad \log t^* = \log a - n \log C$$

where n is the *concentration coefficient* and a is a constant (Watson, 1908). A plot of log C against log t^* therefore gives a straight line whose slope is the value of n (Fig. 4.2). Such plots provide a standard method for deter-

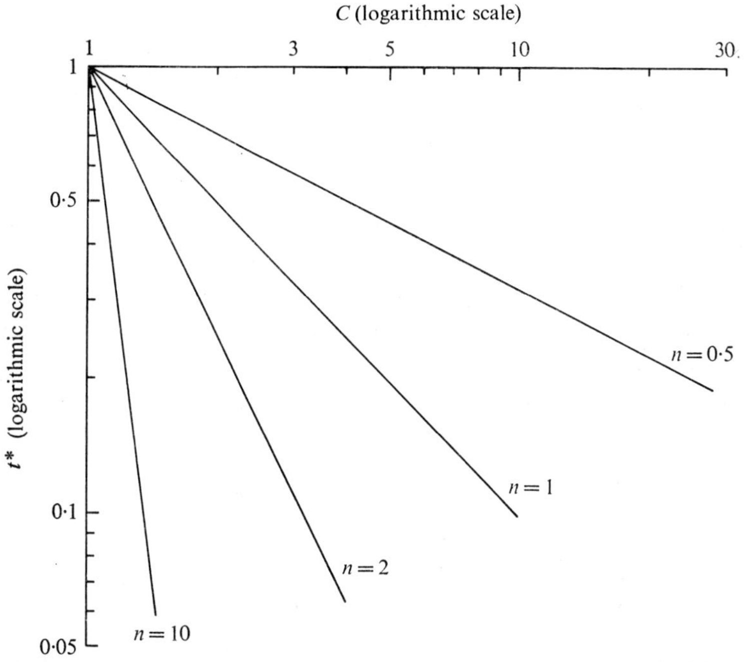

Fig. 4.2. The effect on t^*, the time required to a kill a given fraction of cells, of alterations in C, the disinfectant concentration, for different concentration coefficients, n.

minining the order of a chemical reaction. Consider a true first order reaction, like the decay of an isotope, where the reaction rate and hence t^* is well known to be constant, whatever the concentration of isotope initially. A plot of log initial concentration (on the abscissa) against log t^* then yields a straight line parallel to the abscissa. Thus, a reaction of order 1 yields a line with a slope, n, of 0 and, in general, the order of reaction equals $n+1$:

$$x+y = n+1.$$

When disinfectants act on bacteria, the killing rate and hence the order of reaction may be assumed to be determined solely by the disinfectant concentration. The concentration coefficient, n, then reflects only the properties of the disinfectant. Some observed values of n are shown in Table 4.1 (see Chick, 1930; Clark, 1937; Tilley, 1939; Rahn, 1945). They appear to fall into two groups. The first, $n \sim 1$, giving an order of about 2, which, as it stands, is not necessarily inconsistent with the chemical reaction model since 2nd order reactions are not uncommon. In the second group, $n \geqslant 4$, and here it can be concluded with certainty that the model is invalid since chemical reactions of this order are unknown. For a reaction to occur, the reacting molecules must not only be present but must be in an 'activated' state. The chance of even three molecules being in the activated configuration and bearing the necessary relation to each other is already extremely small, judging from the rarity of 3rd order reactions, so that the chance of witnessing reactions of order greater than 3 is essentially zero. A high value of n is presumably found with certain disinfectants because they act in a manner quite unlike that postulated by the reaction model, as by deterging cell structures (Cowles, 1939). Phenol belongs to this class and, although it cannot be acting through a conventional reaction, it is a curious historical accident that the exponential survival curves obtained with phenol provided part of the early evidence on which the chemical reaction model was based.

Table 4.1. *Concentration coefficients of disinfectants*

	Coefficient	Reference
Hydrogen peroxide	0·5	1
Hydrochloric acid	0·5–1·5	1
Silver nitrate	1·0	1
Mercuric chloride	0·5–3·0	1
Iodine	0·9	2
Thymol	4·0	1
Phenol	4·0–5·5	1
Orthocresol	7·4–7·7	3
Ethanol	11·3–18·9	3

1, Chick (1930); 2, Berg *et al.* (1964); 3, Tilley (1939).

The practical importance of the concentration coefficient is that, when it is small, changes in disinfectant concentration produce relatively little change in killing rate; whereas with a large coefficient, a given change in concentration may produce a dramatic effect (note that n is an exponent), so that it is of especial importance that disinfectants of this type be used at the specified strength.

A rise in temperature usually increases the death rate, whether organisms are killed solely by heat or by a chemical disinfectant whose action is accelerated by an increase in temperature not lethal in itself. The simplest guide to the effect of temperature is the *temperature coefficient*, Q_{10}, but more fundamental parameters can be derived from the theory of absolute reaction rates, described below in the context of heat inactivation.

The Q_{10} is defined as the increase in the killing rate constant per 10° rise in temperature. When the survival curve is exponential, the survival, S, at a given temperature T_1, is related to the exposure, t_1, and the killing rate constant, k_1, by equation (6.5):

$$\log_e S = -k_1 t_1.$$

If the same survival is produced at a higher temperature, T_2, by a shorter exposure, t_2, $-k_1 t_1 = -k_2 t_2$,

$$Q_{(T_1-T_2)} = k_2/k_1 = t_1/t_2.$$

In general, $Q_{10} = Q_a^{10/a}$, where a is the difference in temperature between the tests.

The observed values of Q_{10} for the killing of vegetative bacteria or spores by moist heat are well known to be extremely high; e.g. 10–200 (see Chick, 1930; Rahn, 1945) in contrast to the values of *ca.* 2 observed for many chemical reactions, and have long been attributed to the lethal action of moist heat being due to denaturation of protein, another process known to have an exceptionally large Q_{10}.

On the theory of absolute reaction rates, a reaction is assumed to be possible only when the reacting molecules enter a transient activated state to form an 'activated complex'. The killing rate constant, k, is given by

$$\log_e k = \text{a known constant} + \frac{\Delta S^\ddagger}{R} - \frac{\Delta H^\ddagger}{RT},$$

where R is the gas constant; T is the temperature in °K; and $\Delta S^\ddagger$, the entropy activation, a positive or negative value indicating an increase or decrease in the randomness of the system. A very large positive value of $\Delta S^\ddagger$ therefore indicates a great increase in randomness, as would result from denaturation of a large molecule like that of a protein. $\Delta H^\ddagger$ is the heat of activation, a measure of the energy required for the reactants to become activated. It is clear from the above relationship that a plot of log k against $1/T$ (an 'Arrhenius plot') will yield a straight line which enables the values of $\Delta S^\ddagger$ and $\Delta H^\ddagger$ to be estimated. The larger $\Delta S^\ddagger$, the higher the curve in Fig. 4.3 for a given $\Delta H^\ddagger$; whereas the larger $\Delta H^\ddagger$, the steeper the slope of the curve. $\Delta H^{\ddagger}$ is evidently related to Q_{10} but the latter changes slightly over different ranges of temperature in a system conforming to Fig. 4.3. $\Delta H^{\ddagger}$ of 90 kcal./mole corresponds to Q_{10} of *ca.* 50.

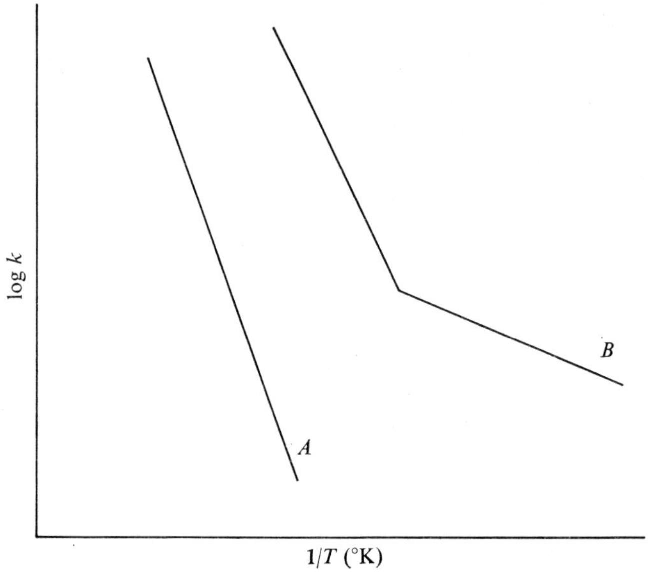

Fig. 4.3. Arrhenius plot for sterilization by heat showing log *k*, the killing rate constant, plotted against 1/*T*, the reciprocal of the absolute temperature. *A*: homogeneous system. *B*: heterogeneous system, often encountered in practice.

Many values of $\Delta S^{\ddagger}$ and $\Delta H^{\ddagger}$ are available for bacteria, viruses, phages and enzymes, both moist and dry (Table 4.2), and consistently reveal marked differences between killing by moist and by dry heat. In almost every case, inactivation by moist heat is associated with large positive values of $\Delta S^{\ddagger}$, indicating extensive molecular disorganization, presumably arising from denaturation of protein and DNA. A notable exception is the single-stranded filamentous DNA phage, Ec9 (Dettori & Neri, 1966). It is striking that the values for purified enzymes are of the same order as those for whole organisms. Inactivation by dry heat, on the other hand, is found with little or no change in $\Delta S^{\ddagger}$, showing that orderliness at the molecular level is unaltered and may even be increased in those cases where $\Delta S^{\ddagger}$ is negative (Ginoza, 1958). The cause of killing by moderate dry heat is not known but a possibility suggested by its potent mutagenic action (Zamenhof, 1961) is the production of multiple lesions in DNA.

The values of $\Delta H^{\ddagger}$ observed with moist heat are extremely large; so

Table 4.2. *Parameters of heat inactivation*

Material		$\Delta H^{\ddagger}$	$\Delta S^{\ddagger}$	$\Delta F^{\ddagger}$	Reference
Phage T 1	(wet)	95,000	207	25,000	1
	(dry)	27,500	0		1
Phage T 2	(wet)	71,700	139		1
	(dry)	18,000	−12		1
Phage T 3	(wet)	105,000	246		1
	(dry)	19,100	−9		1
Phage Ec 9	(wet)	25,400	6	23,200	2
	(dry)	22,900	0	23,200	2
B. stearothermophilus	(wet)	67,700	105		3
Anaerobe	(wet)	72,400	123		3
Invertase	(wet)	53,000	95		1
	(dry)	30,000	0		1
Emulsion	(wet)	44,930	65		4
	(dry)	22,250	−7		4

1, Pollard (1953); 2, Dettori & Neri (1966); 3, Deindorfer (1957); 4, Johnston *et al.* (1954).

large, in fact, that if the rate of killing were governed solely by $\Delta H^{\ddagger}$, the killing rate constant would be extremely small and killing would be almost negligible under ordinary conditions. However, the rate is determined by $\Delta F^{\ddagger}$, the free energy of activation:

$$\Delta F^{\ddagger} = \Delta H^{\ddagger} - T\Delta S^{\ddagger}.$$

The large $\Delta H^{\ddagger}$ found in killing by moist heat is therefore offset by the accompanying large values of $\Delta S^{\ddagger}$, giving a mean value for $\Delta F^{\ddagger}$ of *ca.* 25 kcal./mole which permits killing to take place fairly readily. With dry heat, both $\Delta S^{\ddagger}$ and $\Delta H^{\ddagger}$ are smaller, and $\Delta F^{\ddagger}$ again comes out at a value enabling dry heat to be used as a practical method of sterilization.

Although an Arrhenius plot presumably expresses the most fundamental relation between killing rate and temperature, another relation will also be encountered in the literature. Bigelow (1921) found with bacterial spores that straight lines were obtained if log 'thermal death time' (i.e. the exposure corresponding to a predetermined survival) was plotted against temperature in °C (Fig. 4.9). This implies that Q_{10} is constant (cf. Fig. 6.13*a*), which is not strictly true if the theory of absolute reaction rates is valid; although Q_{10} will appear to be constant over small ranges of temperature, so accounting for Bigelow's observation. The slope of each curve in Fig. 4.9 is the increase in temperature required to decrease the T.D.T. by 10% and is therefore identical with *z*, a parameter commonly used in medical bacteriology and the canning industry (see Ball & Olsen, 1957). If killing is exponential, $z = 10/\log_{10} Q_{10}$. Another parameter met with in this context is F, defined as the T.D.T. in minutes at 121 °C.

Practical implications of kinetic studies

This form of analysis may appear highly academic but it has important practical consequences. First, the smaller values of $\Delta H^{\ddagger}$ (or of Q_{10}) observed with dry as opposed to moist heat means that the killing rate with dry heat is less affected by a given change in temperature (Fig. 4.3). Increasing the temperature in dry heat sterilization therefore does not have the dramatic effects on killing found with moist heat, and it is for this reason that a variety of sterilizing routines has been proposed for ovens and other forms of dry air sterilizer, simply because the influence of temperature is not as clear cut as in autoclaving (Table 4.3). Second, the occurrence of progressive killing, as reflected in the slope of survival curves, implies that there is always a chance of an organism surviving in the culture. It is therefore meaningless, or even worse, misleading, to speak of the 'thermal death time' of a culture as if exposure for this or any longer period invariably produced complete sterility. The rates of sterilization can be expressed in the form: 'the concentration of viable organisms has been lowered to 1 in 1 litre' or '1 in 10^9 litres' or 'the survival is 10^{-9}'. Failure to appreciate the random distribution of surviving organisms amongst replicate cultures given an exposure of 1 'T.D.T.' has led in the past to dangerous overestimates of the efficiency of sterility tests (see pp. 95, 195). The Cutter incident, in which vaccination with formalin-treated poliovirus resulted in a number of cases of paralytic poliomyelitis, provides a good example (Scheele & Shannon, 1955). Third, heat is widely used for the preparation of vaccines; i.e. the method assumes that the antigens of interest resist heat that kills the organism, and it is of obvious practical importance to determine the relative values of $\Delta H^{\ddagger}$ and $\Delta S^{\ddagger}$ for loss of antigenicity as well as viability. This might show, for example, a short exposure at high temperature to be preferable to a longer exposure at a lower temperature, although the latter might intuitively be thought less harmful to the antigen. Moreover, the relative inactivation rates of viability and of antigenicity could well be examined at different pH or in different ionic environments, all of which drastically affect the parameters. An everyday example of this phenomenon is the use of boiling in weak alkali to kill bacterial spores which resist boiling at neutral pH (p. 108).

All this discussion has assumed that the microbial system behaves homogeneously and can be validly described by single values of $\Delta H^{\ddagger}$ and $\Delta S^{\ddagger}$, respectively. This is by no means invariably found in practice. Not only do individual survival curves at a given temperature often indicate a heterogeneous cell population (p. 196) or a complex killing process producing a heterogeneous response (Woese, 1960), but Arrhenius plots are often discontinuous (Fig. 4.3), the implication being that the organism can be inactivated in two different ways. A process with large $\Delta H^{\ddagger}$ often predominates at high temperatures and another with small $\Delta H^{\ddagger}$ at low temperatures (Woese, 1960; Hiatt, 1964; Pollard, 1964).

Comparison of sterilizing procedures

A common practical problem is to compare the action of a given disinfectant acting under different sets of conditions, or the action of different disinfectants which may be dissimilar in nature. Such comparisons can be made in two ways: by determining the entire survival curve; or by making use of a single measurement which may be either the time required to kill a predetermined fraction of the culture or the survival after a predetermined exposure. Finally, an attempt may be made to express the difference by a single index or coefficient so that one agent or treatment can be spoken of as x times as effective as another.

One single point comparative method is the Chick–Martin test which measures the survival after a fixed exposure (Chick, 1930). Another is the Rideal–Walker test (1903), whose result is expressed by the phenol coefficient, defined as 'the dilution of the test agent sterilizing a particular bacterial suspension in a given time/the dilution of phenol sterilizing an exactly similar suspension in the same time'. All procedures of this kind are open to the objection that they provide no evidence that the survival curves under comparison are similar in kind and that the differences between them can be validly expressed by a coefficient. If the two curves differ—for example, if one is convex and the other exponential—no comparison is possible because the difference observed depends on the particular exposure or survival at which the comparison is made (Fig. 6.17). A further objection to the Rideal–Walker or any other test based on measurement of 100% sterility is that the endpoint is never sharp. When the viable count has fallen to *ca.* 1 organism or less in each sample withdrawn from the mixture of bacteria and disinfectant, some samples give growth on subculture although taken later than others which are sterile, because the surviving viable organisms are randomly distributed amongst the samples. Although the mean sterilization time can be estimated from such data

(Mather, 1949), its precision is inevitably limited by this random element in the experiment (p. 195).

The random distribution of surviving organisms also has important consequences for the design of sterility tests since, for many years, the large sampling errors attached to tests on relatively few samples were seriously underestimated. Thus, if 10% of a batch is contaminated, then, on testing 10 samples of a batch of 100, there will be a 35% chance of obtaining 0 contaminated out of the 10 sampled, so that the whole batch of 100 is erroneously judged to be safe. Calculation shows that the information obtained depends far more on the number of samples tested than on the size of the total batch (see Knudsen, 1947).

Provided two survival curves have the same general shape, the difference between them can be validly expressed, with exponential curves, by the ratio of their slopes or inactivation rate constants (p. 198), or, with log-normal curves, by the LT50, the time required to kill 50% of organisms (p. 224). A related index is the 'decimal reduction time', the time required for a 10-fold fall in survival. The D.R.T. is the same for any 10-fold change in survival when the curve is exponential, but otherwise depends on the actual survivals measured. It is related to the inactivation rate constant, k, of an exponential curve as follows:

$$S = e^{-kt}. \tag{6.4}$$

When t = the D.R.T., $S = 0{\cdot}1$,

$$\therefore \quad 0{\cdot}1 = e^{-k\,(\text{D.R.T.})}$$

and the

$$\text{D.R.T.} = 2{\cdot}3/k.$$

Even where two survival curves are of the same kind and can therefore be simply compared in the way that the differences between two exponential curves may be expressed by the ratio of the inactivation rates, it does not follow that the same ratio will hold under different conditions. This would require that all the parameters determining inactivation were the same for the two procedures in question. In particular, that their values of n, $\Delta H^{\ddagger}$ and $\Delta S^{\ddagger}$ were identical. If $\Delta H^{\ddagger}$ differed, for example, raising the temperature of the test would favour the disinfectant possessing the larger value. A particularly well-known case in which comparisons are invalidated because the parameters differ is that of the Rideal–Walker test. Here phenol was chosen as a standard against which disinfectants of all kinds were to be compared. Unfortunately, however, the value of n for phenol is moderately large (Table 4.1), so that the phenol coefficients for disinfectants with different n are likely to be misleading.

Difficulties of this kind are not peculiar to the standardization of disinfectants but recur in bioassays of all kinds. It is commonly found that two systems are not comparable because they differ in kind, not merely in concentration or intensity, so that their behaviour is described by fundamentally different parameters.

METHODS OF STERILIZATION

The usual methods include:

(1) Heat.

(*a*) Moist heat: (i) the autoclave; (ii) intermittent steaming (tyndallization); (iii) boiling; (iv) pasteurization.

(*b*) Dry heat: the hot air oven.

(2) Chemical disinfectants.

(3) Filtration.

(4) Irradiation.

(*a*) Ultra-violet radiation.

(*b*) Photodynamic inactivation.

Table 4.3. *Minimum times required for sterilization by moist and by dry heat at various temperatures*

	Moist heat[1]		Dry heat[2]
Temperature	Time (min.)	Pressure[3]	Time (min.)
121°	15	15	—
126°	10	20	—
134°	3	30	—
140°	—	—	180
150°	—	—	150
160°	—	—	120
170°	—	—	60

[1] Report (1959).

[2] Perkins (1956). The Memorandum (1962) suggests times approximately one-third of these, following Darmady, Hughes & Jones (1958).

[3] I.e. pressure of pure saturated steam in lb./sq. in.

The time-temperature combinations generally accepted as sufficient to kill bacterial spores, the most resistant organisms encountered, are shown in Table 4.3. It is generally true that all organisms, whether vegetative cells or spores, are more susceptible to moist than to dry heat (see Perkins, 1956).

A vegetative bacterium like *Escherichia coli*, which has a high water content, is extremely susceptible even to dry heat, but previous dehydration greatly increases its resistance (Zamenhof, 1960). The times shown in Table 4.3 are not necessarily the times for which the sterilizer should be run in practice, because extra has to be added to allow the load of contaminated material to reach the stated temperature (see Table 4.4). It is essential to remember that for some loads this additional time may be considerably longer than the stated sterilizing time. On the other hand, the surfaces of metal objects like surgical instruments heat up almost instantaneously and it is an open question as to the shortest period of heating that is feasible. Extrapolation of data like those of Fig. 4.9 suggest a 'T.D.T.' of 0·6 sec. at 160°. Exposures of 1 min. at 150° have been successfully used for sterilization in practice (Lane *et al.* 1964).

MOIST HEAT

The autoclave

The working principle of an autoclave is that pure saturated steam at pressures above atmospheric is used to heat the load contained in a sealed chamber. When saturated steam encounters a cooler object, it condenses to form water and, at the same time, liberates its considerable latent heat of condensation so that the temperature of the object rapidly rises. Condensation is also accompanied by a contraction in volume which draws more steam to inner parts of the load. The use of saturated steam diminishes dehydration of materials, like rubber, fabric or paper, which are damaged by dry heat at the same temperature. It is evident that, for proper working of the autoclave, the steam entering the chamber must be saturated with water so that contact with a cooler object immediately leads to condensation and heating, while, at the same time, no moisture is removed from the load.

Properties of steam. The relation of steam, water and ice at different temperatures and pressures is shown in Fig. 4.4. As these are the gaseous, liquid and solid phases of the same substance, Fig. 4.4 is termed a *phase diagram* and the curves the *phase boundaries*, which show the conditions in which the phases are in equilibrium. The line XY joins the points at which steam is in equilibrium with (i.e. is saturated with) water. If water at the pressure A is heated, the system moves in the direction shown by the arrow and the water begins to change into steam when the temperature reaches B. If the temperature is now raised only a fraction above B, all the water will eventually change into steam; conversely, if it is lowered only a fraction below *B*, all the steam will eventually change into water. Hence, at a point on the phase boundary corresponding to the pressure *A* and the temperature *B*, water and steam are in equilibrium and the steam must be therefore saturated, since any water changing into steam will be balanced by the same amount of steam changing into water.

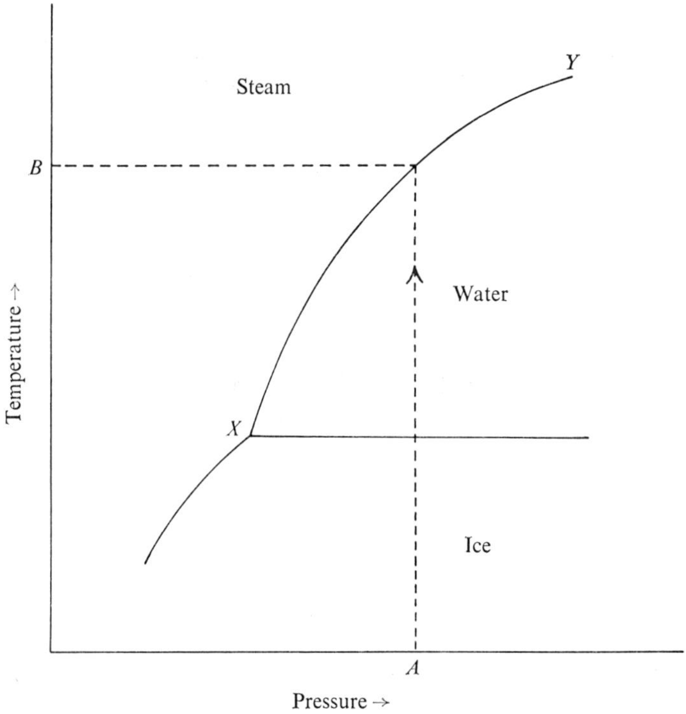

Fig. 4.4. Phase diagram of water. The line *XY* defines the conditions in which steam is saturated, i.e. where steam is in equilibrium with water. The other symbols are explained in the text.

If the system lies above the phase boundary, *XY* in Fig. 4.4, the steam is said to be *superheated*. Although such steam heats any cooler object it encounters, sterilization may not be achieved because the amount of heat liberated by superheated steam on cooling without condensation is far less than the latent heat released by the condensation of saturated steam. At the same time, moisture passes into the atmosphere from the load which may therefore be subject to dry, not moist, heat. Nevertheless, slight superheating does not invariably impair sterilization (Savage, 1937) but moderate degrees are clearly disadvantageous (Fig. 4.5; and also Shull & Ernst,

1962). Steam superheated to a still greater degree is sporicidal but acts effectively as dry heat (uppermost region of Fig. 4.5).

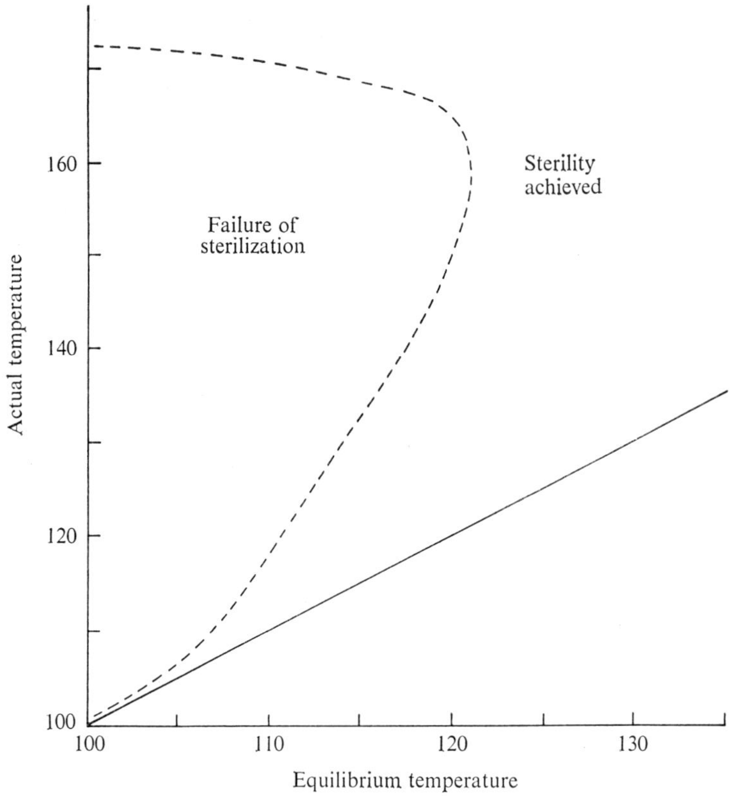

Fig. 4.5. Effect of superheating on sterilization (after Savage, 1937). Samples of earth were exposed to moist heat for 10 min. Ordinate: actual temperature of the steam. Abscissa: equilibrium temperature of the same steam. The continuous line joins corresponding points on the axes and therefore indicates saturated steam. Above this line, the steam is superheated. The dashed line encloses the region in which sterility was not achieved. Sterilization did not invariably fail when superheating occurred, since the lower border of the dashed line and the continuous line did not coincide. Moderate superheating did, however, impair sterilization. Extreme superheating again sterilized the samples but presumably acted as dry, not moist, heat.

Superheated steam may arise from the supply or in the autoclave, and is further considered on p. 107.

The phase diagram shows that pure saturated steam has a temperature of 121° at a gauge pressure of 15 lb./sq. in. (Table 4.7), i.e. at 15 lb./sq. in. above atmospheric pressure (14·7 lb./sq. in. at sea level) or an absolute pressure of *ca.* 30 lb./sq. in. This temperature, and those shown in Table 4.7, hold only for *pure* steam and not for steam mixed with air—for example air trapped in the autoclave during sterilization. This follows from Dalton's Law of Partial Pressures which states that the total absolute pressure of a mixture of steam and air equals the sum of their individual absolute pressures. Thus, the more air present for a given total pressure, the lower

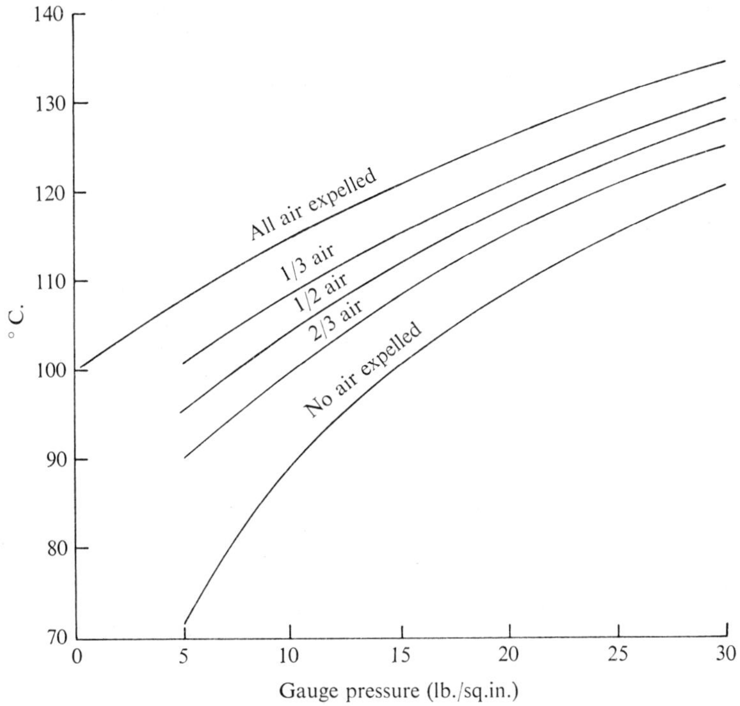

Fig. 4.6. Temperatures of mixtures of air and saturated steam at various pressures. Ordinate: temperature. Abscissa: gauge pressure.

will be the partial pressure of steam, and the lower will be the temperature of the mixture. Suppose, in a simple laboratory autoclave, that the chamber exhaust valve is mistakenly closed before all the air has been expelled and pure saturated steam then admitted until the chamber pressure reaches the desired figure. The fractions on the curves of Fig. 4.6 show the amount of air not expelled by the time the valve is closed. If f is the fraction of air not expelled, its absolute pressure is $f \times$ the absolute atmospheric pressure (say, 14·7 lb./sq.in.). If the absolute pressure of the mixture of air and saturated steam is P, the absolute partial pressure of the steam is $P - 14{\cdot}7f$ lb./sq. in.

The temperature of the mixture corresponds to this pressure of pure saturated steam. Thus, if half the air is not expelled, its absolute partial pressure $= 14{\cdot}7/2 = 7{\cdot}4$ lb./sq. in. If P is 20 lb./sq. in., the absolute partial pressure of the steam in $20 - 7{\cdot}4 = 12{\cdot}6$ lb./sq. in. The temperature of the mixture is therefore that of pure saturated steam at 12·6 lb./sq. in. which Table 4.7 shows to be about 96°.

This necessity for removing air constitutes one of the major drawbacks to the autoclave, since air is only too easily trapped within part of the load which will not be heated to the correct temperature and may not be sterilized.† This difficulty is probably encountered less in laboratory work than in clinical practice where complex loads consisting of bundles of gauze or linen have to be sterilized, but it has always to be borne in mind.

Removal of air. Air can be removed from the chamber and its load either by pumping or by gravity (downward) displacement. In a modern pumped autoclave, as conceived originally, the vacuum pump would rapidly evacuate nearly all the air and then, by using steam at 30 lb./sq. in., a short sterilizing period of perhaps only 3 min. would be feasible, the whole cycle being automated to give strict control of each step. In this way, it was hoped that trapping of air would be abolished while introducing economies in time and skilled manpower. These hopes have not been completely fulfilled. Under routine conditions, pumps proved unreliable and chambers tended to leak (see Knox & Pickerill, 1964), especially since they had to be evacuated to as little as 20 mm. Hg (Knox & Penikett, 1958). More surprisingly small loads were found to be sterilized far less efficiently than large ones. The 'small package' effect is due to residual air re-entering the load when steam is admitted to the chamber (Henry & Scott, 1963; Knox & Pickerill, 1964) and is not encountered with large loads which occupy more of the chamber, partly because there is then less air to be removed and the residual air is distributed over a greater volume of load, and partly because loads of fabric like cotton gauze contribute significant amounts of water vapour to the atmosphere of the chamber which further helps to displace air ('boiling out'). It thus became clear that the critical factor in the autoclaving of surgical dressings and other fabrics was not the final chamber pressure achieved by pumping, which represented air and water vapour, but the corresponding partial pressure of air alone. Problems of air penetration are not encountered with glassware. However, with soda

† The autoclave may be used with no air expelled for coagulating egg or serum media. When run at 5 lb./sq. in. with all air in, the temperature is 72°. At the end of the run, the pressure must be lowered extremely slowly to prevent the medium frothing.

glass, the rapid changes in temperature inherent in short cycles lead to severe losses from breakage. All in all, therefore, pumped autoclaves sterilizing at 134° have certain drawbacks, some of which can be overcome by technical improvements like substituting steam ejectors for vacuum pumps (Knox & Pickerill, 1964), or using separate oil-seal pumps for air and for steam respectively, to avoid contaminating the oil with water vapour (Henry & Scott, 1963). Nevertheless, many of the ancillary features of modern pumped autoclaves, like automated controls and filtration of air admitted at the end of the run, represent valuable improvements.

Downward displacement autoclaves work on the principle that, steam being less dense than air (their densities are 0·598 and 0·946 mg./ml. respectively at 100°), steam introduced at the top of an autoclave displaces air downwards so that it leaves an exit at the bottom of the chamber. For this to occur, it is evident that each part of the load must be arranged so that air can leave. For example, air cannot leave a closed tin or a flask placed upright in the chamber: the interior of either is only heated to the expected temperature by conduction from their outer surfaces. Steam must necessarily be able to enter the load whose wrappings should therefore be permeable. Metallic foil, glazed waterproof cellophane, or manilla paper (which swells on wetting) are unsuitable, whereas Kraft paper or autoclavable plastic film are satisfactory (Hunter, Harbord & Riddett, 1961). Even a suitable wrapping becomes impermeable if its surface is covered with water. Penetration of steam is also delayed by air trapped within material like bundles of dressings (see Perkins, 1956; Report, 1959, 1960*b*).

Running of the autoclave. The essential parts of a downward displacement autoclave are shown in Fig. 4.7. The chamber should be lined with nickel steel, not mild steel, to resist corrosion by broth and saline.

When autoclaving solid objects, like empty bottles or filters, steam is first allowed to enter the jacket, but not the chamber. The walls of the chamber are thus already heated when steam enters which avoids needless condensation and wetting of the load. As steam enters the chamber, air leaves through a discharge line running from the most dependent part of the chamber. This line contains a steam trap (Fig. 4.8) which allows air and water, but not steam, to pass through to waste. After steam has entered the chamber for a time, all air should have left the load, gathered on the floor of the chamber, and finally left through the discharge line which ends over the air break at *W*. The air break prevents backflow of contaminated material from the waste into the chamber and allows water to be seen dropping from the end of the discharge line, indicating that this is clear. An essential part of an autoclave is a thermometer placed just above the steam trap in the discharge line at *T*, for this measures the temperature at the most dependent and hence the potentially coolest part of the chamber. When all air has been expelled, the steam trap closes permanently, the pressure increases to the selected value, and the thermometer in the discharge line should show the temperature expected from the chamber pressure, if air is

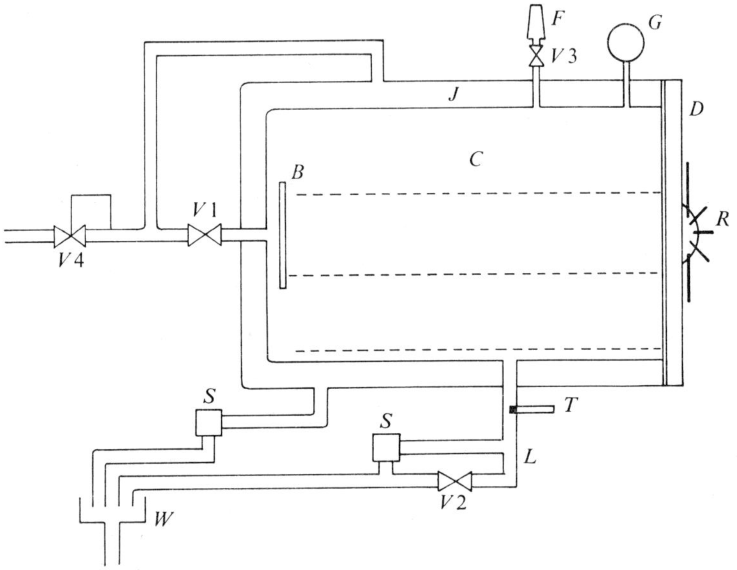

Fig. 4.7. Diagrammatic longitudinal section of a downward-displacement autoclave. *B*: baffle. *C*: chamber containing perforated shelves. *D*: door closed by the capstan, *R*. *F*: air filter, fitted with the non-return valve, *V*3. *G*: pressure gauge. *J*: jacket. *L*: chamber discharge line containing the thermometer, *T*. *S*: steam trap (see Fig. 4.8). *V*1 and 2: valves controlling entry and release of steam from the chamber. *V*4: reducing valve, automatically closed by the feedback when the pressure on the chamber side reaches the intended figure. *W*: waste; the airbreak prevents contaminated liquid being sucked back into the chamber when a vacuum forms.

completely absent (Fig. 4.6). If the temperature is less than expected from the pressure, air is still in the chamber, for example, because the steam trap is defective. However, a correct reading in the discharge line cannot prove that air is not trapped in parts of the load. This can only be checked by direct tests on the load during autoclaving (p. 109).

When the correct chamber temperature is reached, timing begins, and the pressure is maintained long enough for the contents to reach the intended temperature and for sterilization to occur (Tables 4.3, 4.4). At the end of the run, filtered air is admitted to the chamber until the gauge pressure has returned to zero, while the jacket is kept filled with steam to prevent wetting of the load by condensation formed on the walls of the chamber.

Liquids tend to boil out of their containers while the chamber pressure falls at the end of the sterilizing period. This occurs because liquids cool relatively slowly while they are within the chamber, so that their vapour pressure may easily exceed the chamber pressure and boiling results. For

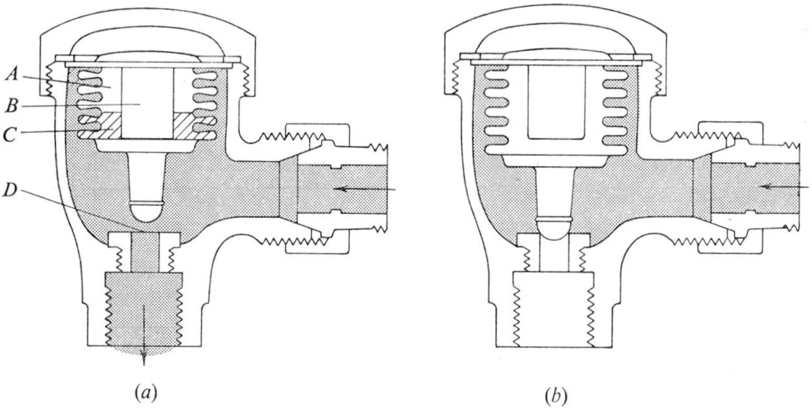

Fig. 4.8. Longitudinal section of a balanced pressure steam trap (by courtesy of Drayton Castle Ltd). *A*: evacuated metal bellows. *B*: stop. *C*: water. *D*: valve. The arrows indicate the direction of flow.

(*a*) As long as air or condensed steam enters the trap, the water in the bellows remains liquid and the bellows remains contracted against the stop. The valve is therefore open, and air and condensate pass through to waste.

(*b*) When steam enters the trap, the water in the bellows vaporizes and the consequent rise in pressure extends the bellows and shuts the valve. Steam is therefore prevented from passing to waste.

the same reason, sealed containers holding liquids may burst as the chamber cools. Cooling is encouraged by not using the jacket, and liquids will then often cool fast enough to allow air to be admitted *slowly* to the chamber through the valve, *V*3, in Fig. 4.7. About 5% of the liquid evaporates during the run.

Some of the damage attributed to autoclaving is probably caused, not during the sterilizing period proper, but during the relatively long time that the load is exposed to high temperatures while air is expelled and, later, during cooling. The first phase is considerably shortened when the air is removed mechanically, while the second phase can be shortened by replacing steam in the chamber with filtered air under pressure at the end of the sterilizing period or, when the load consists of bottles, by spraying with water.

Although this outline of the principles of the autoclave has not covered many important points, such as the quality of the steam supply and the discharge of condensate from the jacket, it provides some criteria for practical work. Much sterilizing equipment in general use obviously lacks many of the features mentioned above, the most extreme example being the domestic pressure cooker, often used in the laboratory. Many laboratory autoclaves have a valve for controlling the flow of steam to the chamber; discharge and safety valves; and a pressure gauge, often inaccurate: but no jacket, steam trap, or thermometer in the discharge line. Such autoclaves are operated by closing the chamber, opening the discharge valve, and allowing steam to run through the chamber for 5–10 min. until all the air is thought to have been expelled. The correct time can be estimated roughly by connecting the discharge line to a bucket of water which bubbles as long as air passes, whereas steam condenses without bubbling. The discharge valve is then closed and the pressure gauge watched until the pressure reaches the required figure. The inflow of steam is cut down until the pressure stays steady and the autoclave then left for a period equal to the time needed to heat the load plus the sterilizing time at that temperature. The steam is then cut off completely and, without opening the discharge valve, the chamber allowed to return gradually to atmospheric pressure when the load is removed. Liquids, including molten agar, usually boil out of their containers if the pressure comes down too quickly. Solid objects, like glassware or metal bowls, are naturally unaffected and are usually drier if the pressure is let down rapidly by opening the discharge valve at the end of the sterilizing period. Needless to say, no attempt should be made to open the door of the chamber while the pressure within is above atmospheric.

Despite the theoretical objections to many laboratory autoclaves, sterilization failures are not common in our experience. This is probably not due, however, to the sterilizing process reaching the correct standard but to the loads being moist and of relatively simple structure so that steam penetration is not hindered. Furthermore, it should be remembered that, although spores are the most heat-resistant organisms encountered, many types die rapidly even in boiling water. This apparent success must not be allowed to disguise the fact that the minimum standards have to be attained consistently to avoid accidents, of which the most serious occur in sterilizing discarded cultures.

Times required to reach sterilizing temperatures. An empty vessel placed with an opening downwards through which air can escape will reach the intended temperature very rapidly owing to condensation on both its inner and outer surfaces. If contained air is not removed, the inner surface will be heated more slowly by conduction from the outer surface.

Table 4.4. *Times taken by various volumes to reach sterilizing temperatures in the autoclave*

Volume of liquid	Time (min.)
100 ml. in bottle	10[1]
568 ml. in bottle	32
5000 ml. in flask	55
50 ml. in conical flask	1[2]
200 ml. in conical flask	3
500 ml. in conical flask	8
1000 ml. in conical flask	12
2000 ml. in conical flask	20

[1] Report (1959). Time required to raise temperature from 20° to 115°.
[2] Perkins (1956, fig. 171). Time required to reach the sterilizing temperature after autoclave is filled with pure saturated steam at 121°.

Liquids are fairly easy to sterilize by autoclaving, for contaminating organisms are already moist and have only to be heated to the correct temperature to be destroyed. Various figures are quoted for different volumes and temperatures (Table 4.4). These figures show that if different volumes of liquid are autoclaved together, the smaller volumes will have to be heated for unnecessarily long periods if the larger are to be sterilized. Ideally, each load should consist of objects of much the same size. The time required for surgical dressings depends largely on the way they are packed and arranged (see Perkins, 1956; Report, 1959).

Sources of superheated steam. The relatively low sterilizing power of superheated steam has already been mentioned. This may be present in the chamber even when the steam supply is saturated, for example if the jacket is hotter than the chamber, either because both are at the same pressure but air remains in the chamber or because the pressure in the jacket exceeds that in the chamber although neither contains air. In either case, the steam in the chamber is heated above the equilibrium temperature and moves above the phase boundary. Another important cause of superheating is the heat released when dry cotton combines with water produced by condensing steam (Henry, 1959).

Steaming and intermittent sterilization

As the temperature reached in steaming does not exceed 100°, a single exposure cannot be relied upon to produce sterility if contaminating organisms are in a neutral environment, though sterilization is greatly enhanced by acid or alkali, a point taken advantage of by boiling in sodium carbonate solution. However, even neutral liquids are sterilized by the method of intermittent steaming or fractional sterilization (tyndallization), introduced by Tyndall (1877) and applied in his refutation of the theory of spontaneous generation (see Bulloch, 1960).

The liquid is brought to 100° and left for 10 min. All vegetative cells are killed, leaving only spores viable. The liquid is then cooled to a temperature favouring spore germination (e.g. 30°) and, after a few hours, is steamed again. When this cycle has been repeated 2–3 times, all the spores present initially will have germinated only to have been killed in the subsequent steaming. The efficiency of the method is increased by the fact that heating often activates the germination of spores which otherwise might have remained dormant. This method depends on the occurrence of germination, and is therefore likely to fail with non-nutrient or inhibitory liquids or on holding the liquid at an unsuitable temperature between steamings.

Steamers are run on piped steam or are self-contained. The Koch and Arnold steamers are of the latter type which consists of an insulated container with boiling water placed at the bottom under a grid which supports the load. They differ in the design of the lid which is arranged so that condensate runs back inside the steamer.

Boiling

Boiling cannot be relied upon to kill spores suspended in neutral solutions and it is therefore used only occasionally. Spores are said to be killed by boiling for 10 min. in 2%, w/v, sodium carbonate solution, but some undoubtedly survive.

Pasteurization

Heating to temperatures of less than 100° is the basis of the various pasteurizing processes used in food preservation (see Ball & Olsen, 1957). The only laboratory use is in killing bacteria in stocks of heat-resistant phages, like the T-even coliphages and the A1 and A2 phages of *Salmonella typhimurium*, by heating to 56°–60° for 40 min.

DRY HEAT

Dry heat is considerably less lethal to organisms than moist heat at the same temperature (Table 4.3), and causes more damage to many materials like rubber, fabric, and paper. It is therefore generally used in sterilization of items not penetrated by steam, such as closed tins of Petri dishes, assembled syringes, and anhydrous hydrophobic material like greases and powders. Glassware and other heat-resistant material is also sterilized by this method.

The usual apparatus is some form of oven. The main practical difficulty is to heat the load uniformly to the predetermined temperature, and experience shows that this cannot be done unless the load is spaced out on perforated shelves and the air within the oven is forcibly circulated by a fan. Unless these precautions are taken, some parts of the load may never reach the nominal oven temperature within the time allowed for the run (Darmady *et al.* 1961).

Accepted time-temperature combinations for hot air sterilization are given in Table 4.3. As in autoclaving, these are minimum times for the run as a whole, since extra time must be allowed for the load to reach the sterilizing temperature. The length of this extra period depends largely on the oven and the load, but is unlikely to be less than 1 hr. after the oven thermometer reaches the sterilizing temperature. At the end of the run, the oven should be allowed to cool with the door shut to avoid violent mixing of contaminated cold air with the sterile contents of the oven.

The temperature of a sterilizing oven should never rise above 160°, for tarry compounds and fatty acids inhibitory to bacteria may distil out of cotton wool and some kinds of paper (p. 56). They are also released in smaller quantities at lower temperatures and are deposited on the walls of the oven as it cools, where, after a time, quite large amounts accumulate only to spoil future loads if the oven is not regularly cleaned.

CONTROLS ON THE EFFICIENCY OF STERILIZATION

Since spores are the most heat-resistant organisms, it might be thought that sterilizers should be judged by their ability to kill spore suspensions. This is not so, however, for the heat-resistance of spores depends so much on incidental factors, like the environment in which sporulation occurred and the cultural conditions after heating, that reproducible results are difficult to obtain (see Curran, 1952; Kelsey, 1958). If spores are to be used, standardized preparations are sold (Kelsey, 1961 *a*), or contaminated soil or hay should be used since spores formed under natural conditions are usually more heat-resistant than those produced in the laboratory.

The alternatives to spores include direct measurement and the chemical indicators described below. Both show immediately if the sterilizer is working correctly, whereas heated spore suspensions require at least 3 days' incubation before they can be accepted as sterile.

Physical and chemical indicators

Thermocouples can be placed within different parts of the load with leads attached to recorders outside the sterilizer. This is the most satisfactory method, but is probably too complex for routine use.

Table 4.5. *Times in minutes required for Browne's tubes, Types* I–IV, *to change from red to green at various temperatures (manufacturers' data)*

	Type of tube[1]			
Temperature	I (black spot)	II (yellow spot)	III (green spot)	IV (blue spot)
120°	16	9	—	—
125°	10	6	—	—
130°	7	4	—	—
134°	5	3	—	—
160°	—	—	60	45
170°	—	—	31	23
180°	—	—	16	12
190°	—	—	8	6·5

[1] Individual tubes are identifiable by coloured spots.

Chemical indicators. Browne's tubes (Albert Browne Ltd, Chancery Street, Leicester) are sealed glass phials enclosing a red liquid which changes on heating to amber and finally, when sterilization is complete, to green. The liquid can be made to any specification but there are four standard types. With each, the time taken for the green colour to appear varies directly with the temperature (Table 4.5, Fig. 4.9). Type I is designed for autoclaves working at less than 126°; Type II is for pumped autoclaves working at over 126° (the designation, Type II, was previously used for a tube more sensitive than Type I—see Kelsey (1958)—but this has been superseded); Type III is for hot air ovens at 160°; and Type IV is for infra-red sterilization. Types I, III, and IV can be stored at room temperature for about a year; Type II can be stored at not more than 21° for up to six months. If these precautions are ignored, the results may be erroneous (Brown & Ridout, 1960).

Patent indicator papers placed within the load or stuck to its outside show a colour change after heating for the correct time. They are probably best considered as indicators that the object has been heated, rather than as accurate controls of the sterilizer (Brown & Ridout, 1960). One type is the

'Klinter' paper made by Robert Whitelaw (Newcastle) Ltd, Klinter House, Newcastle-upon-Tyne. Bowic, Kclscy & Thompson (1963) recommend the 'type 1222' paper made by the Minnesota Mining and Manufacturing Co. for autoclaving at 134° and 126°.

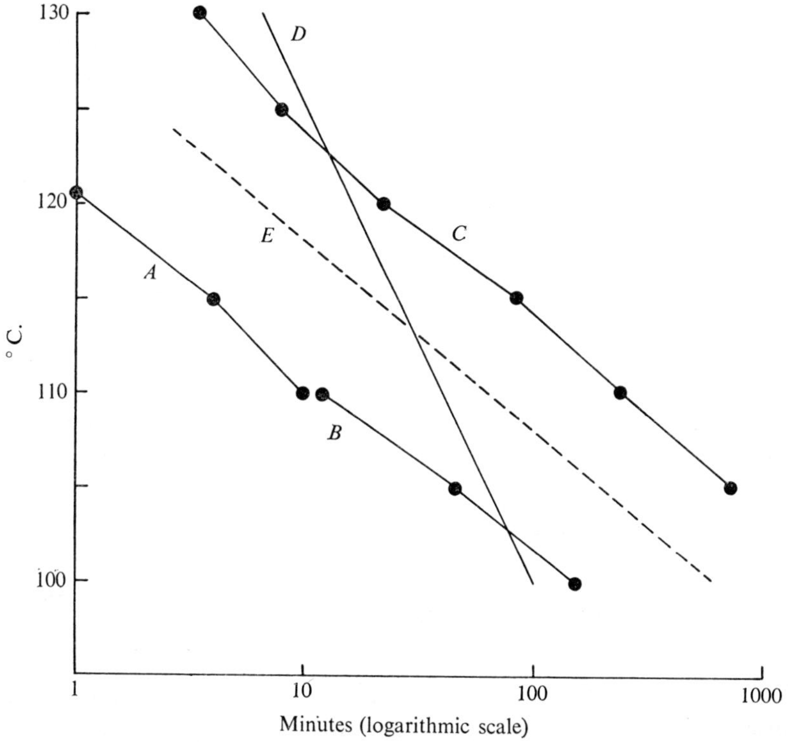

Fig. 4.9. 'Thermal death times' of spores exposed to moist heat at various temperatures (after Kelsey, 1958). Ordinate: temperature. Abscissa: log T.D.T. in min. (see p. 93). *A*: *Clostridium oedematiens*. *B*: *Cl. sporogenes*. *C*: an unspecified thermophile, a type unlikely to cause contamination in the laboratory or in medical work. *D*: Browne's Type I Tube (Table 4.5). *E*: ideal indicator, which would have the same slope as curves *A* and *B* but would be displaced slightly to the right for safety. The value of Q_{10} for the spore shown here is about 10 whereas for Browne's Type I Tube it is only about 2·5.

The ideal chemical indicator would be slightly more heat-resistant than spores at all temperatures, the difference representing a margin of safety. The behaviour of such an indicator is depicted by the dashed curve, *E*, in Fig. 4.9, but when this is compared with the curve for Browne's Type I tubes, they are seen to coincide only at 114° because of differences in the slopes of their curves (i.e. in their respective values of z or Q_{10}: p. 94).

At lower temperatures the tubes turn red too soon, which overestimates the efficiency of the autoclave; and at temperatures higher than 114°, the tubes change colour too slowly. Despite this disadvantage, Type I tubes are the best simple indicator available (Kelsey, 1958).

Whichever test is used, the indicator must occupy the position least favourable for sterilization. For example, the bottom of an autoclave where air is most likely to persist; the centre of a crate of tubes rather than the edge; or immersed in a liquid rather than placed on the outside of its container.

Failures in sterilization. If the load should be contaminated after sterilization, the contaminating organisms must be identified for their type may show their origin. Spore-forming organisms, like *Bacillus*, are consistent with poor running of the sterilizer; but a non-sporing species, like a *Micrococcus*, suggests contamination following sterilization, as virtually any sterilizer may be relied upon to kill vegetative cells. Sources of recontamination include influx of contaminated air into the autoclave after the sterilizing run, prevented by passing the air through a filter (Report, 1960*b*); wet or dislodged cotton wool plugs in bottles of medium (prevented by covering plugs with Kraft paper during autoclaving); and, possibly, streaks of dried medium between a plug and the mouth of the container down which air-borne organisms eventually grow.

CHEMICAL DISINFECTANTS

Chemical disinfectants are rarely used in preparative sterilization, as the majority cannot be removed from the material, but they are generally chosen for discarded objects like plastic centrifuge tubes that melt in the autoclave or pipettes which become difficult to clean after autoclaving because broth cakes on the inner surface. Of the enormous range of compounds available (see McCulloch, 1945; Sykes, 1965; Report, 1965; Whittet, Hugo & Wilkinson, 1965) only lysol, hypochlorite, formaldehyde, mercurial compounds, and certain volatile agents are considered here. Detergents should generally be avoided as they often adsorb to glassware and are difficult to remove.

Non-volatile disinfectants

Lysol and other phenolic disinfectants. The phenolic disinfectants isolated from coal tar are poorly soluble in water and are therefore either dissolved in soap solution to give 'black fluids' including lysol (Liquor Cresolis

Saponatus, B.P., U.S.P., N.F.) or emulsified with a colloid like gelatin or dextrin to give 'white fluids'. In lysol, the main bactericidal component is cresol emulsified by the detergent activity of soap, which, like any surface-active agent, may form micelles if its concentration is too high, i.e. aggregates of the surface-active molecules arranged with the hydrophilic groups on their surface. Since the cresol molecules may be enclosed in such micelles, increase in concentration may lower the antibacterial activity of these disinfectants.

Concentrated solutions of lysol are clear and dark brown in colour but turn milky on adding water. The solutions are highly corrosive and should not come in contact with the skin. For the same reason, the solutions must not be autoclaved, for the chamber becomes filled with vapour which may injure the operator when the door is opened. The usual working concentration of 3–5%, v/v, rapidly kills vegetative cells but is not very effective against spores. Its activity is little diminished by organic matter.

A common use of lysol is in disinfection of water-repellent surfaces like waxed floors or animal fur. As it adheres to glass, it is less suitable than hypochlorite for sterilizing contaminated pipettes.

Hypochlorite. This is often sold as an acid solution of sodium hypochlorite (NaOCl) with a trace of permanganate (e.g. Chloros, I.C.I.). The bactericidal power of hypochlorite rapidly falls in alkaline solution, probably because its action is due to undissociated hypochlorous acid, HOCl, a strong oxidizing agent formed in the reaction

$$OCl^- + H_2O \rightleftharpoons HOCl + OH^-.$$

A rise in pH therefore decreases the proportion of acid present in the active form. Permanganate acts as a visible indicator of the state of oxidation of the solution, for when the hypochlorite is totally reduced to chloride and is no longer bactericidal, the permanganate is also reduced and becomes colourless. Chloros is therefore only effective as long as it remains red. Its action is lowered by the presence of organic matter, with which it combines.

Vegetative cells and spores of certain species, including *Bacillus anthracis*, are rapidly killed by hypochlorite solutions diluted to 200 parts per million of available chlorine.

The chief use is in disinfection of pipettes. If these are subsequently to be autoclaved before removal from hypochlorite, it must first be inactivated by adding thiosulphate, for, unless this is done, the autoclave becomes filled with gaseous chlorine. Chloros corrodes metal and is unsuitable for the sterilization of instruments.

Formaldehyde. The usual form is the aqueous solution although the gas is used for disinfection of rooms (Report, 1958; Wilson *et al.* 1958). 'Formalin' is the saturated aqueous solution containing 37–41 %, v/v, formaldehyde and a stabilizer such as methanol to prevent the formation of insoluble polymers on standing.

A 3·8 %, v/v, solution of formaldehyde (i.e. 10 %, v/v, formalin) kills both spores and vegetative organisms in a short time, even in the presence of organic matter. A concentration of 0·1 %, v/v, formaldehyde kills vegetative cells in several hours. Its action at low concentrations is said to be bacteriostatic and to be reversible by sodium sulphite. A concentration of 0·5 %, v/v, formaldehyde is used to sterilize suspensions of enterobacteria whose flagellar antigen is to be preserved.

Mercurial compounds. These owe their antibacterial activity to their ability to combine with sulphydryl groups. As the union is weak, their action is easily reversed by dilution or by the addition of sulphydryl compounds like thioglycollate, which compete for Hg^{2+}. Considerable amounts of mercurial ion are taken up by the organisms so that the effective concentration of the disinfectant is partly determined by the bacterial concentration.

Mercurial compounds are now largely used as preservatives. $HgCl_2$ is effective in a concentration of 1/1000–1/5000, w/v, but organic compounds with some surface activity are now more common. One is sodium ethyl mercurithiosalicylate ('Merthiolate', Eli Lilly and Co.), used at a concentration of 0·01 %, w/v, for maintaining the sterility of sera.

Volatile disinfectants

A reliable disinfectant that could be removed simply by gentle warmth or in a vacuum would be extremely valuable for sterilizing solutions of heat-labile compounds (see Opfell, 1965). At present, however, only liquid ethylene oxide or β-propiolactone appear in the least practicable, and even these undergo spontaneous hydrolysis in aqueous solution to yield stable products exerting anti-microbial effects (Toplin & Gaden, 1961). Neither is therefore likely to be useful for culture media, but may find a place in the sterilization of concentrated solutions (e.g. of sugars or vitamins) that are considerably diluted before use. Chloroform and a few other compounds are used to prevent the growth of contaminants in solutions that have to be stored before sterilization by autoclaving.

Ethylene oxide. The liquid boils at 10·7° but can be used for sterilizing material like serum or milk in the cold. 1 volume of liquid ethylene oxide, cooled to 0°–4° in a sealed vessel, is transferred with a *chilled* pipette to 100–200 volumes of the material, also at 0°–4°, and left for 1 hr. in the cold. The oxide is then removed in a fume cupboard by warming for 24 hr. at 37° or for a shorter time at 45° (Wilson & Bruno, 1950; Judge & Pelczar, 1955). Hydrolysis produces ethylene glycol.

The gas is now widely used for sterilizing heat-susceptible articles, like plastic Petri dishes, but as it is explosive and toxic, special equipment is needed (Bruch, 1961; Kelsey, 1961*b*).

β-Propiolactone. The liquid is sold as 96 % and 99 % pure. It boils at 155° and hydrolyses in aqueous solution to give β-hydroxy-propionic acid: the half-life is 210 and 20 min. at 25° and 50° respectively. The gas is a powerful lachrymator (Hoffman & Warshowsky, 1958). The liquid has been used to sterilize solutions in a final concentration of 0·2–0·5 %, v/v (see Toplin & Gaden, 1961). Himmelfarb, Read & Litsky (1961) used 0·2 %, v/v, to sterilize 20 % sugar solutions by incubation at 37° for 2 hr., the mixture being shaken every 10–15 min. The sterilized solutions were diluted 1/40 in culture medium on the following day.

Chloroform. Organic solvents, like chloroform, toluol, xylol, acetone or ether, kill enterobacteria almost instantly, though their action on other organisms is less consistent (Bray, 1945; Hutner & Bjerknes, 1948). Chloroform has the advantage of not being inflammable and is often used in a final concentration of 1·5 %, v/v, in preparing broth. Phage stocks grown on Gram-negative bacteria are often sterilized in the same way, but it is not invariably inert as it inactivates filamentous phages; moreover, it reverses the mutant host-range phenotype of phage λ*h* (Newcombe & Rhynas, 1958) and increases the sensitivity of phage T2*r* to X-rays (Cotton & Lockingen, 1963). Chloroform vapour is also bactericidal, so that bacterial colonies or phage plaques can be preserved by putting a few drops of the liquid in the lid of a glass petri dish and leaving it for 30 min. at room temperature. Plastic dishes dissolve. Exposure for up to a minute is sometimes useful for killing swarming organisms carried on to the surface of overlays or pour plates.

Other volatile compounds. Hutner & Bjerknes (1948), after testing a large number of possibilities, found that solutions remained sterile if they contained 1 %, v/v, of a mixture of 1 vol. *o*-fluorotoluene, 1 vol. 1:2-dichlorethane and 2 vol. *n*-butyl chloride. These are subsequently driven off by autoclaving at 121° for 15 min.

FILTRATION

Bacteriological filtration is almost always concerned with freeing liquids from unwanted suspended particles. In coarse filtration, as in preparing broth, this is achieved by sieving-out the particles on the filter. But in sterilization, where the particles are far smaller, a large part of the filter's action is due to adsorption of particles to the large area presented to the liquid by the walls of the filter pores. The mean size of particle retained is then often smaller than the average pore diameter (a.p.d.) of the filter (see Ferry, 1936; Elford, 1938). Another common finding inexplicable by a sieving mechanism is that the performance of filters of all kinds is largely

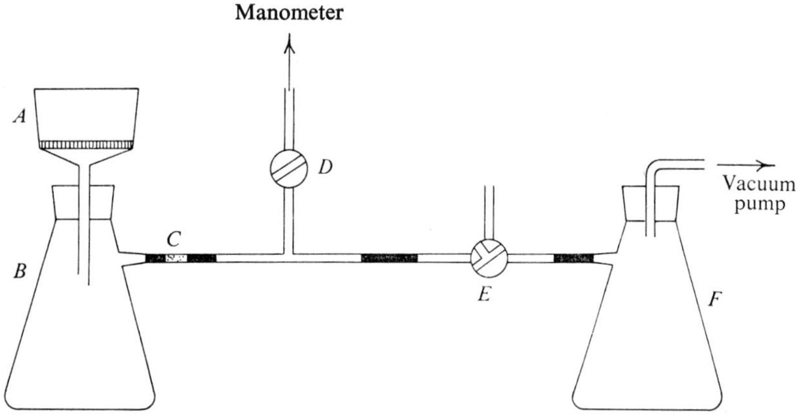

Fig. 4.10. Filter and vacuum line arranged for filtration under negative pressure. *A*: sintered glass filter mounted in Buchner funnel. *B*: filter flask as receiver. *C*: air filter of non-absorbent cotton wool held between constrictions in glass tubing. *D*: two-way glass stopcock leading to a manometer (e.g. to one limb of a U-tube half-filled with mercury with the other limb open to the air). *E*: three-way glass stopcock. *F*: filter flask acting as trap to catch water coming back from the filter pump.

The filter, receiver, and air filter are usually autoclaved together after being assembled loosely and closed by Kraft paper. The paper is removed when the receiver is connected to the vacuum line: the air filter prevents contaminated air entering the receiver. When drawing the vacuum, tap *D* is open while tap *E* is closed to the air. When the necessary vacuum is present, tap *E* is turned so that the manometer and receiver are isolated from the pump, which then sucks air through the sidearm and can be turned off without a backrush of water into the trap.

This vacuum line is also used for partial evacuation of anaerobic jars before filling with H_2 (p. 80) or CO_2 (p. 74). The lids of these jars carry two taps, one which is attached to the vacuum line in place of the filter, *C*, while the other is attached to the manometer (previously removed from *D*). The jar is then evacuated to a known amount after closing tap *D*. Tap *E* is adjusted as in filtration. If the jar is simply connected to the line in place of the receiver shown in the figure, the manometer often grossly overestimates the vacuum within the jar, presumably because the taps on its lid are of sufficiently small bore to produce a marked pressure gradient across them.

determined by the composition of the liquid phase, a point taken advantage of in freeing virus or phage suspensions from contaminating bacteria. Thus, a filter such as the Seitz filter has a finite capacity to absorb virus particles and may remove all the virus from a low titre stock though not from one of higher titre. However, all the virus may be recovered, even from a low titre stock, without affecting the retention of bacteria, if broth is first passed through the filter. The accepted explanation is that broth protein saturates the adsorptive capacity of the filter which can then no longer take up virus (Galloway & Elford, 1931).

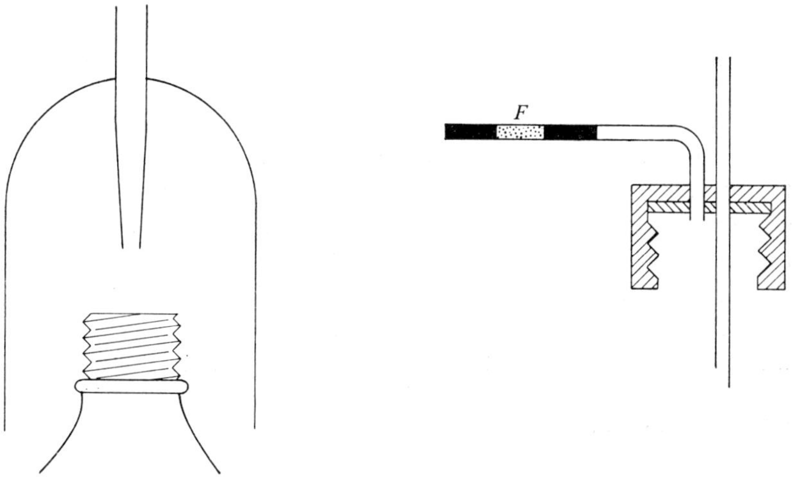

(*a*) (*b*)

Fig. 4.11. Two devices for the distribution of filtrates. (*a*) In pressure filtration, the filtrate can be collected from a pipette fused into a glass hood, as shown, which allows receivers to be interchanged without disturbing the filter. (*b*) In vacuum filtration, the filtrate can be collected in a screw-capped bottle closed by a screw cap fitted with a rubber liner, shown here in section, which carries a straight tube coming from the receiver and an angled tube leading to the pump via the cotton wool air filter, *F*.

Three types of sterilizing filter are commonly used for microbiological work although these represent only a fraction of the types now available, many of which like sintered stainless steel discs or plastic-impregnated paper for air filtration are likely to be of considerable value.

(1) Membrane filters (porous discs of cellulose esters);

(2) Seitz filters (discs of an asbestos–cellulose mixture). These and membrane filters can be set up to take any volume between a few ml. and many litres;

(3) Sintered (fritted) glass filters, composed of glass fragments fused into a disc, and usually used for relatively small volumes (e.g. 5–100 ml.).

For other filters and related apparatus, see Cummins & Hale (1956). Filter candles of unglazed ceramic are relatively cheap and are useful for large volumes.

Filtration is usually encouraged by a drop in pressure across the filter, produced either by applying a positive pressure to the unfiltered liquid or by connecting the receiver to a vacuum line (Fig. 4.10). Electric vacuum pumps are greatly to be preferred to water pumps whose performance fluctuates. Pressure filtration has the advantages compared to vacuum filtration that the receiver can be changed without disturbing the filter (Fig. 4.11 *a*) and that protein-containing filtrates do not froth. The receivers for negative-pressure filtration range from conical filter flasks or thick-walled test tubes fitted with sidearms, to screw-necked bottles connected to the filter by a special screw cap (Fig. 4.11 *b*). In Fig. 4.10 the bevel of the filter funnel faces away from the sidearm to prevent filtrate being sucked down the vacuum line. Contaminated air in the vacuum line is prevented from reaching the filtrate by a simple air filter attached to the sidearm of the receiver, usually a wad of non-absorbent cotton wool about 2 cm. long held in a glass tube between two constrictions. The end of the tube is covered with Kraft paper during autoclaving to keep the wool dry.

Membrane filters

These are made in a wide range of diameters and pore sizes (e.g. 5 μ–10 nm.) and have the great advantage of relatively high flow rates. The two principal makers are: the Membranfiltergesellschaft, Göttingen, Germany, and the Millipore Filter Corp., Bedford, Mass., U.S.A. Both supply several types of filter, the main varieties being for aqueous and for air filtration. For removal of bacteria from aqueous suspension, an a.p.d. of 450–200 nm. is recommended (M.F.G., MF 15 or 30; Millipore grade HA). The filters are thin membranes of cellulose esters, *ca.* 100 μ in thickness and of refractive index 1·5, that are resistant to water, dilute acid or alkali, but susceptible to methanol, ethanol, ketones, esters and ether.

Oxo Ltd supply a similar filter of rather larger a.p.d. whose refractive index is 1·47.

The filters are not always supplied sterile. All makers suggest, amongst other methods, placing the membranes in water or between moist absorbent pads in a container which is then autoclaved for 15 min. at 121°. For use, the sterilized membrane is transferred to a filter holder previously sterilized by moist heat. Membranes can also be autoclaved *in situ* in the holder but the makers suggest that this method is less reliable. Millipore state that the membrane may be damaged by direct contact with steam.

If unsterilized membranes are used for separation, not sterilization, they should be soaked in distilled water before use.

Seitz filters

The filter consists of an asbestos–cellulose pad with a floccular upper surface supported by a woven base (see Webb, 1946). The porosity increases, the higher the proportion of cellulose. Various grades and sizes are made, for example, by T. B. Ford Ltd and sold by Gallenkamp Ltd under the trade name of Ford's 'Sterimats'. Grade GS will retain *Serratia marcescens* (syn. *Chr. prodigiosum*): grade SB is finer.

Many types of filter holder are available from retailers or manufacturers (e.g. John C. Carlson, Ltd, Ashton-under-Lyne, Lancs., England, or Seitz-Werke, Kreuznech, Rheinland, Germany), extending from those taking one pad to filter presses taking numerous pads arranged in parallel, with possibly pre-filters as well. The usual laboratory holder has two halves held together by bolts and intersected by a perforated plate which supports a single filter pad. The liquid to be filtered is placed in the upper half. The lower half leads the filtrate to the receiver. The Uhlenhuth pattern is the most useful as it can be used for either pressure or vacuum filtration. Most filter holders are made of silver-plated metal or stainless steel, but a glass holder with stainless steel supporting disc for vacuum filtration is sold by Gallenkamp Ltd.

For sterilization, the holder, filter pad, and receiver are assembled with the bolts loose, and then wrapped in Kraft paper and autoclaved for 30 min. at 108° (5 lb./sq. in.). Before use, the pad is saturated with sterile saline and only then should the two halves of the holder be firmly screwed together by gently tightening each screw in turn, so that the two halves of the holder grip the pad symmetrically.

The drop in pressure across the pad should be kept as constant and as small as possible, and must not exceed 26 cm. Hg. (5 lb./sq. in.).

Sintered glass filters

Several grades are available, the finest grade with a.p.d. $\leqslant 2\mu$ being used for sterilization (No. 5 Pyrex). As this grade is not uniform, each filter is supplied with a certificate stating its individual a.p.d. Filters are available in many forms: for example, as Buchner funnels where the disc is fused into a complete funnel, or as Gooch crucibles which are fitted to a funnel by means of a rubber cone. The latter are convenient as they lessen the risk of breaking the whole funnel when it is removed from the receiver for cleaning. New filters should be rinsed with hot hydrochloric acid followed by distilled water to remove traces of glass and dust.

Used filters should be cleaned by sucking distilled water *back* through the filter, for example by inverting the filter on a wide bung perforated by a tube leading to a vacuum line. It can then be immersed overnight in concentrated sulphuric acid containing a little potassium nitrate and potassium perchlorate, followed, next day, by repeated washing with distilled water, recently boiled to remove dissolved air. Filters must *not* be cleaned by forcing water back through the disc by connection to the mains, as the pressure is too great and may damage the union between disc and funnel. Chromic acid should be avoided as Cr absorbs to glass and may damage subsequent filtrates. Strong alkali may increase the pore size.

Side effects of filtration

All filters may grossly alter the properties of the filtrate, quite apart from removing suspended particles. The inhibition of bacterial growth caused by passing broth through cotton wool, a source of fatty acid, is mentioned on p. 56; and inhibition has even been caused by filter paper impregnated with formaldehyde during manufacture.

Seitz pads raise the pH of unbuffered solutions by as much as 4 pH units, owing to their content of alkali which includes sodium silicate present for bonding (see Webb, 1946). Virus and phage particles are absorbed by pads of relatively high porosity, but this can be prevented, without affecting the retention of bacteria, by first filtering broth, gelatin, or, if a non-antigenic agent is needed, 0·04–0·05 %, w/v, sodium alginate (Hyslop, 1961). Enzymes are likewise absorbed by untreated pads. Citrated plasma usually clots after passing a Seitz pad, for, although the first fraction of the filtrate remains liquid, clotting subsequently occurs more rapidly, the more plasma has been filtered. This is because the pad removes fibrinogen and prothrombin from the first fraction, and then ceases to take up fibrinogen and begins to release thrombin formed by activation of absorbed prothrombin. Activation, and hence clotting, is prevented by bringing the plasma to pH 10·6 before filtration (Bushby & Whitby, 1942) or by treating it with kaolin (Maizels, 1944).

Sintered glass and membrane filters are by no means inert. Some phage stocks are inactivated by passage through glass or membrane filters, especially when the stocks are made in salts medium, not broth. Indeed, sintered glass filters have been used as model receptors in studies of phage adsorption (Puck, Garen & Cline, 1951). Some membrane filters contain detergent, which may be toxic (Cahn, 1967), and also glycerol, removable by boiling in three changes of distilled water, which enters the filtrate in appreciable amounts. The grid markings on these filters may inhibit the growth of *Pasteurella tularensis* to such an extent that the colony count is related to the heaviness of the imprint (Gasper & Leise, 1956). The colony count on membrane filters may also be lowered if the organisms are suspended before filtration in a solution of potentially toxic cations (e.g. 5×10^{-6}M-Cu^{2+}) which are adsorbed and concentrated in the filter (Shipe & Fields, 1954). On the other hand, low concentrations of cations such as Fe, Mg or Mn, raise the colony count of phenol-treated *Escherichia coli* on membrane filters to the figure obtained by direct inoculation of nutrient agar (see Harris, 1963).

IRRADIATION

Irradiation is used in the laboratory as much for its incidental effects like mutagenesis or phage induction as for its ability to sterilize. Of the variety of means available, only u.v. irradiation and photodynamic inactivation are practicable without taking considerable precautions. X-irradiation has been used relatively widely, as in phage work (see Stent, 1958), while gamma radiation by ^{60}Co sources is used on an industrial scale for sterilizing plastics and food (see Bridges, 1964; Whittet, Hugo & Wilkinson, 1965).

Ultra-violet radiation

The ultra-violet region of the spectrum covers wavelengths of less than 400 nm., but those less than 180 nm. are not in general use due to their absorption by atmospheric oxygen. All the common effects of u.v. light on micro-organisms, such as killing, mutation, or the induction of phage growth in lysogenic bacteria, are most efficiently produced by radiation of wavelength *ca.* 260 nm. (see Kleczkowski, 1951). This wavelength is also that maximally absorbed by nucleic acid bases, suggesting that absorption by these compounds is responsible for at least part of the action of u.v. light. The original interpretation, that the target for u.v. light is nucleic acid which also adsorbs maximally at this wavelength, has since been strongly supported by chemical and genetic studies with bacteria (see Moseley, 1968).

Intensity of irradiation is usually expressed in ergs/sec./sq. mm. (10^7 ergs/sec. = 1 joule/sec. = 1 watt). A simple biological method for measuring intensity is to determine the survival curve of phage T2 plated on *Escherichia coli*, strain B. Independent physical calibration shows that the ultra-violet dose which gives 1 % survival is 200 ergs/sq. mm. (Latarjet, Morenne & Berger, 1953).

The usual u.v. source in the laboratory is a commercial germicidal tube consisting of an electric arc operating in low-pressure mercury vapour. This emits at several discrete wavelengths to produce a line spectrum, *ca.* 80 % of the total u.v. emission being at 253·7 nm. which has 85 % of the biological activity of the most effective wavelength, 260 nm. In contrast, the high-pressure mercury lamps used for lighting have relatively little biological effect as their emission is spread more evenly throughout the u.v. region. The apparatus usually consists of a glass tube about 45 cm. long and 2 cm. in diameter mounted between electric contacts and run off the mains with a choke. A 15 watt tube is suitable for most laboratory work and gives a dose-rate of the order of 4 ergs/sec./sq. mm. at a distance of 91·5 cm. (3 ft.). The glass of the tube is chosen so as to transmit essentially all 253·7 nm. radiation, while absorbing all radiation of less than 200 nm. This prevents formation of ozone which is toxic to man in concentrations of *ca.* 0·05 p.p.m.: the minimum concentration detectable by smell is *ca.* 0·015 p.p.m.

Ultra-violet light must not reach the eyes, either directly or by reflection, as it produces severe corneal irritation with characteristic symptoms of lachrymation and photophobia coming on shortly after exposure. The symptoms disappear spontaneously within 24–36 hr. and no treatment is needed beyond rest with closed eyes. A safe and convenient arrangement is to enclose the tube in a deep box of non-reflecting material, open downwards, that is suspended from adjustable brackets mounted on a wall. This prevents scatter of u.v. light into the room while allowing the tube to be placed at different distances from the bench.

The intensity of irradiation received by the object depends on several factors besides the wattage of the tube:

(1) The age of the tube. The intensity falls rapidly in the first 100 hr. of use, and new tubes are usually calibrated as of 100 hr. age. Thereafter, the fall is less rapid, for example from 100 % to 80 % between 100 and 3000 hr.

(2) The surroundings of the tube. Appreciable reflection occurs from polished aluminium or chromium plate, and even from white paper.

(3) The distance of the tube from the object. For distances greater than

the length of the tube, the intensity is approximately inversely proportional to the square of the distance, while, for distances less than one third of the tube length, it is inversely proportional to the distance (Buttolph, 1955).

(4) The suspending medium of the organisms. Ultra-violet light is heavily absorbed by protein as well as by nucleic acid and free bases. Organisms grown in broth must therefore be washed or diluted at least 1/300 in buffer before irradiation. Dilution 1/100 is probably too little: a few measurements suggest that tryptic digest broth, 1 %, w/v, peptone water, and 1 %, w/v, acid-hydrolysed casein each diluted 1/100 in water absorb about 65 %, 25 % and 10 % of irradiation at 2530 Å.

(5) The concentration of the organisms. Dilute suspensions (e.g. 10^6 *Escherichia coli*/ml.) can be irradiated in shallow layers in watch glasses, but concentrated suspensions (e.g. 10^9 cells/ml.) should be rocked in glass Petri dishes during irradiation to prevent shielding of some of the organisms. Plastic dishes are useless as their surfaces are non-wettable and the suspension does not flow evenly.

Many organisms are inactivated exponentially (p. 196; for the radiobiological characteristics of many phages, see Stent, 1958). The apparent survival of bacteriophage, however, may depend on the number of inactivated particles inoculated on each plate, owing to the occurrence of multiplicity reactivation. Thus, if a series of plates are inoculated with consecutive dilutions of irradiated phage, the plaque counts from successive dilutions decrease disproportionately rapidly (Luria & Dulbecco, 1949: cf. dotted curve in Fig. 6.3). In general, the effects of irradiation are strongly modified by the state of the organisms before exposure and by their later treatment (see de Serres, 1961). In particular, exposure to fluorescent light or strong sunlight, or plating on a medium different to that in which the cells were grown before irradiation, can partly counteract its effects. The survival of bacteria may also be affected by their increased liability to inactivation by the diluent used for viable counts, as shown by the toxic effect of isotonic saline on irradiated *Escherichia coli* (Hollaender, 1943).

Although the killing effects of all types of irradiation are usually expressed in terms of loss of colony-forming ability, its results are not all-or-none as this method of measurement might suggest. Microscopy of irradiated cells placed on nutrient agar shows that, while some do not change in shape, others elongate without dividing to form 'snakes' (Payne *et al.* 1956; Hartman, Payne & Mudd, 1955). Diploid yeasts may divide several times after X-irradiation and then fail to divide further (Gunter & Kohn, 1956).

Photodynamic inactivation

Photodynamic effects comprise a wide range of biological and chemical phenomena, all resulting from the action of light on a substrate exposed to a sensitizing agent (Blum, 1941; Clare, 1956). Inactivation is a typical example in which bacteria or viruses exposed to a dye like methylene blue become susceptible to killing by visible light. Photodynamic effects also occur readily in non-living systems, as in the formation of toxic amounts of peroxide in eosin–methylene blue plates exposed to sunlight (Liebert & Kaper, 1937), while, with whole organisms, they include not only inactivation but less drastic effects like the induction of vegetative phage growth in lysogenic bacteria (Freifelder, 1966).

Sensitization is produced by many compounds, all of which are fluorescent, and the absorption spectrum of the compound concerned appears to determine the relative efficiencies with which different wavelengths produce their photodynamic effect. Oxygen is essential, and one model reaction postulates that the sensitizer is excited by light and transfers its energy to the substrate which then becomes able to combine with O_2. Although the course of the reaction might be expected to be first order, since it may be presumed to be initiated by single quanta, dose-response curves often have shoulders or tend to flatten out as the exposure increases.

Extremely small concentrations (10^{-4}–10^{-6}M) of sensitizing agent are effective. Convenient light sources are domestic 'daylight' fluorescent tubes or photoflood lamps (the latter also give out considerable heat). Masking occurs, as with ultraviolet irradiation, and the suspensions should be agitated during irradiation. However, protein in moderate concentration does not diminish photodynamic effects, which has the important practical consequence that the method can be used successfully with organisms suspended in broth or serum. It has therefore proved valuable for preparing killed virus vaccines (see Turner & Kaplan, 1965) and, since unrelated species of organisms differ considerably in resistance, contaminating viruses or bacteria can be killed without inactivating the virus of interest (Hiatt, 1960). The sensitizing agent is readily removed by passing the irradiated suspension through a column of exchange resin (Wallis, Melnick & Phillips, 1965). Despite its technical simplicity and fundamental interest, the microbiological aspects of photodynamic action have so far received relatively little attention although it would seem to have many applications.

Table 4.6. *Methods of sterilization for laboratory materials*

Material	Container	Method
Clean pipettes or petri dishes	Bulk: copper or aluminium boxes[1] with overlapping lids[2]	Oven
	Individual: Kraft paper wrapping	Oven
Glass centrifuge or test tubes closed with metal caps, cotton wool plugs, or foil	Wire racks	Oven
Glass flasks, beakers or bottles closed with all-metal caps or foil	Wire crates	Oven
Assembled all-glass syringes[3]	Glass boiling tubes with wool plugs; aluminium tubes	Oven
Empty glass bottles with rubber-lined screw caps[4]	Wire crates	Autoclave
Filter holders on flasks	Kraft paper wrapping	Autoclave
Most liquid and solid culture media, solutions of salts, sugars, etc.	Screw-capped bottles;[5] closed tubes, incl. ampoules	Autoclave
Heat-labile solutions	Screw-capped bottles;[5] closed tubes, incl. ampoules	Tyndallization; filtration; volatile disinfectants
Serum	Screw-capped bottles;[5] closed tubes, incl. ampoules	Volatile disinfectants; filtration
Serum-containing media	Screw-capped bottles; closed tubes	Tyndallization;[6] volatile disinfectants; filtration
Used petri dishes, tubes, flasks, filters, etc.	Metal discard bins with lids[7]	Autoclave[8]
Used pipettes	Jars[9]	Hypochlorite;[10] lysol; autoclave

[1] Boxes for pipettes should have glass wool at the base to prevent tips breaking.

[2] To prevent contamination by incoming air (Barton-Wright *et al.* 1936).

[3] See Memorandum (1962). Metal-glass joints may melt, even at 115° in the autoclave.

[4] Silicone rubber stands dry heat.

[5] Containers can be closed (as the contents are already moist and have only to be heated to the correct temperature by conduction) or open, to avoid breakages.

[6] These media coagulate on heating if the pH is less than *ca.* 7·4.

[7] Boxes must be sound and free from holes. For this reason, no acid solutions, for example hypochlorite, should be placed in them. Lids with deep overhanging rims are left on when the box is not in use to prevent entry of insects attracted by culture medium. Nothing should be touched or removed before autoclaving after being placed in a discard bin.

[8] 20 min. at 121° (15 lb./sq. in.) for safety.

[9] Made of polythene, rubber or glass. Certain rubber jars withstand repeated autoclaving. Glass jars need a pad of glass wool at the bottom to avoid breaking the tips of pipettes. Cotton wool is unsuitable as it reduces hypochlorite.

Table 4.6 (*cont.*)

Material	Container	Method
Used slides, broken glass, cotton wool, etc.	Jars	Lysol[10]
Clean nylon and cellulose nitrate[11] centrifuge tubes and other heat-sensitive plastics	—	Ultra-violet irradiation[12]
Used nylon and cellulose nitrate[11] centrifuge tubes and other heat-sensitive plastics	—	Lysol; HCl
Polypropylene and other heat-resistant plastics	—	Autoclave
Air[13]	—	Heat; filtration; irradiation

[10] The disinfectant must be deep enough to completely cover the objects, which must be left immersed for at least 12 hr.

[11] Cellulose nitrate tubes may explode on autoclaving (Silver, 1963).

[12] U.v. light does not penetrate glass, plastic, or solutions of protein or nucleic acid bases. The objects must therefore be exposed directly to the lamp and afterwards stored in sterile containers.

[13] See Humphrey (1960) and Cherry, Kemp & Parker (1963).

Table 4.7. *Temperature of pure saturated steam at various pressures*

A. Pressures below atmospheric

Absolute pressure[1] (lb./sq. in.)	Temperature °F.	°C.
1	102·1	38·94
2	126·3	52·39
3	141·6	60·89
4	153·1	67·28
5	162·3	72·39
6	170·2	76·78
7	176·9	80·50
8	182·9	83·83
9	188·3	86·83
10	193·3	89·61
11	197·8	92·11
12	202·0	94·45
13	205·9	96·61
14	209·6	98·67

B. Pressures above atmospheric

Gauge pressure[1] (lb./sq. in.)	Temperature °F.	°C.
0	212·0	100·0
1	215·4	101·9
2	218·5	103·6
3	221·5	105·3
4	224·4	106·9
5	227·1	108·4
6	229·6	109·8
7	232·3	111·3
8	234·7	112·6
9	237·0	113·9
10	239·4	115·2
11	241·5	116·4
12	243·7	117·6
13	245·8	118·8
14	247·8	119·9
15	249·8	121·0
16	251·6	122·0
17	253·4	123·0
18	255·4	124·1
19	257·0	125·0
20	258·8	126·0
21	260·4	126·9
22	262·0	127·8
23	263·7	128·7
24	265·3	129·6
25	266·7	130·4
26	268·3	131·3
27	269·8	132·1
28	271·2	132·9
29	272·7	133·7
30	274·1	134·5

[1] Absolute pressure = gauge pressure + atmospheric pressure (14·7 lb./sq. in.). Subatmospheric pressures are sometimes expressed by the vacuum, i.e. atmospheric pressure − gauge pressure (see legend to Fig. 4.6).

5

EXAMINATION OF BACTERIA BY MICROSCOPY

The study of bacterial morphology has been advanced so rapidly by electron microscopy that the examination of organisms by visible light might be thought outdated. This is by no means so, for although traditional methods show relatively little bacterial structure, they are so convenient that they remain of great value in many experiments as well as for routine identification. Bacterial nuclei and other organelles are easily counted, an important point with heterogeneous populations when the state of individual cells has to be determined, as in measuring the distribution of flagella in clones (Kerridge, 1961) or the mean number of nuclei per cell in an asynchronously dividing culture (Maaløe, 1960). Moreover, the scope of conventional microscopy is continually being extended by improvements in method: notably autoradiography, the use of suspending media of high refractive index that allow the nuclei of living cells to be seen, and fluorescence microscopy using fluorochromes or labelled antibody, as in recent outstanding studies of cell wall synthesis in bacteria (see Cole, 1965).

The successful examination of stained films or living organisms requires correct adjustment of both microscope and lamp. This is not to say that preparations cannot be examined with a microscope set up empirically—simply, that far better results are obtained if a little more trouble is taken. Practice is more closely related to theory than elsewhere in this book, but, quite clearly, a full account of microscopical theory is impossible here (see Françon, 1961; Martin, 1961, 1966. For general introductions to microscopy, see Barer, 1968*a*; Hartley, 1962; Cosslett, 1966). We therefore give a short account of image formation while emphasizing that what is seen in the microscope is not the object but an image of the object and that the two are by no means identical, a point discussed for bacteria by Ross (1957).

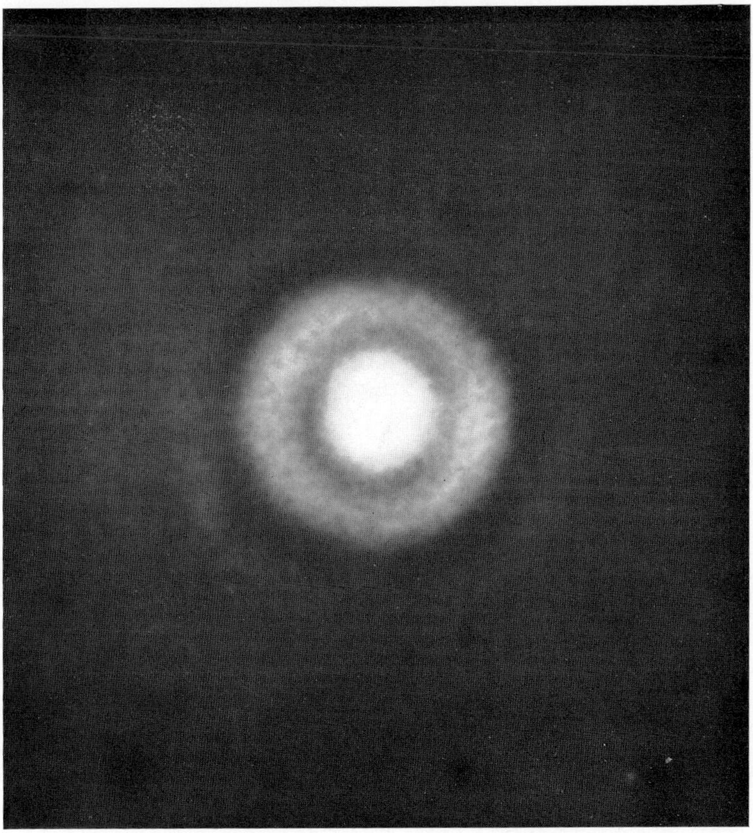

Plate 5.1. Image of a luminous point: Airy pattern.

[*Facing page 128*]

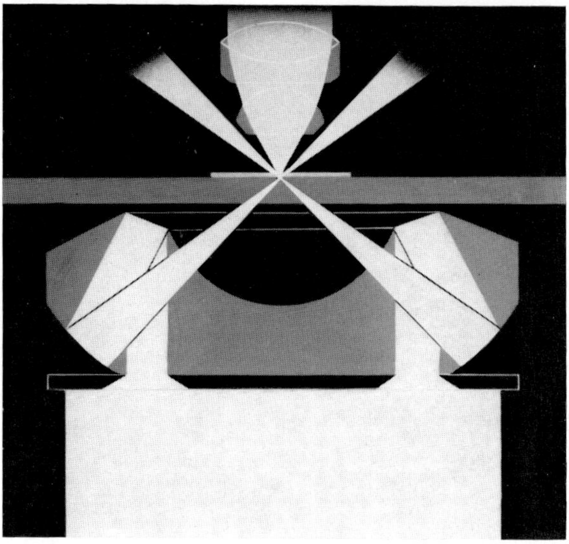

(*a*) Dry dark-field condenser.

(*b*) Oil-immersed dark-field condenser.

Plate 5.2. Ray paths in high power dark-field microscopy. Each figure shows a hollow cone of light focused on the slide with some light scattered into the objective by the object.

THE COMPOUND MICROSCOPE

Image formation

In geometrical optics, the objective, *O*, forms a real inverted image, *A′*, of the object, *A*, at *Z* (Fig. 5.1). *Z* lies within the focal point of the ocular, *E*, so that the observer views an enlarged virtual image of *A′* at *X* which is usually assumed to be 25 cm. from the eye, the normal distance for comfortable close vision. If the image were formed in this way, then the image of a point would clearly be a point, and that of a line, a line. That this is not so in reality is easily seen by viewing a pinhole in an opaque screen, or, even more simply, by quickly drying a thick film of nigrosin on a slide so that a network of fine cracks is formed. When either preparation is viewed under oil immersion, its image is found not to correspond to the object.

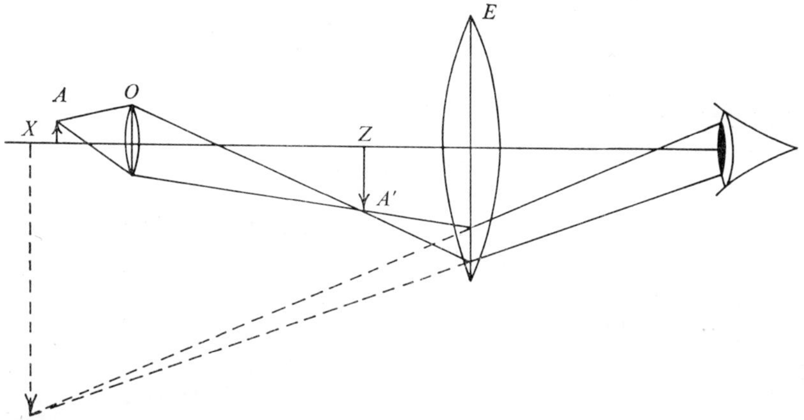

Fig. 5.1. The compound microscope. The objective, *O*, forms a real image, *A′*, of the object, *A*, at *Z*. *Z* lies within the focal point of the ocular, *E*, so that the observer views a virtual image of the object at *X*.

The pinhole appears as a white disc surrounded by alternating dark and light rings (an Airy pattern: Plate 5.1) while each fine crack is bordered by dark and light bands although wide cracks have sharp borders. It is clear, therefore, that geometrical optics, although giving the proper position of the images on the axis, does not correctly describe their structure.

This discrepancy between the structure of object and image becomes increasingly obvious, the smaller the object, and arises from the wave nature of light. The incident light can be regarded as a uniform succession of waves, resembling ripples passing across water, whose arrangement is disturbed by the object. Some light continues on its original path but the remainder diverges and is diffracted. It is the interaction between the diffracted and undiffracted rays that gives rise to the image, including the surrounding bands like those shown in Plate 5.1.

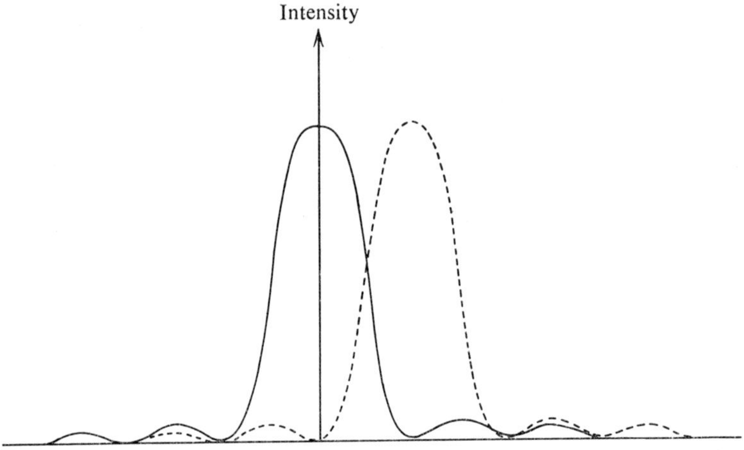

Fig. 5.2. Airy pattern, showing distribution of intensity in a plane perpendicular to the optical axis. By definition, two similar patterns are just resolved when the maximum of one coincides with the first minimum of the other.

Resolution and magnification. The complex structure of an image can be described quantitatively by the distribution of light intensity in a plane perpendicular to the optical axis. Thus, an Airy pattern gives a central peak surrounded by alternating troughs and peaks which diminish in amplitude on passing away from the axis (Fig. 5.2). Since the image of a point clearly has no sharp boundary, it becomes a matter of definition as to when two points are regarded as distinct. The usual criterion is that the maximum of one distribution shall coincide with the first minimum of the other. The smallest distance between two points like *A* and *B* in Fig. 5.3 that is thus resolvable equals

$$\frac{\lambda\ 0{\cdot}61}{n \sin u}, \tag{5.1}$$

where λ is the wavelength of the light; n is the refractive index of the medium in contact with the lens; and u is half the effective acceptance angle of the lens (Fig. 5.3). The resolving power of a lens is expressed by the product, $n \sin u$, termed its *numerical aperture*. The N.A. of any lens in air cannot be greater than 1, since the refractive index of air $= 1$ and the angle u cannot be greater than 90°, giving $\sin u = \sin 90° = 1$. Resolution is therefore increased either by increasing n, the refractive index of the medium ($n = 1$

for air, 1·33 for water and *ca.* 1·5 for immersion oil); increasing u by increasing the acceptance angle of the lens; or decreasing λ by using blue or ultra-violet light. In practice, the maximum N.A. attainable in air is *ca.* 0·95 and, in oil, *ca.* 1·4. The respective resolving powers therefore equal 0·39 μ and 0·26 μ approximately, taking k as 0·65 and the wavelength of green light as 0·55 μ.

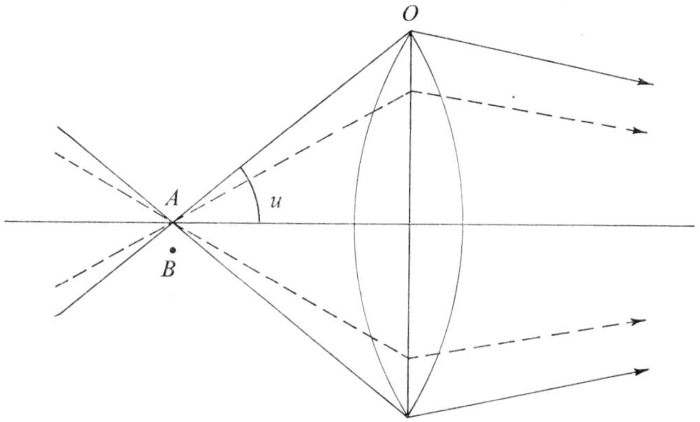

Fig. 5.3. Numerical aperture. Rays from a point, *A*, pass to the objective, *O*, at the maximum angle, *u*, to the optical axis. *B* is another point near *A*. The dashed lines show how the angle, *u*, and hence the working N.A. of the objective, is decreased if an iris to the left of *A* is partly closed.

The two functions of the microscope—resolution and magnification—must be distinguished as they vary independently. Resolution is the formation of a distinct image, whereas magnification is the production of an enlarged image which may or may not be distinct. The overall magnification produced by the microscope equals the magnification of the objective multiplied by that of the ocular. Mere increase in the magnification of the ocular will not increase resolution, since the ocular cannot improve the quality of the image presented to it by the objective. Conversely, resolution can be decreased without change in magnification. Suppose that the optical system shown in Fig. 5.3 is unaltered except for reducing the aperture of the iris of a substage condenser through which light reaches the object. This reduces the cone of light falling on the objective, so reducing its working N.A., with the result that resolution is also reduced according to equation (5.1). The magnification must nevertheless be the same since the object, objective, and ocular are unchanged. This example incidentally shows that the setting of the substage condenser and its iris both play a critical part in determining the quality of the image.

The total magnification required depends partly on the structure of the human eye, which restricts the maximum resolution possible. Two discrete points are usually said not to be separated if they subtend an angle of less than *ca.* 1 min. to the eye, that is, they cannot be separated if they are less than 0·07 mm. apart at a distance of 25 cm. from the eye. The magnification required for an object whose finest detail is s mm. is therefore $0{\cdot}07/s$. This is a minimum estimate based on the assumptions that the object has high contrast and that the eye is giving its maximum performance. In bacteriology, the objects are usually small and the maximum resolution is usually sought by using an objective of N.A. 1·3–1·4 immersed in oil. The maximum resolution is *ca.* 0·2μ so that the minimum magnification needed is *ca.* $0{\cdot}07 \times 10^3/0{\cdot}2 = 350\times$. Although the total magnification is the product of the magnifications of the ocular and objective and does not depend on their individual powers, it is preferable to use a high-power objective and a relatively weak ocular, rather than the reverse. The usual rule is that the minimum and maximum magnifications for a given objective are 500 and 1000 times the N.A. Apart from failing to improve resolution, the higher the power of an ocular, the smaller the beam entering the observer's eye which lowers its acuity and increases distortion due to visual defects like astigmatism or vitreous opacities.

It was seen above that resolution is governed not only by the objective but also by the substage condenser and its iris. Ideally, the condenser N.A. should be as large as the N.A. of the objective, or the latter is effectively reduced and resolution impaired. In practice, however, the condenser N.A. has usually to be slightly lessened by closure of its iris in order to avoid glare. The lamp iris is usually arranged so that its image exactly fills the field and is neither larger nor smaller. Neither the condenser nor the condenser iris or the lamp iris should be used to control the brightness of the field, or resolution suffers. Intensity of illumination must only be controlled by reducing the brightness of the lamp filament, or by interposing neutral density filters between the lamp and microscope.

The optical system

Although all methods ideally require perfect centration of the optical system, the degree to which this is possible in practice is often restricted by the construction of the microscope. It is nowadays uncommon to be able to centre the objectives or the substage iris separately and, if any of these components is seriously off centre, it is usually necessary to consult the makers.

Centring the light source on the lamp lens. Adjust the lamp housing until its axis is perpendicular to a distant wall. Note the position of a point on the wall that is on the axis of the housing. Focus the lamp filament on the wall. Adjust the centring screws of the bulb holder (this is usually adjustable, the lamp lens being fixed in its housing) until its image is centred over the marked point. The light source should then be on the same axis as the lamp lens.

Centring the illuminating system on the axis of the microscope. Without altering the focus of the lamp lens, place the lamp squarely about 10 in. before the microscope and adjust its housing so that the light falls on the centre of the plane mirror. Remove the condenser, objective and ocular, and hold a piece of paper against the top of the drawtube. Adjust the mirror until the paper is evenly illuminated.

Centring the condenser. Put in a low-power objective and a $\times 10$ ocular. Focus on a slide. Rack up the mount of the substage condenser, insert the condenser with its iris fully open, and focus it on the object plane by producing an image of the filament when the slide is viewed through the ocular. (The condenser sometimes appears to come to the upper limit of its range of movement before a sharp image is obtained. When its mount is in the form of a circular collar, the usual reason is that the condenser has not been pushed home.) Remove the ocular, half close the substage iris, and observe its image in the back lens of the objective. Adjust the substage iris until it has approximately the same diameter as the back lens of the objective. Adjust the substage centring screws to centre the image of the iris in the back lens. The condenser should then be central. If necessary, the process can be repeated with a high-power objective.

Systems of illumination. With low-power objectives (N.A. $\leqslant 0{\cdot}1$), the object can be adequately illuminated by a concave mirror alone. However, with higher powers, a condenser is needed since the mirror delivers neither sufficient light to illuminate the object nor a cone of light that is sufficiently wide to fill an objective of high N.A., with the result that resolution falls, as in Fig. 5.3. The widest cone is obtained if the apparent light source is focused by the condenser on the object. This can be verified by removing the ocular and inspecting the back lens of the objective with the condenser focused on the object. The back lens of the objective should be evenly filled with light and, as the condenser is moved away, it becomes steadily less evenly illuminated until eventually only a central spot of light is seen. For light from the condenser to be able to fill the objective, the angle u in Fig. 5.3, and hence the N.A. of the condenser, must be at least as great as that of the objective. Since the N.A. of a condenser is defined as for an objective, it cannot exceed 1·0 in air, so that a condenser used with a high-power oil immersion objective (N.A. $\simeq$ 1·4) must be oiled to the under-surface of the slide. It is also essential to use thin slides or their thickness will prevent the condenser approaching sufficiently close to the object to be focused upon it.

Two systems of illumination are described, critical illumination having largely been superseded by the Köhler system.

In *critical illumination* parallel rays emerge from the lens, *B*, placed at its focal length from the light source, *S* (Fig. 5.4). The rays fall on the substage

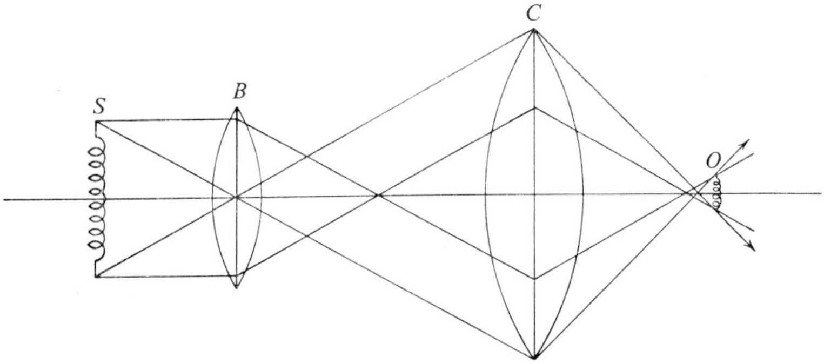

Fig. 5.4. Critical illumination. The filament, *S*, is at the focus of the lamp lens, *B*. Parallel rays therefore enter the substage condenser, *C*, and are focused on the object, *O*. Consequently, an image of the filament appears in the object plane.

condenser, *C*, which focuses an image of the light source in the plane of the object, *O*. The main disadvantage of this method is that an image of the light source appears in the object plane, which is therefore unevenly illuminated unless the source is unusually homogeneous. Furthermore, a very large light source is needed unless the field of view is very small. The Köhler system avoids both these difficulties.

In *Köhler illumination* the lamp lens, *B*, is placed so as to form an image of the light source on *I*, the iris of the substage condenser (Fig. 5.5*a*). This iris is separated from the condenser by a distance approximately equal to the condenser's focal length; therefore, the image of the light source formed by the condenser is at infinity and each point of the light source gives rise to a bundle of parallel rays leaving the top lens of the condenser. These bundles have next to be focused on the object. Note that in Fig. 5.5*a* the rays *QZ* and *PX* can be thought of as arising from the point *Y* at the centre of the lamp lens *B*. Hence, if *Y* is now brought into focus in the object plane, the continuations of *QZ* and *PX*, namely, *ZW* and *XW*, must intersect in the object plane and, at the same time, their bundles of rays are made to converge on the object. In practice, the edge of the lamp iris, not the lens *B*, is focused on the object since the iris and lens are close together.

In this system, the light source seen in the object plane by the observer is not the real source, *S* (i.e. the lamp filament), but the lens, *B*, which has

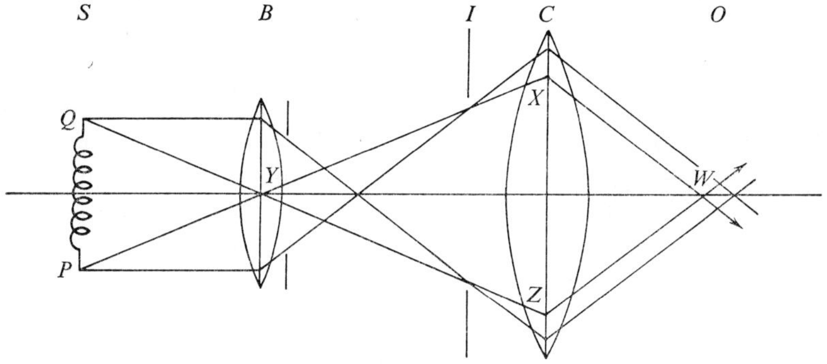

(*a*)

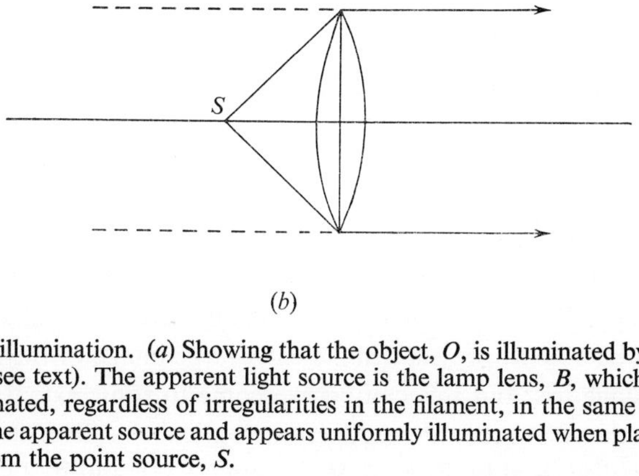

(*b*)

Fig. 5.5. Köhler illumination. (*a*) Showing that the object, *O*, is illuminated by bundles of parallel rays (see text). The apparent light source is the lamp lens, *B*, which appears uniformly illuminated, regardless of irregularities in the filament, in the same way that the lens in (*b*) is the apparent source and appears uniformly illuminated when placed at its focal distance from the point source, *S*.

become the apparent source. This occurs because the observer, whether he inspects the object plane by eye or by means of an objective and ocular, receives bundles of parallel rays from each point of the real source; i.e. it is as if the filament in Fig. 5.5*a* was viewed through a convex lens at a distance equal to its focal length so that it appeared at infinity and its apparent size was that of the lens. The whole lens therefore appears to be evenly filled with light however small and uneven the real source, namely the lamp filament. Thus, the advantages of Köhler illumination are, first, that although the real light source may be small and irregular, the apparent source, the lamp lens, *B*, can readily be made large enough for the condenser and objective to be filled completely with light; and, second, that the irregularity of the real source is largely obliterated since beams from different points of the filament overlap to produce an even intensity of illumination in the object plane.

The practical steps in obtaining Köhler illumination are as follows. Close the substage iris and open the lamp iris. Adjust the lamp lens (*B* in Fig. 5.5*a*) so as to focus an image of the filament as sharply as possible on the substage iris which it should cover completely. The image is more easily seen if the undersurface of the iris is viewed in a small mirror placed on the bench in front of the microscope. If the image is smaller than the iris, place the lamp further from the microscope. If this fails, the lamp lens is probably unsuitable for Köhler illumination.

Open the substage iris and almost close the lamp iris. Focus a low-power objective on the object plane. For oil-immersion microscopy, oil the top lens of the condenser and bring it up to the undersurface of the slide until a seal forms. Then adjust the condenser until an image of the lamp iris is seen in the object plane viewed through the ocular.

Should the object plane not be uniformly illuminated, a sheet of ground glass can be placed between the lamp bulb and the lamp lens as close to the bulb as possible to avoid unnecessary loss of light. The ground glass must not be put between the lamp iris and the microscope for the iris then ceases to be effective.

The Köhler system is preferable for most purposes, and can be used for dark-field and phase-contrast as well as for ordinary transmitted light microscopy. The only difficulty that may be met with is that the lamp lens may not be able to project a sufficiently large image of the light source on the substage iris; but this can be overcome to some extent by placing the lamp further from the microscope. The optical characteristics of a suitable lamp are discussed by Barer & Weinstein (1953), and by Dade (1958) who gives a design based on stock lenses.

Adjustment of the irises and illumination. Open the lamp iris fully. Remove the ocular and, while looking down the tube, adjust the substage iris until its margin just appears within the boundary of the evenly illuminated back lens of the objective. The working N.A. of the condenser is then just less than the maximum N.A. of the objective. If the substage iris is opened further, excess light is scattered into the objective to cause glare, and discrimination is reduced. In theory, resolution suffers if the maximum N.A. of the objective is not used (p. 131). In practice, however, the clearest images are usually obtained when the substage iris is closed until 10–30 % of the area of the back lens of the objective is occluded. Finally, the lamp iris is closed until the image of its margin coincides with that of the field seen through the ocular.

At this stage, the light may be too strong for comfort and will need to be cut down, for example by altering the brightness of the filament by means of a resistance or multi-step transformer or, better, by restricting the light passing to the microscope by interposing neutral density filters. If the filament is dimmed, the light usually becomes very yellow whereas with filters, its colour is unaltered. The intensity must not be altered by varying the position of the substage or lamp iris, or the substage condenser, for this will decrease resolution.

Difficulties in obtaining distinct images are commonly due to dirt throughout the optical system and often the only remedy is to have the instrument cleaned by an expert. Wooden microscope cases usually allow lenses to become covered with dust within a few weeks and it is well to cover every microscope with a plastic bag, whether it is on the bench or in its case. Even then, a yearly overhaul is worthwhile when a microscope is used to even a moderate extent.

Oil should be wiped from oil-immersion objectives (and substage condensers) with lens tissue after use, for, although modern oils do not set, dirt gathers on the glass and may damage it in time. Dried oil is removed with lens tissue moistened with benzene or xylol. Either may damage the lens mounting, but benzene is safer as it evaporates quickly.

Streaming and marked distortion usually follow if different makes of immersion oil are mixed, for their refractive indices are usually slightly different.

Light sources

Incandescent filaments such as those of tungsten bulbs emit a continuous spectrum like any hot body, whose maximum is at progressively shorter wavelengths, the hotter the filament. Filaments are usually characterized by their 'colour temperature', that is, the temperature at which the same spectrum is emitted by liquid platinum. The maximum temperature attainable by a tungsten filament is limited to 3700 °K by the melting point of tungsten, whereas the colour temperature of daylight is 5600–6500 °K.

Any tungsten lamp will therefore be redder than daylight and, for this reason, the colour balance is generally restored by bluish 'daylight filters'. The scope of filament lamps has been considerably extended by the introduction of quartz-iodine (tungsten-iodine) bulbs running at 3300 °K compared to 2800 °K for a conventional filament lamp. These bulbs work on the principle that tungsten evaporating from the filament will combine with a gaseous halogen in the bulb, typically iodine, to form tungsten

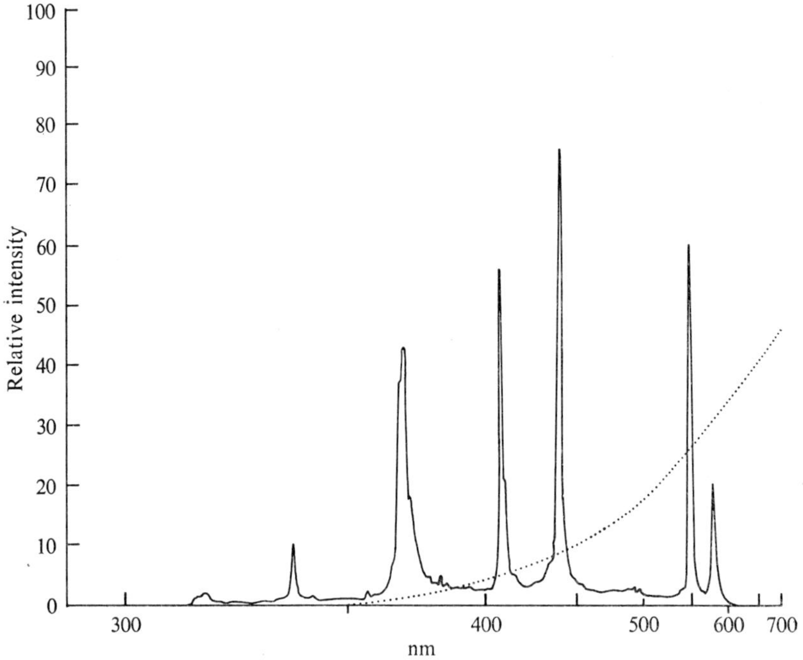

Fig. 5.6. Emission spectra. The discontinuous spectrum of a mercury discharge source (the Osram HBO 200 W/4 lamp) and the continuous spectrum (shown dotted) of a 12 V. tungsten-iodine bulb. Note that the two curves are drawn to different scales on the ordinate.

iodide, which decomposes on coming into contact with the filament with deposition of tungsten and release of iodine, Not only are these bulbs more intense than previous filament lamps, but they are compact, reasonably priced, can be turned on and off repeatedly, and emit sufficiently at wavelengths $\leqslant$ 400 nm., to be useful in fluorescence microscopy (p. 150).

Discharge lamps differ from filament lamps in emitting a discontinuous line spectrum characteristic of the vapour in the bulb. A medium-pressure mercury lamp, like the widely used Osram HBO200, gives a series of prominent lines at $\leqslant$ 435 nm. superimposed on a continuous low background (Fig. 5.6) and, for this reason, is widely used in fluorescence microscopy. Its excellent performance outweighs its disadvantages of cost and need for cooling it before it can be re-used as well as the hazard of u.v. light accidentally reaching the eyes. Obviously, no u.v. source should be used without an efficient u.v. filter to protect the user.

Filters

Two classes are available: absorption filters of coloured gelatin film (e.g. Kodak or Ilford) or glass (Kodak, Ilford, Chance, Schott & Gen) which absorb unwanted wavelengths; and interference filters, which remove the unwanted wavelengths by interference (e.g. Balzer; Grubb Parsons). In general, absorption filters are cheaper, more robust, come in wider ranges and are perfectly adequate for most purposes. Interference filters provide sharper cut-offs: the peak transmission shifts to shorter wavelengths if the filter is at an angle to the optical axis. Solid filters of these types have superseded liquid filters once commonly used in optical troughs (for these, see Withrow & Withrow, 1956).

Colour filters. Gelatin filters provide a wide choice (see the makers lists or the *Handbook of Chemistry and Physics*). Although they are damaged by moisture or heat, it is often sufficient to mount them in the 2 in. glass squares sold for transparencies; many are available already sealed. A green filter like the Wratten no. 66 is often comfortable and gives improved images for transmitted light microscopy.

Daylight filters are bluish colour filters that preferentially absorb red and yellow light and, when used with a tungsten source, produce illumination approximating to daylight.

Neutral density filters. These absorb all visible wavelengths equally, and are used to produce a uniform reduction in intensity of visible light (p. 137). Several filters are needed. Some are characterized by their densities (log I_0/I) as on p. 15, for example 0·6, 0·9, 1·2, where the transmission of each differs from its neighbour by a factor of 2.

Heat filters are used to prevent overheating of colour filters and of the object by absorbing wavelengths greater than *ca.* 700 nm. An effective alternative is an optical trough containing water or Mohr's solution (50 g. ferrous ammonium sulphate dissolved in 250 ml. water + 1·3 ml. of 33 %, v/v, H_2SO_4).

MICROMETRY

Measurement requires (*a*) a graticule or scale ruled in arbitrary units printed on a glass disc that can be placed on the stop of the ocular beneath its top lens; and (*b*) a scale in known units, for example 1 mm., 0·1 mm., etc., printed on a slide. The calibrated slide is placed on the stage and brought into focus. The rulings of the slide and those of the graticule should then appear superimposed. The ocular scale is next calibrated for each objective and thus enables the size of the field or of an object on the stage to be measured. A more precise method, using an image-splitting eyepiece, is discussed by Powell & Errington (1963*b*).

A different approach is to photograph a stained film of bacteria and, after preparing a print of known magnification, to measure individual organisms directly.

All methods have difficulties. The graduated scale in the ocular may not be as sharply focused as the organisms. If phase-contrast is used, organisms do not present a sharp outline, especially if they are viable (Powell & Errington, 1963*b*). Nor do they appear distinct by ordinary bright-field microscopy, owing to diffraction (Ross & Galavazi, 1965; Plate 5.1). If stained preparations are measured, the apparent size of the organisms depends on the degree of distortion caused by the fixative and stain, and on whether a cytoplasmic or a cell wall stain is used.

DARK-FIELD MICROSCOPY

The basic ray diagram is shown in Plate 5.2. No rays leave the centre of the condenser and the inner angle of the hollow cone of light thus formed exceeds the acceptance angle of the objective. Therefore, in the absence of an object, no light enters the objective and the field appears black. An object diffracts light into the objective and appears white, and will be visible even if its actual diameter is less than the resolving power of the objective. Plate 5.2 also shows that the N.A. of the condenser must be greater than that of the objective which it usually exceeds by 0·2–0·4.

Dark-field condensers are of two general types: (*a*) where the central rays of an ordinary condenser are blocked by a central stop placed in the sub-stage condenser ring; this is only suitable for low-power work with objectives of N.A. $\leqslant$ 0·65; and (*b*) where the hollow cone of light is produced by reflecting surfaces within the condenser (Plate 5.2). This type is suitable for high-power work. If the N.A. of the objective is 0·95, that of the condenser must be appreciably greater than 0·95 and must therefore by oiled to the slide to obtain its maximum N.A., usually *ca.* 1·3 (p. 131). The working distances of such condensers are so small that only thin slides can be used.

Low-power microscopy. First, focus and centre the condenser without the stop in the usual way. Then put in a central stop and dark-ground should be obtained.

High-power microscopy. Köhler illumination is used. Centring of the substage condenser depends partly on its type. Some have a ring engraved

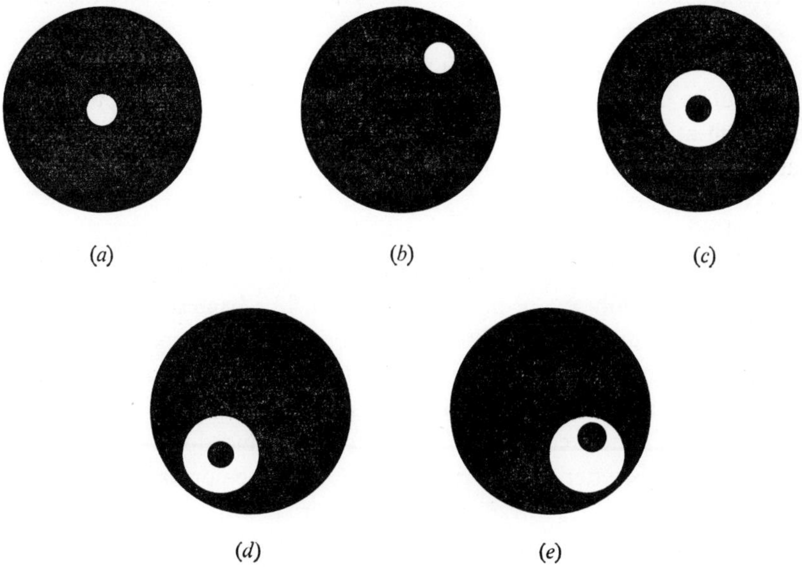

Fig. 5.7. Appearances seen in a film of indian ink while centring a dark-field condenser. (*a*) The condenser is centred and focused on the object plane. The apex of the cone of light is seen as a central bright disc. (*b*) The condenser is off centre but is focused on the object plane. (*c*) The condenser is centred but not focused. The unilluminated centre of the cone of light shows as a central dark spot in the bright disc. (*d*) The condenser is both off centre and not focused. (*e*) The condenser is again off centre and not focused, but, in addition, the light source is not centred on the condenser. The dark spot is therefore not centred in the bright disc.

on the centre of the upper lens which is visible before oil is added. The ring can then be centred in the field. When no ring is present, focus a low-power (16 mm.) objective on a suspension of indian ink before putting in the condenser. On inserting the condenser and focusing it up and down, a series of images will be seen in the indian ink (Fig. 5.7). When appearance (*a*) is obtained, the condenser is central and focused on the object. Plate 5.2 shows that the effective acceptance angle of the objective must not exceed the angle of the dark cone from the condenser. Since the N.A. of an oil-immersed condenser is *ca.* 1·3, the N.A. of the objective has to be reduced to 0·9–1·0 to obtain dark-field illumination. This is done either by using a special objective fitted with an iris diaphragm, which is partly closed, or by dropping a funnel stop into the upper opening of an ordinary objective after unscrewing it from the objective carrier. Makers state the precise apertures needed for their own equipment.

The light source has to be more intense than for bright-field microscopy, as only a small fraction of the light enters the objective. Neither the lamp iris nor the substage iris should be closed.

Two points soon become obvious when using dark-field microscopy. First, unlike bright-field microscopy, centration must be accurate to obtain any result at all, and second, cleanliness is essential as the method reveals particles that usually go unnoticed.

PHASE-CONTRAST MICROSCOPY

Many unstained objects of biological interest, like bacteria suspended in broth or water, are almost indetectable by bright-field microscopy because their optical properties closely resemble those of their surroundings. A simple method for rendering such objects visible is to close the substage condenser iris almost completely and to focus up and down with the fine adjustment. Far better results are obtained by phase-contrast microscopy (see Bennett *et al.* 1951; Françon, 1961; Barer, 1966).

An object is invisible by conventional bright-field microscopy if it fails to alter the wavelength (colour) or amplitude (intensity) of the light which encounters it. Nevertheless, such an object often produces changes in phase which are proportional to the product of its thickness and the difference between its refractive index and that of its surroundings. Rays like those in Fig. 5.1 are each assumed to consist of an advancing front of independent waves. As they are in phase, no interference occurs between them. When a wave enters a region of higher refractive index, like a bacterium, its velocity falls so that, on emerging, it is retarded relative to, and is therefore out of phase with, waves on either side. Each point in the object is to be regarded as an independent source (Huyghen's principle) from which waves travel forwards as from a stone dropped in water. The retarded and unretarded waves therefore cross and, as they are now out of phase, interference occurs to produce a new class of waves (the 'diffracted' waves) which appear to originate from the object. Thus, there are three classes of waves: unretarded (class *a*); retarded (class *b*); and the result of their interaction, the diffracted waves (class *c*).

The retarded and unretarded waves are shown by curves *B* and *A* in Fig. 5.8*a*. The difference between them is responsible for the diffracted

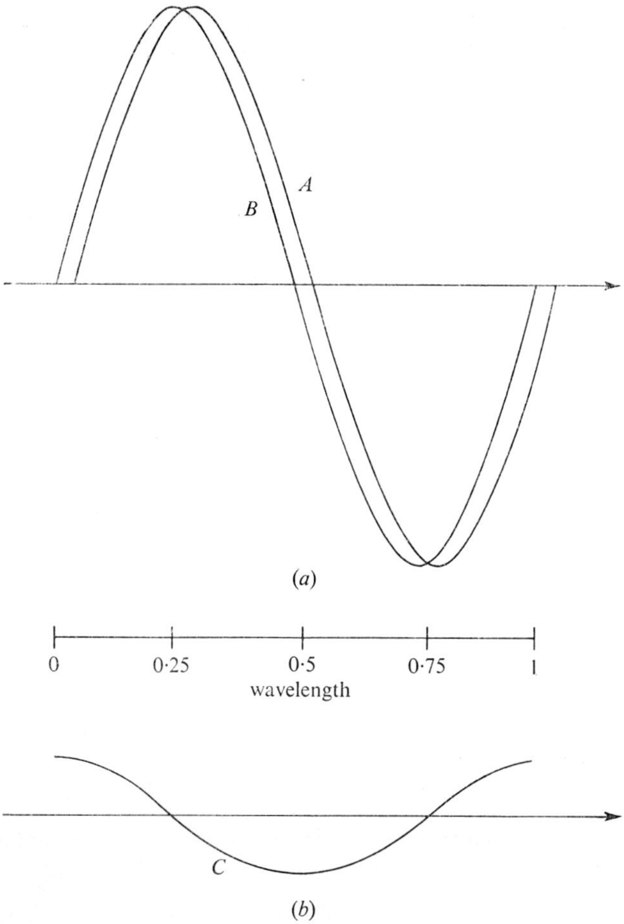

Fig. 5.8. Characteristics of diffracted and undiffracted rays. The refractive index of the object is assumed to be greater than that of the surroundings.

(*a*) The class of rays which traverse the object (*B*) are slightly retarded compared to those not encountering the object (*A*), i.e. the two sets of rays are not in phase. Their amplitudes and wavelengths are unchanged by the type of object considered. The difference between the rays gives the nature of the rays diffracted by the object which are depicted by curve *C*.

(*b*) Note that *C* is approximately one-quarter of a wavelength behind *A*.

waves which can therefore be depicted by curve *C* of Fig. 5.8*b*, constructed by subtracting curve *B* from curve *A*. This method of determining the properties of the diffracted waves is equivalent to that using vectors (e.g. Martin, 1947) since, in each case, the properties of the diffracted waves are determined solely by those of the other two classes of rays. Fig. 5.8 brings out the important fact that the diffracted waves are *retarded by approximately one-quarter of a wavelength* relative to the unretarded waves of class *a*. They are also of smaller amplitude since the light encountering the object is only slightly retarded.

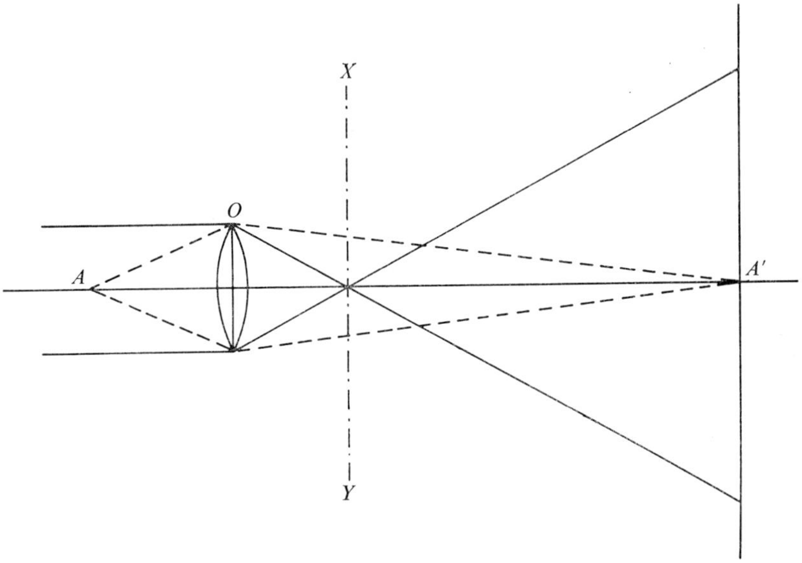

Fig. 5.9. Differing paths of diffracted and undiffracted rays. Undiffracted, shown by continuous lines, come to a focus in *XY*, the rear focal plane of the objective, *O*. Diffracted rays from the object, *A*, are shown by dashed lines and come to a focus at *A′*.

It should be stressed that diffracted waves do not invariably have these characteristics which depend solely on the sort of transparent object, such as a bacterium in water, considered here. If the refractive index of the object differed greatly from that of the medium so that the waves of class *b* were considerably retarded, the diffracted waves would necessarily have quite different properties.

Paths of diffracted and undiffracted waves. Their differing paths can now be represented geometrically. Fig. 5.9 shows an object placed in a microscope set up for Köhler illumination, so that the light leaving the condenser consists of bundles of parallel rays. Those not encountering the object (class *a*) therefore come to a focus in *XY*, the rear focal plane of the objective, while diffracted rays originating in the object come to a focus at *A′*. At this point, *A′*, the intensity of illumination of the image is the resultant of the *squares* of the amplitudes of the undiffracted and diffracted rays (Fig. 5.10) but, owing to the small amplitude of the latter (class *c*), the difference in intensity between *A′* and the background is too slight to be

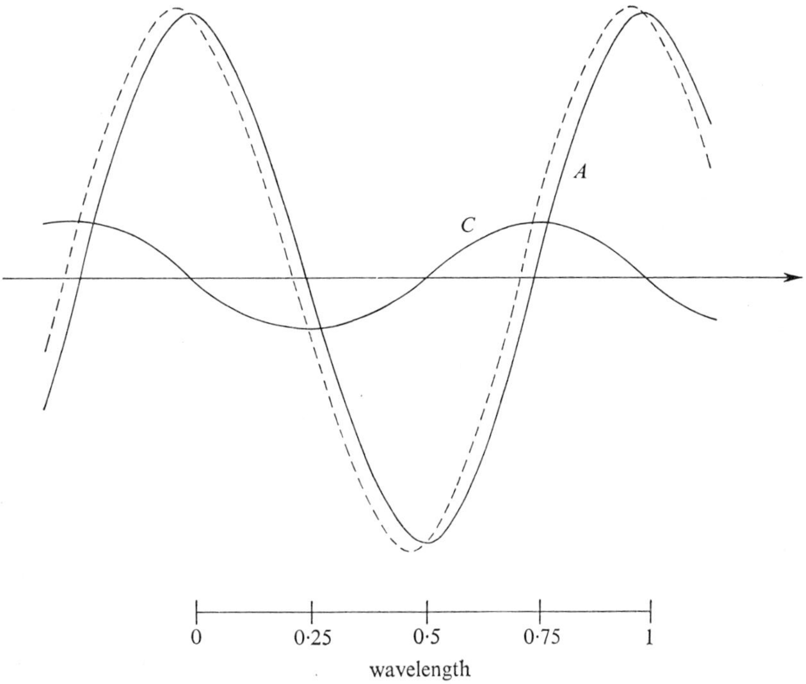

Fig. 5.10. The relation between undiffracted rays (that is, class *a*, depicted by the curve *A*) and diffracted rays (*C*) at the point *A′* in Fig. 5.9. The two sets of rays interact, but their product, indicated by the dashed curve, does not differ sufficiently from the background (*A*) to be detectable by eye.

apparent to the eye. However, the difference in intensity at *A′* can be increased to a degree that is easily visible by retarding the *diffracted* waves by a quarter of a wavelength as they pass through *XY* in Fig. 5.9 (positive phase-contrast: Fig. 5.11). The wave formed by the algebraic sum of the diffracted (*c*) and undiffracted rays (*a*) then has a markedly lesser amplitude than the background formed by the undiffracted waves alone, with the result that *A′* appears to the observer as a dark spot on a brighter field.

Alternatively, *A′* can be made to appear as a bright spot in a dark field by the reverse manœuvre of retarding the *undiffracted* waves (class *c*) by a quarter of a wavelength (negative phase-contrast: Fig. 5.12).

In this example, the zero-order waves were retarded by the object because its refractive index was greater than that of the medium, so it appeared darker or brighter than the field by positive or negative phase-contrast respectively. On the other hand, if the object's refractive index had been less than that of the medium, its appearance would have been reversed by these methods.

The apparatus

The optical system consists of: a diaphragm, often a translucent ring in an otherwise opaque plate, placed under the substage condenser; and a diffraction plate (the phase plate) placed in the rear focal plane of the objective at the level corresponding to *XY* in Fig. 5.9. The microscope and lamp are set up for Köhler illumination so that the object is illuminated by a hollow cone of light formed by the ring in the diaphragm. All the undeviated rays pass through a ring (the *conjugate* area) of the diffraction plate corresponding to that of the substage translucent ring. However, diffracted rays pass through the whole of the diffraction plate. It can now be seen that, if the diffraction plate is constructed so that diffracted rays are retarded by one-quarter of a wavelength in relation to undiffracted rays, sets of waves will interact to produce a visible image at *A′* as shown in Fig. 5.11. Positive phase-contrast is therefore obtained.

The contrast of the image is further improved by equalizing the amplitudes of the two sets of rays so that they cancel completely, and in Fig. 5.13 the object will then appear black instead of just darker than its background as in Fig. 5.11. Since less light is diffracted than undiffracted, the diffraction plate is therefore made with the conjugate area less translucent than the remainder of the plate. The usual general-purpose diffraction plates transmit *ca.* 25 % of the undiffracted rays. However, with objects such as bacteria that retard only slightly, contrast is improved by use of a diffraction plate that transmits less than 25 % of undeviated light, for example 7–14 % (see Bennett *et al.* 1951).

Contrast is dramatically increased by appropriate treatment of the mounting medium and of the object. The refractive index of the medium is increased by adding 30 %, w/v, gelatin or 15 %, w/v, polyvinyl-pyrollidone of mol. wt. 10,000 which enable structures like the nuclei of living cells to be seen (p. 166; Mason & Powelson, 1956; Schaechter & Laing,

1961). Contrast may even be reversed (Barer, Ross & Tkaczyk, 1953). A technique of far greater potential importance is the use of antibody, for this combines specifically with localized regions of the bacterial surface whose contrast is then altered, so revealing unsuspected differentiation in structure (p. 165; Tomcsik, 1956*a*, *b*, 1961; Tomcsik & Guex-Holzer, 1954).

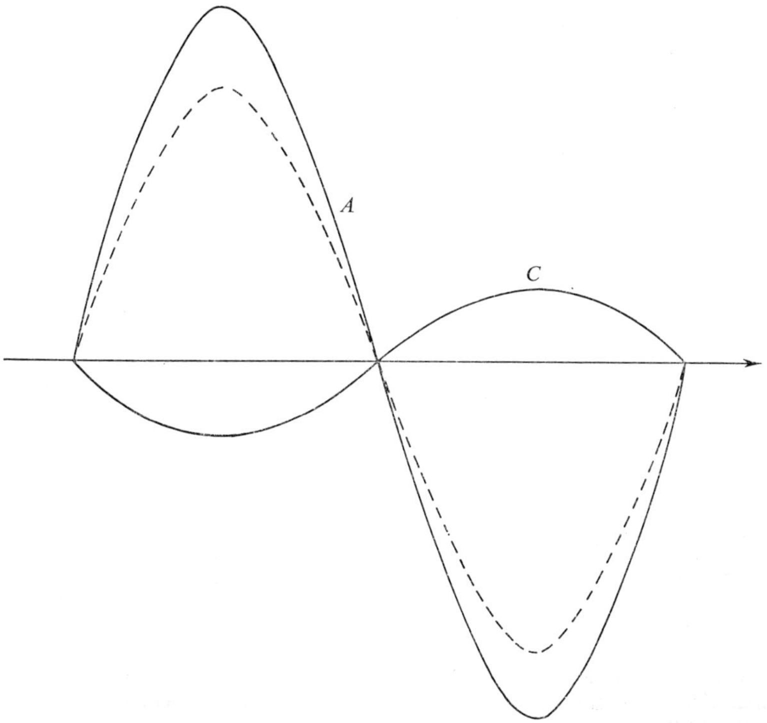

Fig. 5.11. Positive phase-contrast. The diffracted rays (*C*) have been retarded by a further quarter of a wavelength relative to the undiffracted (*A*). Consequently, the product of their interaction at *A′* in Fig. 5.9 (shown by the dashed curve) now has a markedly smaller amplitude than the background (*A*). The object therefore appears darker.

The practical details of phase-contrast microscopy depend largely on the position of the phase diaphragm and are usually supplied with the equipment. The microscope is first set up as for bright-field microscopy with Köhler illumination without putting in the phase diaphragm. It should be noted that objectives fitted with phase plates can be used for ordinary transmitted light microscopy. The objective is focused on the object and the corresponding diaphragm brought into position, usually in the substage condenser, and centred on the phase plate by its centring screws

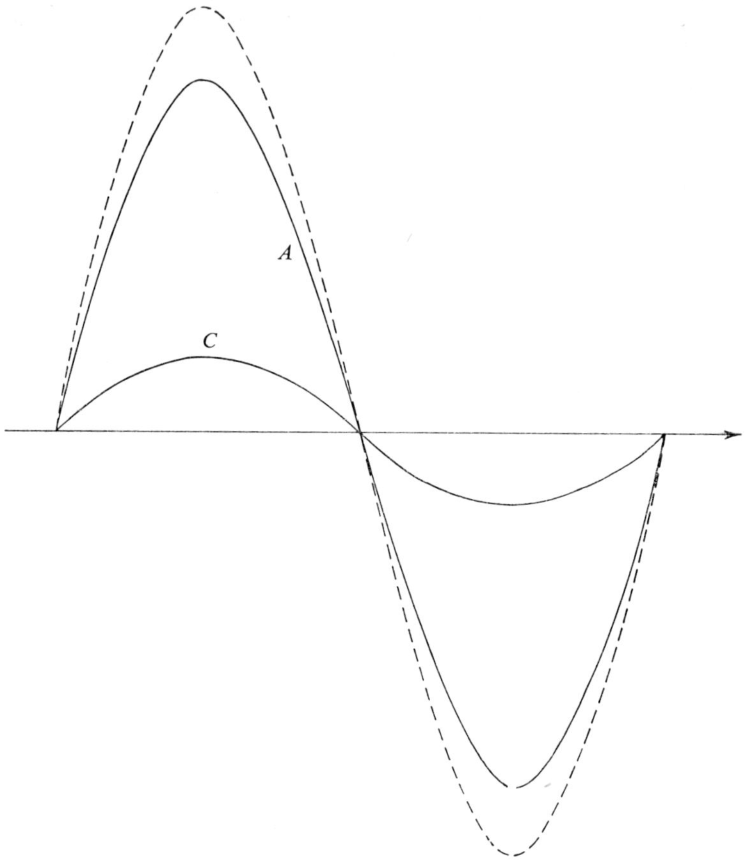

Fig. 5.12. Negative phase-contrast. The undiffracted rays (*A*) have been retarded by a quarter of a wavelength relative to the diffracted rays (*C*). The object therefore appears lighter than the background.

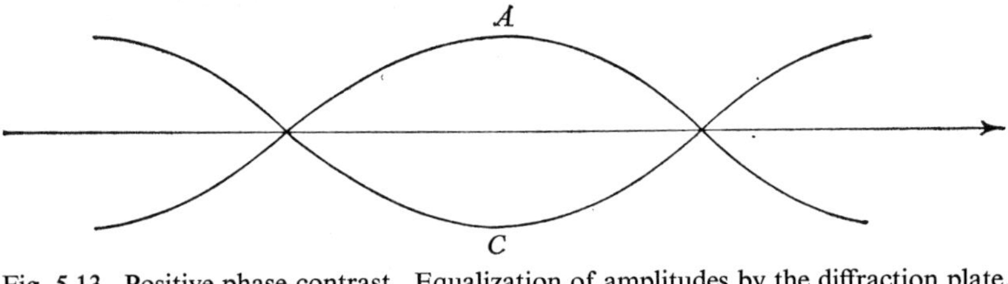

Fig. 5.13. Positive phase-contrast. Equalization of amplitudes by the diffraction plate. In positive phase-contrast, considered here, the diffracted and undiffracted rays are half a wavelength out of phase. However, in Fig. 5.11, the object is merely darker than the background. Contrast is improved by reducing the amplitude of undiffracted rays (*A*) as they pass through the conjugate area of the diffraction plate, so that their amplitude equals that of diffracted rays (*C*), as shown here. The two sets of rays then cancel and the object therefore appears black.

while inspecting its image through a telescope inserted in the drawtube of the microscope in place of the ocular. When the images of the phase plate and diaphragm coincide, the telescope is removed and, on replacing the ocular, the object should be seen by phase-contrast.

FLUORESCENCE MICROSCOPY

Many compounds fluoresce when irradiated by ultra-violet or visible light, in that they emit light distributed over a characteristic spectrum. The physical basis of the phenomenon is that energy absorbed from the irradiating light causes the substrate to enter an excited state, from which it returns to its ground state with concomitant emission of energy as light. Since only a fraction of the absorbed energy is emitted, the mean wavelength of the emitted light is always greater than that of the radiation used for excitation (see Williams & Bridges, 1964).

Fluorescence microscopy of micro-organisms is concerned with three situations: (1) Autofluorescence, when unstained organisms fluoresce. This phenomenon is not marked but needs to be borne in mind when interpreting unstained controls. (2) Organisms stained with fluorescent dyes (fluorochromes), notably acridine orange for DNA and RNA and to a far lesser extent, in vital staining; and auramine for acid-fast bacilli. (3) Immunofluorescence, using an immunological reagent, almost invariably antibody, previously conjugated to a fluorescent marker like fluorescein. Maxima for the absorption and emission spectra of the usual labels are given in Table 5.1 (see Porro *et al.* 1963, 1965).

Table 5.1. *Fluorescent agents*

Fluorescent agent	Absorption maxima (nm.)	Emission maxima (nm.)
Acridine orange	267, 492	540
Auramine	440	550
Fluorescein isothiocyanate	280, 325, 495	520
Lissamine rhodamine B	350, 575	595, 710
DANS	310–370	525

There is no bar, in principle, to using fluorescence microscopy for studying the reaction of micro-organisms with any test reagent, provided it is either autofluorescent or can be conjugated to a fluorescent label. Thus, the inherently fluorescent trypanocidal agent, trypaflavin, can be seen to stain sensitive but not resistant strains of trypanosomes. In the same way, the union of labelled polymyxin (Newton, 1955) or labelled β-glucosidase (Pital *et al.* 1967) with bacteria can be demonstrated directly.

Whichever system is used, the basic requirements of the method are the same:

1. A light source emitting sufficiently strongly at the exciting wavelength.
2. An exciter (or primary) filter placed between the light source and the object which transmits the exciting but not the emitted wavelengths.
3. A barrier (or secondary) filter placed between the object and the observer's eye which absorbs all the exciting light passed by the primary filter while transmitting the fluorescent light emitted by the object. The primary and secondary filters are therefore complementary so that, in the absence of fluorescence, the field should appear black. The compatibility of different pairs of filters can thus be checked by reversing their positions: any light seen under these conditions must arise from imperfections in the system, like fluorescent lenses or mountants, or stray light. No system is perfect in practice, and a compromise has generally to be made in matching filters, sources and reagents.

Two types of light source are available: mercury discharge lamps for 'ultraviolet fluorescence' and incandescent filaments for 'blue light fluorescence'. The Osram HBO 200 W/4 mercury lamp emits a continuous spectrum from 300 nm. upwards with strong maxima superimposed at 365 and 435 nm., and, although these maxima do not correspond to the absorption maxima of the commonly-used fluorescent agents, the overall intensity of the lamp is so high as to give excellent excitation. Moreover, its spectrum is sufficiently distinct from that emitted by the object to allow suitable filters to be devised (these must include a u.v. filter to protect the eyes). A 12 V. quartz-iodine lamp, on the other hand, gives a continuous emission whose intensity diminishes rapidly at less than 500 nm. to become negligible at 350 nm. and below. Its emission is therefore closer to many emission spectra than is that of the HBO 200 lamp and, for this reason, and because its intensity is less, adequate pairs of filters are harder to select, although this difficulty is likely to become less as new interference filters are introduced (Richards & Waters, 1967). There is no doubt that the quartz-iodine lamp is adequate for many purposes and, considering its advantages, it should be considered before opting immediately for a mercury source.

Filters for the HBO 200 W/4 lamp are supplied by the appropriate optical firms. Various combinations have been suggested for quartz-iodine lamps used with fluorescein conjugates: Corning C.S.Y./59, no. 5850 and Ilford no. 107 (Young & Armstrong, 1967), and Wratten no. 32 and 38/A (or

Wratten 38/A sandwiched between two Schott BG23) with Wratten no. 12 (Tomlinson, 1967, and personal communication) as primary and secondary filters, respectively. No special microscopical equipment is needed with either light source. In fact, the simpler the optical system the better, since most fluorescent images, although of high contrast, are relatively weak and are only further diminished by introducing unnecessary lenses and prisms. Fluorite objectives are thus preferable to apochromats, particularly plan-apochromats, and, with difficult objects, a straight monocular tube is preferable to an inclined binocular tube. Quartz optics are never required, as ordinary glass transmits wavelengths greater than 300 nm. It is obviously important that no component of the optical system should itself be fluorescent, whether it be lens, immersion oil or mounting medium. Dark-ground illumination has the advantage over conventional transmitted light microscopy that only the fluorescent light emitted by the object enters the objective, so that a greater proportion of the exciting spectrum can be used to illuminate the object and secondary filters of higher transmission are feasible. With transmitted light, on the other hand, the object is illuminated more intensely, but highly efficient filters are needed. Optically, the system differs fundamentally from the microscopy of stained specimens, since a fluorescent object constitutes a point source which emits independently in all directions. There is thus nothing to be gained in matching the apertures of the substage iris and the objective, and the iris is kept fully open to obtain maximum illumination of the object. For the same reason, the condenser should be oiled to the slide. Whichever system of illumination is chosen, the condenser should be focused in u.v., not visible, light, since the focal length differs appreciably at different wavelengths. With thick specimens, a substantial fraction of the light emitted is absorbed by the object; vertical illumination may then be more effective than transmitted light and also allows weaker barrier filters to be used since most of the exciting light passes away from the eye. A combination of vertical and conventional illumination necessarily gives the maximum intensity of fluorescence (Barer, 1968*b*). Remembering that fluorescent images are usually faint, it is important to work in a dark room and for the observer to be dark-adapted. A dim photographic safelight is useful in allowing adjustments to be made without adaptation being lost.

Fluorochromes

Acridines. Acridine orange and related fluorochromes provide a simple but specific means for demonstrating the localization and characterization of DNA and RNA within bacteria and other cells (see Hicks & Matthaei, 1955; Mayor, 1963). Modifications of the technique also enable the type of nucleic acid to be determined by a macroscopic test (Bradley, 1966).

1. Dry slides in air and fix for 5 min. in Carnoy's solution (p. 158).
2. Rinse in absolute ethanol and dry in air.
3. Stain for 5 min in 10 ml. citric acid-phosphate buffer (p. 66) containing 0·1 ml. of 1 %, w/v, aqueous acridine orange. The optimal pH should be determined by trial but is usually 3·5–5·0. Mayor (1963) recommends pH 4·0.
4. Rinse in buffer. Examine wet in buffer or dry under oil.

DNA or double-stranded RNA appear yellow-green, and single-stranded DNA or RNA are reddish-orange (see Mayor & Hill, 1961; Mayor, 1963; Bradley, 1966).

RNA and DNA can be selectively removed before staining by enzymic digestion between stages 2 and 3. Suitable conditions are 30 min. at 37° in either 0·05 %, w/v, RNase dissolved in citric acid-phosphate buffer pH 4, or 0·01 %, w/v, DNase dissolved in 0·025 M veronal buffer pH 7·5, containing 0·003 M-$MgSO_4$ (Mayor & Hill, 1961). DNA animal viruses may resist DNase unless they are first treated with 0·02 %, w/v, pepsin at pH 2 for 1·5 hr. at 37° (Anderson, Armstrong & Niven, 1959). Fixed slides covered by drops of enzyme solution are incubated in a closed petri dish, supported on glass rods over moist filter paper.

Acridine orange staining can also be applied to drops of concentrated virus suspension dried on a slide and, used with empirical after-treatments, enable the type of nucleic acid to be assessed macroscopically (for details, see Bradley, 1966). The identification is easily confirmed with RNA phages if RNase (50 μg./ml. medium) abolishes plaque formation.

Fluorochromes can also be used for vital staining, and, in this respect, euchrysine 2GNX is preferable to acridine orange since it gives differential staining of living as well as of fixed cells at pH 7 (Young & Smith, 1963). Acridine orange staining has been used as an index to the viability of individual cells, notably by Strugger (1949), but the significance of the test is probably no greater than that of the many techniques using non-fluorescent dyes (p. 23). Live and dead cells are said to appear green and orange-red, respectively, when suspended in acridine orange, 1/30,000, in saline.

Auramine staining of acid-fast bacilli. The method provides a valuable alternative to conventional stains, since individual organisms fluoresce brightly against a black background and small numbers are readily detected, even in thick films (p. 162).

Immunofluorescence

Highly specific fluorescent staining is obtained by conjugating a fluorescent compound to a serological reagent, almost invariably antibody, though complement has also been used (see Steiner & Edelhoch, 1962; Nairn, 1969; Goldman, 1968). The site of fluorescence in the object therefore provides a sensitive means for demonstrating the position or presence of the homologous antigen. The method has found wide applications in diagnostic microbiology (see Cherry & Moody, 1965) and histology (see Pearse, 1968), while in experimental work, the method has been used to demonstrate flagellar as well as somatic antigens (Fleck, Minck & Kirn, 1962; Caldwell, Stulberg & Peterson, 1966) and the appearance of new antigens during sporulation and germination (Walker & Batty, 1965). Its unique contribution has been in elucidating the nature of cell wall growth in various bacteria (see Cole, 1965). Whatever the field, however, the same criteria apply. The method depends essentially on serological specificity and, in every system, it is essential to demonstrate that specificity is maintained, by showing that neither non-specific labelling nor quenching of specifically localized fluorescence occurs. Specificity is also demonstrable by 'blocking': that is, by specifically inhibiting labelling by previous exposure to sufficient unlabelled antibody to saturate the antigen.

The antibody itself is labelled ('direct method') or the specimen is treated with unlabelled antibody, washed, and then exposed to the appropriate anti-globulin previously coupled to the fluorescent agent ('indirect method'). The indirect method has the advantage that fluorescent anti-globulin is available commercially and one sample, as of fluorescent anti-rabbit globulin, can be used with a variety of different rabbit antisera.

Labelling in itself is, perhaps, a misleadingly simple technique since the usual methods evidently produce a highly heterogeneous population of globulin molecules, some lightly, some heavily, labelled, of which the latter are responsible for the non-specific staining removed in the past by various empirical methods like absorption with tissue powder. In the future, practical immunofluorescence is therefore likely to involve, not only the use of more highly fractionated globulins and fluorescent agents, but more precise characterization of the globulin present after labelling (e.g. Wood, Thompson & Goldstein, 1965).

The most widely used label is fluorescein isothiocyanate (FITC), which is greenish-yellow; followed by lissamine rhodamine B (RB200) and

1-dimethylaminonaphthalene-5-sulphonic acid (DANS) which are orange and green, respectively. Labelling with FITC requires:

(1) Immune serum, from which the globulin can be separated by simple salt fractionation using Na_2SO_4 rather than $(NH_4)_2SO_4$, since residual NH_4^{2+} will interfere with conjugation.

(2) Solid FITC.

(3) 0·5 M-carbonate-bicarbonate buffer, pH 9·0, containing 3·7 g. $NaHCO_3$ and 0·6 g. anhydrous Na_2CO_3 dissolved in distilled water to 100 ml. and stored in well-closed bottles to exclude CO_2.

(4) A fine glass rod and 100 ml. beaker. Magnetic stirrer at 4 °C. 0·15 M saline. Neutralized saturated $(NH_4)_2SO_4$. Sephadex G-25 in a simple column.

The method is as follows:

(1) Weigh out FITC in 1 ml. beaker (20–50 mg. FITC/mg. protein).

(2) Mix in 100 ml. beaker, 1 ml. serum or globulin, 0·7 ml. bicarbonate buffer and 5·3 ml. of 0·15 M saline. Cool to 4°.

(3) Add 1 drop of bicarbonate buffer to FITC in 1 ml. beaker and work into a paste with a fine glass rod. Continue adding drops of buffer, while bringing the FITC into an even suspension.

(4) Very gradually, with continual stirring, add the FITC to the diluted serum. Put in chaser of magnetic stirrer. Cover with foil and leave stirring at 4° for at least 18 hr. The pH should be checked during the initial 2 hr. and corrected if necessary.

(5) Following conjugation, unreacted FITC must be removed by one of several methods:

(*a*) Passage through Sephadex G-25 or G-50. The labelled globulin is easily seen as a yellow band passing rapidly down the column.

(*b*) Precipitation of globulin by adding an equal volume of neutralized saturated $(NH_4)_2SO_4$ followed by washing of the precipitate with half-saturated $(NH_4)_2SO_4$, until the supernatant is colourless, and dialysis against phosphate-buffered saline, pH 7·2, for a few hours at 4° to remove $(NH_4)_2SO_4$.

(*c*) Dialysis alone.

(6) If the labelled serum is inconveniently dilute, it is readily concentrated by ultrafiltration; with Sephadex; by holding it in dialysis tubing suspended *in vacuo* or laid against polyethylene glycol (Howe, Groom & Carter, 1964).

Staining is equally simple, in principle. Smears are dried in air and then fixed either by gentle heat or with methanol, acetone or 95 % ethanol for

1 min. and allowed to dry. For staining, slides are placed on a platform inside a petri dish containing moistened paper. In the direct method, a drop of labelled serum is placed on the smear and left for 15–30 min. Excess serum is then tipped off and the slide gently washed in three changes of 0·15 M saline buffered to pH 7·2 with 0·01 M phosphate in the course of 10 min. The usual mountant is a mixture of 9 parts glycerine and one part 0·01 M phosphate, pH 7·2. In the indirect method, slides are treated with unlabelled globulin and washed before applying homologous fluorescent anti-globulin.

GENERAL METHODS FOR EXAMINING BACTERIA

Living organisms

Wet mounts. A drop of bacterial suspension is placed on a slide and covered with a coverslip. Such preparations are best sealed to avoid turbulence from convection and from drying at the edges of the film, for example with paraffin wax; 'vaspar', a mixture of equal parts of paraffin wax and petroleum jelly; beeswax; or nail varnish. Motile organisms are sometimes trapped on the glass and motility is then more easily examined in a hanging-drop preparation. A drop of culture is put on a coverslip which is covered by a slide bearing either a concavity surrounded by a thin film of petroleum jelly or a ring of petroleum jelly or paraffin wax. The slide is pressed gently against the coverslip and, if it is turned over quickly, the drop remains in the centre of the coverslip, especially if the glass is first made water-repellent by rubbing with silicone-impregnated lens tissue.

Many motile species travel too fast for accurate observation. Methylcellulose added to the suspension slows their rate of travel and also allows flagellar movement to be seen (Pijper, 1947). The centre of wet preparations rapidly becomes oxygen-deficient and, in fact, anaerobic metabolism was discovered by Pasteur who observed that an obligate anaerobe remained motile at the centre but not at the edges of a wet preparation (Bulloch, 1960). Contrast by phase-contrast microscopy is enhanced by adding polyvinylpyrollidone to the suspending medium or by exposing the organisms to antibody (pp. 146, 165).

Organisms in dilute suspension. These first require to be concentrated, as by centrifugation. A quicker method is collection on a membrane filter where they can be examined *in situ* after staining (Ehrlich, 1955; Lumpkins & Arveson, 1968). The background may itself be stained to some extent and is often not homogeneous, and it is better to resuspend the organisms in a concentrated suspension that is examined by the usual methods.

Colonies. Organisms can be grown on thin layers of nutrient agar placed on sterile slides (Ørskov, 1922), kept moist by incubation in a closed chamber or by enclosure with a coverslip sealed at the edges (e.g. Knaysi, 1957). The positions of microcolonies are easily noted if the coverslip is first layered with a formvar replica of a numbered grid (Taubeneck, 1959). Such methods can be extended by the use of vertical or oblique microscopy using special chambers (Pearce & Powell, 1951), some of which enclose circulating medium so that growth occurs in a constant environment (Powell, 1956*c*).

Full-sized colonies are usually examined under a plate microscope. The most revealing method is to use oblique illumination, with a mirror placed horizontally between lamp and microscope so that the light strikes the undersurface of the plate at an angle. The structure of rough and smooth colonies or the presence of sectors is then seen far more clearly than by ordinary transmitted light or by vertical illumination (Henry, 1933).

Fixed specimens

The main steps of fixation, staining, and mounting are similar to those used for tissues. Occasionally, films have to be cleared before staining or dehydrated before mounting. It should be remembered that staining is a form of biochemical test whose results often depend as much on the culture medium (Duguid & Wilkinson, 1961) and on the growth phase (Wade, 1955) and viability of the organisms (p. 23) as on more obvious factors like the species and details of technique.

In diagnostic bacteriology, films are usually made on slides and examined under oil without a coverslip. For more precise work, coverslips are preferable as they can be conveniently stained in small dishes and are needed in any case for correct performance of the objective. Coverslips $\frac{3}{4}$ in. square are a convenient size to mount on 3×1 in. slides. The surface bearing the film is best marked with grease pencil to aid in focusing.

New glass tends to be greasy so that films do not spread evenly and organisms lie in clumps, instead of being evenly dispersed. The grease can be removed in several ways: by immersion in dichromate and sulphuric acid, followed by water and, finally, ethanol; by heating in a Bunsen flame until it turns yellow; or, most simple of all, by dipping in detergent, followed by drying on a clean cloth and flaming on each side.

Preparation of films. (*a*) Growth on solid medium is smeared direct on the coverslip or is first suspended in a drop of water. Undue pressure will grossly damage the cells (Porter & Yegian, 1945; Bartholomew & Mittwer, 1952; Murray & Robinow, 1952).

(*b*) Blocks of agar (e.g. 0·5 cm. each face) carrying microcolonies are cut from a nutrient agar plate and placed on the coverslip with the bacteria downwards, the whole being fixed overnight in Bouin's fixative. The agar is removed next day when fixation is complete.

(*c*) Impression films are made by momentarily pressing a previously inoculated block of agar face downwards on the coverslip, followed by immediate fixation (Robinow, 1943–4).

(*d*) Broth cultures give poor films as the organisms become coated by broth protein precipitated during fixation. This is avoided by inoculating blocks of agar (not containing broth) with the broth culture and by then making impression films immediately. Protein can obviously be eliminated completely by using a medium such as acid-hydrolysed casein, or by washing the organisms.

Inoculated coverslips are most easily fixed and stained in watch glasses or, better, in staining blocks which do not swivel when the coverslips are removed. Slides are stained in jars or on parallel glass rods laid across a sink.

Heat fixation. After drying in air, the film is fixed by passing the slide or coverslip rapidly through a Bunsen flame. This is only satisfactory for crude work. This degree of heating will not kill many species, especially spore-formers (Rudat, 1955), and heat-fixed films should be regarded as infectious.

Chemical fixation. (*a*) *Osmium tetroxide*, OsO_4 ('osmic acid'). This consists of greenish yellow crystals that dissolve in water (solubility in water = 6·2 g./100 ml. at 25°). The tetroxide is extremely volatile and is commonly sold in sealed ampoules containing, for example, 0·1 g. crystals which are dissolved in water to give a 1–2 %, w/v, solution. *The fumes are highly toxic and may injure the eyes.* Solutions are easily reduced to the brownish compound, $OsO_2.2H_2O$, and are best kept in glass-stoppered bottles in darkness.

Wet or dry films are fixed by exposure for 3 min. to the vapour of a 2 %, w/v, solution poured into a wide-mouthed brown glass bottle half-full of glass beads which form a platform for the coverslips. Dried or impression films are fixed by immersion for the same time. After osmic fixation, cells will not stain regularly with simple stains (p. 159), unless treated by another fixative as well, for example Schaudinn's fixative for 1–2 min. at 45°–50°.

(*b*) *Bouin's fixative*

Saturated aqueous picric acid	75 ml.
40%, v/v, formaldehyde in water	25 ml.
Glacial acetic acid	5 ml.

Solubility of picric acid in water at 25° = 1·4 g./100 ml.

(*c*) *Schaudinn's fixative*

Saturated aqueous $HgCl_2$	60 ml.
Absolute ethanol	30 ml.

Solubility of $HgCl_2$ in water at 20° = 6·9 g./100 ml.

(*d*) *Carnoy's fixative*

Glacial acetic acid	10 ml.
Chloroform	30 ml.
Absolute ethanol	60 ml.

These fixatives keep indefinitely.

Before staining, the coverslip is washed in water. After osmic fixation, the film may be dark brown but the colour goes on washing.

(*e*) *Electron microscopy* (see Kay, 1965). Opinions as to optimal conditions for fixation are always changing but it seems that in emergency no great harm is done if samples are fixed in formalin or glutaraldehyde diluted 1/40, v/v, in 0·05M-phosphate buffer, pH 7·4.

Clearing smears before staining. Broth or serum frequently obscure detail and often have as great an affinity for stains as the organisms. Gram-negative organisms are particularly difficult to see in preparations stained by Gram's method, but become readily detectable after the following treatment. Following fixation, immerse the smear in 2%, w/v, aqueous acetic acid for 2 min. Rinse with 95%, v/v, ethanol and leave the film covered with ethanol for 2 min. Wash with tap water. Then stain as usual.

Fat is troublesome in films made from milk or egg but can be removed before staining by immersing the fixed film in a fat solvent like ether or chloroform for 2 min. (see Levine & Black, 1948). Otherwise, the untreated film can be stained by a modified periodic-acid-Schiff technique which stains bacterial nucleic acid and polysaccharide but not milk fat or protein (Moats, 1959).

Temporary mounts. Films are often mounted in water under a cover slip, and last for several days if properly sealed like wet mounts. More lasting preparations are made by sealing the coverslip to the slide by one of the many adhesive mounting media that are sold.

Permanent mounts. (*a*) *Farrant's medium*. Dissolve 10 g. gum arabic in

10 ml. distilled water and then add 5 g. glycerol. This medium has the advantage that films can be mounted directly from water.

(*b*) *Canada balsam* dissolved in an organic solvent, usually xylol. Films must be either dry or dehydrated by passage through increasing strengths of ethanol and xylol (which may decolorize the cells) before mounting.

(*c*) *Neutral mounting media*, such as DPX (Distrene-Plasticizer-Xylol: Kirkpatrick & Lendrum, 1939) consisting of polystyrene dissolved in xylol with a plasticizer, tricresyl phosphate, to prevent retraction of the mount during drying.

General stains

Positive stains like methylene blue or Gram's stain colour the organisms and other material fixed on the slide, whereas negative stains fill the interstices between the organisms which therefore appear unstained in silhouette (see Maneval, 1934).

Positive stains. Any of these solutions will stain the majority of species in 30–60 sec. at room temperature. After staining, rinse in tap water.

Methylene blue (Löffler)

Saturated alcoholic solution of methylene blue	30 ml.
0·01 %, w/v, potassium hydroxide	100 ml.

Solubility of methylene blue in 95 %, w/v, ethanol at 20° = *ca.* 1·6 g./100 ml.

Carbol fuchsin (Ziehl)

Basic fuchsin	1 g.
Absolute ethanol	10 ml.
5 %, w/v, aqueous phenol	100 ml.

Dissolve the dye in ethanol and add to the phenol.

The following form 1 %, w/v, solutions in distilled water: methyl violet, crystal violet, malachite green.

Gram's stain. This is only one of many methods suggested for this stain.

(1) Immerse the fixed film in 1 %, w/v, aqueous methyl violet for 30 sec. Rinse in tap water.

(2) Immerse in Burke's iodine solution for 1 min.

(3) Pour off the iodine. Decolorize by adding acetone and, almost immediately, wash under the tap since acetone acts very rapidly. A slower decolorizing agent is 95 %, v/v, ethanol, applied until stain ceases to leave the film.

(4) Counterstain with 0·5 %, w/v, safranin, 1 %, w/v, neutral red, or Ziehl's carbol fuchsin diluted 1/20, v/v, in distilled water.

(*a*) *Iodine solution*

Potassium iodide	2 g.
Iodine	1 g.
Distilled water	100 ml.

Take a mortar able to contain 30–50 ml. liquid and place in it 2 g. KI, previously dried in a desiccator over P_2O_5, and 1 g. iodine. After grinding for a few seconds, add 1 ml. distilled water. Continue grinding, adding 5 ml. water after a short while, and the KI and iodine should dissolve. Then add 10 ml. water, mix, and transfer to the stock bottle. Rinse the mortar and pestle, put the rinsings in the stock bottle, and make up to 100 ml. with water. The solution is stable for months (Bartholomew, 1962).

Both Gram-negative and Gram-positive organisms are stained initially by methyl violet, but, after treatment with iodine solution, Gram-positive species are not decolorized by acetone. Gram-positive species therefore appear dark violet-black whereas Gram-negative organisms take up the colour of the counterstain, which, in this case, is red. The mechanism of the stain is still debatable (see Salton, 1963).

All of the methods described in the literature may give unexpected results (see Bartholomew & Mittwer, 1952). The difficulty is that until the mechanism of the stain is understood, there is no independent method for determining how a particular bacterial culture should react. Gram-negative organisms appear Gram-positive if the film is too thick and hinders decolorization. Gram-positive organisms may be over-decolorized, especially if they come from stationary phase or dying cultures, or the methyl violet may be masked or displaced by the counterstain. The safest rule is to become thoroughly familiar with one method and the results it gives in different circumstances. Known species can be included on the slide as controls.

A modification found to give reproducible results also uses acetone for decolorization but slows its action by mixing it with iodine (Preston & Morrell, 1962). The timing of each stage is not critical.

(1) Flood the film with Hucker's crystal violet for 30 sec.

(2) Remove by flooding with Burke's iodine solution which is then left for 30 sec.

(3) Wash off the iodine with a mixture of acetone and iodine (v.i.) and leave it covering the slide for 30 sec.

(4) Wash well with tap water.

(5) Counterstain as usual.

(*b*) *Hucker's crystal violet.* Dissolve 2 g. crystal violet in 20 ml. of 95 %, v/v, ethanol. This 20 ml. of solution is added to 80 ml. of 1 %, w/v, aqueous ammonium oxalate to give the stock solution which should be allowed to stand for 48 hr. before use (Bartholomew, 1962).

(*c*) *Acetone–iodine solution* (Preston & Morrell, 1962). This contains 96·5 ml. acetone + 3·5 ml. Liquor Iodi Fortis, B.P. (iodine, 10 g.; KI, 6 g.; methylated spirits, 74 O.P., 90 ml.; and distilled water, 10 ml.).

Sandiford's counterstain is used in detection of Neisseria in films of exudates, preferably after staining with Hucker's crystal violet and decolorization with acetone.

Malachite green	0·05 g.
Pyronine G	0·15 g.
Distilled water	100 ml.

This keeps for about one month. Gram-positive and -negative bacteria are stained purple and red respectively, but the background is bluish green instead of red as with the usual counterstains. Animal cells have blue-green nuclei and cytoplasm. Gram-negative species other than Neisseria may stain green if the cells are in the late stationary phase of growth (Sandiford, 1938).

For staining organisms in histological sections by Gram's method, see Ollett (1947).

Giemsa's stain. This is perhaps the best known of the large class of Romanowsky stains, each consisting of complex mixtures formed by the interaction of methylene blue and eosin. The chemical nature of the method is not fully understood (see Conn, 1961) and, as the precise nature of the stain evidently depends on details of preparation, it is preferable to buy it from commercial sources. Even so, differences between batches are the rule.

All stains of this type produce a polychrome effect: i.e. the colour of different structures extends from red in the case of so-called acidophilic material, through purple, to blue for basophilic material. Considerable differentiation is obtained and these stains are therefore widely used for smears containing a mixture of cells, for example bacteria and phagocytes, malarial parasites in blood cells, or viral inclusion bodies in host cells. Giemsa is also used for Robinow's method for bacterial nuclei.

Films of exudates are stained as follows:

(1) Fix air-dried films in absolute methanol for 3 min.

(2) Stain for 1 hr. in a staining jar with 10–20 %, v/v, Giemsa in 0·01 M phosphate buffer, pH 7·0. A weaker concentration of stain can be used for a longer time, such as 1–2 % for 24 hr.

(3) Wash with buffer and leave the film in buffer for 30 sec. to differentiate.

(4) Blot gently and dry in air.

Stains for acid-fast species. Certain organisms, notably tubercle bacilli and other mycobacteria, are not stained by ordinary positive stains unless they are used either with phenol and heat (Ziehl–Neelsen method) or with detergent at room temperature. The usual stain is carbol fuchsin but the fluorescent dye, auramine, is frequently used for fluorescence microscopy. Once stained, these species cannot be decolorized with acid: *Mycobacterium tuberculosis* also resists alcohol, unlike saprophytic acid-fast organisms, and both acid and alcohol are therefore used in diagnostic work.

For the *Ziehl–Neelsen method* (Cruickshank, 1965), (1) place a slide carrying a heat-fixed film on a slide carrier over a sink. Flood with carbol fuchsin solution and heat until steam rises. Leave for 5 min., heating occasionally to keep the stain steaming.

(2) Wash with water.

(3) Pour on 20 %, v/v, H_2SO_4, wash off with water, and repeat several times until the film is a faint pink. Finally, wash well with water.

(4) Treat with 95 %, v/v, ethanol for 2 min.

(5) Wash with water.

(6) Counterstain with 0·2 %, w/v, malachite green. Wash and blot dry. Acid- and alcohol-fast organisms are red; other organisms are green.

The carbol fuchsin solution contains

Basic fuchsin	5 g.
Phenol (crystalline)	25 g.
95 % or absolute ethanol	50 ml.
Distilled water	500 ml.

Dissolve the fuchsin in phenol by placing them in a 1 l. flask over a boiling-water bath for 5 min., shaking the contents from time to time. When the dye is completely dissolved, cool, add the ethanol, and mix thoroughly. Then add the distilled water. Filter before use.

Staining at *room temperature* requires detergent. Dissolve 3·5 g. basic fuchsin in 12·5 g. pure phenol at 80° or place in a boiling-water bath. When dissolved, cool and add 25 ml. of 95 %, v/v, ethanol and mix well. Make up to 300 ml. with distilled water. While stirring, slowly add 30 drops of Tween 80. To use, filter the stain on to heat-fixed slides. Stain for 3 min. and then wash off with water. Decolorize with acid and ethanol, and counterstain with malachite green (Auber, 1950).

Auramine can be used alone as a fluorescent stain for acid-fast organisms but is usually combined with rhodamine for better contrast (e.g. Silver, Sonnenwirth & Alex, 1966).

The stain contains

Auramine O	1·5 g.
Rhodamine B	0·75 g.
Glycerol	75 ml.
Phenol crystals liquefied at 50°	10 ml.
Distilled water	50 ml.

1. Smears are air-dried and fixed by flaming.
2. Filter the stain through Whatman no. 2 paper.
3. Stain slide on a rack in a 37° incubator for 30 min. Staining troughs must never be used with *Mycobacteria* since organisms can enter the stain and be deposited on subsequent slides to give false-positive results.
4. Decolorize with a mixture of concentrated HCl, 0·5 ml., NaCl, 0·5 g., 70 %, v/v, ethanol, 99·5 ml.
5. Counterstain for 2 min. with a solution of 0·5 %, w/v, aqueous $KMnO_4$. This greatly decreases non-specific fluorescence of the background but, if applied too long, will decolorize the organisms.

Suitable excitation and barrier filters for the HBO200 lamp are the BG-12 and the OG-1, respectively.

Mycobacteria typically appear as distinct slender slightly curved rods, often beaded, that are stained reddish-yellow on a black background. *Myco. leprae* is less alcohol-fast than *Myco. tuberculosis* and smears should be treated after staining with 25 % H_2SO_4 for 20–30 min. or with 0·5 % HCl without alcohol.

Negative stains. The background is rendered semi-opaque by the stain and the organisms, which are not penetrated by the stain, appear as light objects on a homogeneous dark ground. Organisms and structures, like spirochaetes and extracellular slime, that are difficult to demonstrate by positive stains are easily seen by this method. Spores appear as refractile areas in the bodies of vegetative cells.

The stain is either applied to a dried fixed film and itself allowed to dry; or a drop of stain and a drop of bacterial suspension are mixed at one edge of a slide and spread lengthwise before they dry. The thickness of the film then usually decreases along its length so that at least one part should be of suitable opacity.

The common negative stains are indian ink (p. 165) or nigrosin, an aqueous solution of a mixture of a yellow and a blue dye which forms a black background. Nigrosin, as supplied, is usually too concentrated and better contrast is often obtained with a 1/50 dilution in water. Negative staining can be combined with positive staining, although nigrosin often decolorizes other stains. It can, however, be used successfully with phloxine B, by dissolving 1 g. phloxine in 98 ml. distilled water and adding 2 ml. aqueous nigrosin (Meynell & Meynell, 1966).

STAINS FOR THE PRINCIPAL CELL STRUCTURES

For other methods, including those for polysaccharide, lipid, and polymetaphosphate (volutin), see the Symposium (1956), Wilkinson & Duguid (1960), Murray (1960) and Robinow (1960*a*, *b*).

Cell walls

(1) Inoculate an agar plate with the organisms and cut out a block of agar. Transfer it face downwards to a coverslip. Fix the whole block by immersion in Bouin's fluid overnight. Remove the agar next day, leaving the organisms on the glass.

(2) Mordant with 5–10 %, w/v, aqueous tannic acid for 20–30 min. by floating the coverslip face downwards on the surface of the acid. Another mordant is 1 %, w/v, phosphomolybdic acid for 5 min. at 20° (Hale, 1953).

(3) Wash in water.

(4) Stain the cell wall by floating on 0·02 %, w/v, aqueous crystal violet for 5–10 sec. Cell walls and septa appear black or dark blue, while the cytoplasm is unstained (Robinow, 1949).

There are considerable differences in the ease with which the walls of different genera are stained, those of *Bacillus* being most easily shown. A variation of the above method is to stain in step 4 with 0·5 %, w/v, aqueous crystal violet for 1·5–2 min.; wash; decolorize with 0·5 %, w/v, aqueous congo red for 2–3 min.; and to wash and blot dry (Webb, 1954). With *Bacillus*, cell walls and nuclei are demonstrable simultaneously. Fix in osmium vapour for 1·5–2 min. Hydrolyse in 1 N-HCl for 7 min. at 60° and wash in water. Stain with 0·1 %, w/v, basic fuchsin for 5 min.; wash; mordant in 10 %, w/v, tannic acid for 5 min.; wash; and finally stain again with basic fuchsin for 5–10 sec. (Cassel, 1951).

Capsules and slime

These components, being composed of highly hydrated gummy or gelatinous material, are easily distorted so that fixed specimens are of value only for knowing whether extracellular material is present. To obtain a realistic idea of its size or structure, wet preparations are needed. In the usual indian ink methods capsules appear as undifferentiated haloes surrounding the cells but after treatment with specific antibody, the capsules of certain *Bacillus* species are seen by phase-contrast microscopy to possess septa extending laterally from the wall which intersect the capsule proper (see Tomcsik, 1956*a*, *b*, 1961). These changes produced by combination with antibody are now known to be responsible for the phenomenon of so-called 'capsular swelling', in which capsules become visible by conventional transmitted light microscopy after mixing organisms with antibody. Non-specific reactions also occur between capsules and various proteins like casein, apparently by co-precipitation at the isoelectric point (Tomcsik & Guex-Holzer, 1954). Such studies can be usefully extended by initially treating the organisms with lysozyme which removes the capsular septa (Tomscik, 1956*b*). These immuno-cytological methods are potentially of great value but have so far been applied to relatively few species.

Wet indian ink method (Duguid, 1951). The ink must be dense and contain only fine particles. Dilute inks can be concentrated by evaporation. Coarse particles can be disintegrated by shaking 4 vol. of ink with 1 vol. of grade 12 glass beads (0·2 mm. diam. ballotini) for 60 min. in a shaker.

Place a large loopful of undiluted ink on a clean slide. To this add a small part of a colony or the pellet of a centrifuged broth culture. Place a clean coverslip on the mixture and press it down gently under a pad of blotting paper. Capsules appear as clear light zones between the refractile cell outlines and the dark background of the ink.

Protoplasts

With the majority of stains, the boundary of the protoplast normally merges with the cell wall and cannot usually be distinguished by light microscopy. However, Victoria blue B stains the cell membrane preferentially and the protoplast can then be seen, especially if it is first shrunk by one of the methods described by Robinow & Murray (1953). The cells may be plasmolysed by suspension in a drop of 5 %, w/v, KNO_3, dried in air, and fixed in 70 %, v/v, ethanol. Some shrinkage occurs with growing but not stationary phase *Bacillus* when unfixed organisms are dried in air; fixation is for 5 min. in saturated aqueous picric acid or in Bouin's fixative. Alternatively, *Bacillus* growing on agar is exposed to ether, acetone or chloroform vapour for no more than 5–10 min. followed by immediate fixation with Bouin's solution before lysis occurs. For staining, the fixed films are rinsed in water and stained for 20–40 sec. in 0·05 %, w/v, Victoria blue B. Exposure to saturated aqueous $HgCl_2$ before staining prevents diffusion of the stain from the membrane into the cytoplasm when preparations are examined in water mounts.

Spheroplasts derived from whole cells are readily seen by phase-contrast or dark-field microscopy and can be stained with conventional cytoplasmic or nuclear stains (e.g. Spiegelman, Aronson & Fitz-James, 1958).

Colonies of *L* forms or *Mycoplasma* growing on agar can be fixed *in situ* with Bouin's fixative (method *b* on p. 157) followed by staining with Giemsa, or agar slices can be examined (Dienes, 1967).

Nuclei

The three accepted methods for demonstrating nuclei are phase-contrast microscopy of unstained organisms suspended in media of high refractive index (p. 146); fluorescence microscopy of films stained with fluorochromes (p. 151); and bright field microscopy of organisms stained with Giemsa. For counting nuclei, the last method is preferable as the contrast between nucleus and cytoplasm is far higher than by the other methods. Before staining, the organisms are treated with warm HCl. The rationale is that, although Giemsa stains the nuclei of cells not exposed to HCl, the RNA of the cytoplasm is also stained so deeply that nuclei are barely visible (Wade, 1955). Indeed, the nuclei may appear as relatively unstained areas in the heavily stained cytoplasm (Murray, 1960). Treatment with HCl or ribonuclease hydrolyses RNA, so preventing staining of the cytoplasm, while acid hydrolysis also increases the affinity of the nuclei for the stain (see Robinow, 1956, 1960*a*; Kellenberger, 1960).

Robinow (1943–4) gives two methods. The simplest is to dip unfixed air-dried impression preparations in boiling 0·2 N-HCl for 5 sec. Rinse in water and mount in 0·1 %, w/v, aqueous crystal violet. The second and more usual method is as follows. Cut a small block of agar out of an inoculated agar plate and place face upwards on a coverslip. Fix in osmic acid vapour for 2–3 min., and then place face downwards on a clean dry coverslip. Alternatively, the block of agar is placed face downwards on the coverslip and fixed by allowing Bouin's fluid to diffuse through the agar block overnight. In either case, the fixed films are dried in air, treated with Schaudinn's fixative for 1–2 min. at 45°–50°, rinsed, and stored in 70 %, v/v, ethanol. In staining, first place the films in 1 N-HCl for *ca.* 10 min. at 60°. Rinse in tap water and two changes of distilled water, and float the coverslips face downwards on Giemsa's stain added to 0·01 M phosphate buffer, pH 7·0, in the proportions 2–3 drops/ml. If films are to be examined in water mounts, staining for 30 min. at 37° is sufficient; several hours are needed if films are to be dehydrated and mounted in balsam.

Intracellular inclusions

Localized accumulations of various polymers appear in stationary phase cells grown in the appropriate medium (see Duguid & Wilkinson, 1961) and are presumably a manifestation of 'unbalanced growth'.

Polyhydroxybutyrate ('lipid'). Fix an air-dried film by gentle flaming. Flood with alcoholic Sudan Black (0·3 g. stain dissolved in 100 ml. 70 %, v/v, ethanol; shaken at intervals; and allowed to stand overnight before use). Leave at room temperature for 5–15 min. Drain. Clear with xylol. Blot. Counterstain with 0·5 %, w/v, aqueous safranin for 5–10 sec. (Burdon, 1946).

Fat is seen as black deposits within the cells. These should be removed specifically by previous treatment with lipase. Considerable internal structure is evident by electron microscopy (Ellar *et al.* 1968).

Polyphosphate ('volutin' or 'metachromatic' granules). Fix an air-dried film by gentle flaming. Apply stain for 3–5 min. Wash in tap water. Blot. Flood with iodine for 1 min. Wash and blot dry. Stain: toluidine blue, 0·15 g., malachite green, 0·2 g., glacial acetic acid, 1 ml., 95 % ethanol, 2 ml.; distilled water, 100 ml. Iodine solution: iodine 2 g., potassium iodide, 3 g., distilled water, 100 ml.

Granules are bluish-black and cytoplasm is green (Laybourn, 1924). The biochemical genetics of polyphosphate formation are reviewed by Harold (1966).

Polysaccharide. Many species develop inclusions that stain by the periodic acid-Schiff method (e.g. Hotchkiss, 1948) but the method is not entirely specific (see Vidra, 1956). In some cases, starch granules are formed and are easily stained blue merely by suspending the organisms in Gram's iodine (Carrier & McCleskey, 1962).

Flagella

The flagella of most species are only 10–30 nm. in thickness and are therefore beyond the limits of resolution of ordinary bright field microscopy. However, they are visible by dark field microscopy, especially when their motion is slowed by suspending the organisms in 0·5 %, w/v, methylcellulose (Pijper, 1947; Mitani & Iino, 1968). Staining reveals flagella by coating them with a precipitate of stain, so that they are enlarged as well as rendered opaque.

Para-rosaniline method (Leifson, 1951). (1) Slides must be absolutely clean and free from grease. Just before use, flame each slide briefly and allow to cool. Draw a transverse line across the slide with a grease pencil and continue the line round the edges of half the slide so as to form a rectangle.

(2) The bacteria to be stained must be freed from broth, for example by centrifugation twice with resuspension in distilled water. Flagella may be damaged while resuspending the pellet of deposited organisms: this is best done by adding water to the pellet, leaving it to stand for a minute or two, and by then gently tapping the tube. Bacterial growth on solid medium can be gently suspended in distilled water.

Place a loopful of the suspension at one edge of the rectangle marked on the slide, and let it run to the opposite edge by tilting the slide. Dry in air at room temperature.

(3) Place exactly 1 ml. stain on the smear without allowing any to flow over the wax line. Separate fixation is not needed owing to the tannic acid in the stain. Leave at room temperature for 7–15 min.; the best period has to be determined by trial and error. Otherwise, view the slide against a dark background and illuminate it from beneath by a beam of light. As soon as a precipitate forms throughout the film, remove the stain (Leifson, 1958).

(4) Rinse gently under the tap without pouring off the stain.

(5) If the bacterial bodies do not stain red, they can be counterstained with a positive stain, like methylene blue.

The stain is prepared from the following solutions:

(1) 1·5 %, w/v, NaCl in distilled water.

(2) 3 %, w/v, tannic acid in distilled water, choosing a sample that forms a light yellow solution.

(3) 0·9 %, w/v, para-rosaniline acetate and 0·3 %, w/v, para-rosaniline hydrochloride in 95 %, v/v, ethanol. These take several hours to dissolve, even with shaking. Basic fuchsin certified for flagellar staining can be used at 1·2 %, w/v, in 95 %, v/v, ethanol in place of the rosaniline salt.

Mix equal volumes of the three solutions and store in a tightly stoppered bottle. The mixture lasts several weeks at 4°, and months or years at −10° to −20°.

The precipitation of stain on flagella in the above method is accompanied by considerable deposition on the slide, and the price of heavily stained flagella is usually a very dirty background. Flagella can often be demonstrated simply by mordanting air-dried films with 10 % aqueous tannic acid for 5 min., followed by staining with Ziehl–Neelsen's carbol fuchsin for 30 sec. The flagella are distinct but very thin, while the background is clean. Note that formalin fixation alters flagellar morphology in some species (Leifson, 1961).

Spores

Spore formation is now arbitrarily divided into seven stages covering a period of about six hours, largely as a result of electron microscopy of thin sections of *Bacillus* species (see Murrell, 1967): I, condensation of bacterial nuclei to form an axial thread; II, formation of a membranous cross septum at one end of the bacterium; III, encirclement of the spore protoplast by further membrane growth; IV, deposition of a cortex outside the membrane; V, coat formation; VI, 'maturation', the onset of heat-resistance; and VII, liberation of the free spore from the vegetative cell. Although electron microscopy is essential for a detailed examination of structure, the stages of sporulation are distinguishable by staining and by phase or interference microscopy (see Robinow, 1960*b*). The nuclear changes of stage I are visible after staining with nuclear stains like the osmic acid-HCl method. Septum formation in stage II and enclosure of the spore protoplast in stage III can be visualized in *Bacillus cereus* by mounting live cells in a drop of 0·03 %, w/v, aqueous crystal violet (Gordon & Murrell, 1967). When the coat appears in stage V, the spore ceases to be stained by simple stains, including nigrosin. Thereafter, it becomes increasingly refractile, as its refractive index increases, until the end of stage VI. Stages I–III cannot be separated by phase-contrast microscopy but II and III may be identified using interference optics (Hitchins, Kahn & Slepecky, 1968).

Germination is accompanied by a reverse of these changes in which the spores rapidly become non-refractile (Levinson, Hyatt & Holmes, 1967).

Traditional spore stains were devised for relatively mature spores in stages V onwards, which resisted positive stains like methylene blue. Either the stain, usually carbol fuchsin, was mixed with phenol and heated, as in the Ziehl–Neelsen method for acid-fast bacilli (p. 162); or conventional stains were used at room temperature after exposing the spores to acid, heat or mechanical pressure (Lechtman *et al.* 1965). In the first method, the spore remains stained after the film is treated with acid, although the vegetative part of the sporing cell is decolorized and can be counterstained.

(1) Place a slide carrying a heat-fixed film on a slide carrier placed across a sink. Flood with carbol fuchsin solution (p. 162) and heat until steam rises. Leave for 3 min., heating occasionally.

(2) Wash with water.

(3) Pour on either 5 %, w/v, Na_2SO_3 or 0·25–0·5 %, v/v, H_2SO_4. Leave for 1–5 min. and then wash well with water.

(4) Counterstain with methylene blue or 0·2 %, w/v, aqueous malachite green. Wash and blot dry. The spores are red and the rest of the organism takes the colour of the counterstain.

AUTORADIOGRAPHY

The autoradiographic localization of isotopically-labelled compounds in bacteria and phage is largely confined to studies of tritium (^{3}H)-labelled DNA. Two typical applications are in measuring the segregation of labelled DNA at bacterial division (Forro, 1965; Lark, 1966) and the amount of donor DNA transferred to the recipient during conjugation (Gross & Caro, 1966). Note that only thymine-requiring strains incorporate exogenous ^{3}H-thymidine efficiently unless a high concentration of a nucleoside is present, e.g. deoxyadenosine, 250 μg./ml. (Boyce & Setlow, 1962).

In general, the resolving power of the technique is determined by the type of particles emitted by the isotope and by the distance between the source and the emulsion. The higher the energy of the particles, the greater the area over which a given number of tracks are distributed, so that resolution is diminished. Separating the emulsion and source has the same effect. The advantages of ^{3}H as a label are not only that it is readily introduced into a wide variety of substrates but that the β particles emitted are of low energy compared to those arising from ^{14}C or ^{32}P and therefore travel only short distances ($\leqslant 3\ \mu$). This in turn simplifies quantitative analysis, since the thickness of emulsion obtainable in practice invariably exceeds 3 μ so that it must contain within it all the β particles arising from the source. The efficiency with which β particles are detected as grains has been determined for several emulsions: the value for Ilford L-4 Nuclear Emulsion, a particularly fine grain emulsion also suitable for electron microscopy, is 1·31 grains per decay (Caro, 1967). It should be appreciated that the shape of the grains revealed by electron microscopy does not indicate the particle's path through the emulsion. Tortuous filamentous grains are produced by the usual 'chemical' developers like Kodak Microdol-X or Ilford ID-19, whereas smaller comma-shaped grains are formed with a 'physical' developer like the paraphenylenediamine formula of Caro & van Tubergen (1962). The practical details that follow are intended as a guide to only one of the techniques described (see Kopriwa & Leblond, 1962; Rogers, 1967).

Microscope slides should have a ground writing area and be 0·8–1·0 mm. thick. After cleaning, they are 'subbed', i.e. coated with a thin layer of gelatin to improve adhesion of the emulsion and to prevent its distortion on drying. *Subbing solution* contains 0·5 %, w/v, purified gelatin (e.g. Difco) dissolved in distilled water to which is added chrome alum ($KCr(SO_4)_2 . 12H_2O$) to a final concentration of 0·01 %, w/v. Slides are dipped in the molten solution, dried at room temperature in a clean atmosphere and stored in slide boxes. The bacterial suspension, previously fixed with 1 %, v/v, formalin and washed free of unassimilated label on a membrane filter, is suspended in dilute broth which acts as a wetting agent and a drop gently spread on the slide without scratching the gelatin film. Fixation is either in Carnoy, followed by rehydration through alcohols, or in 5–10 %, w/v, aqueous trichloracetic acid for 10 min., followed by washing in two lots of distilled water for 2 min. Slides are then dried in air.

Working in the darkroom with an Ilford 902 safelight, 30 g. of L-4 emulsion (supplied in the form of rubbery lumps) are dissolved in 20 ml. or less of distilled water at 45°. Initially, the surface of the solution will be covered with bubbles but if it is stirred carefully for about 20 min. while it is dissolving and then left to stand for an hour, most of the bubbles disappear and the remainder can be taken off by dipping a few clean slides.

Each of the slides is dipped in turn, held vertically for a few moments while wiping the back and then placed face upwards on a horizontal surface to dry slowly. Rapid drying distorts the gelatin and it is often convenient to leave the slides for 3–5 hr. in a light-proof box at room temperature. Finally, the dried slides are stored at 4° in light-proof boxes (Clay Adams, black plastic, from Arnold Horwell Ltd) well away from radio-active sources. A small packet of self-indicating silica gel is included to take up condensation.

Developers are dissolved and used according to the maker's instructions. After slides are developed, they are rinsed successively in distilled water, 1 % aqueous acetic acid and distilled water and, finally, are fixed for 1 min. before being washed in three changes of distilled water.

For subsequent light microscopy, bacteria can be stained with dilute (0·3 %, w/v) aqueous methylene blue for 20 sec., rinsed and dried. Alternatively, films may be stained before dipping but, if this is done, the preparation must first be separated from the emulsion, e.g. by twice dipping in 0·25–1 %, w/v, collodion dissolved in equal parts of ether and absolute ethanol. The collodion is smeared with a little egg-albumin to allow the emulsion to adhere. The extra thickness somewhat lowers the sensitivity of the method for ^3H (Kopriwa & Leblond, 1962). All preparations are mounted under a coverslip before examination (DPX is suitable, p. 159).

The results obtained with bacteria possessing ^{3}H-DNA necessarily depend in part on the generation time of the culture, since this determines the DNA content per rod. Using ^{3}H-thymidine at a specific activity of 2 C./mM and assuming 1000 μg. DNA per rod, an average of about 6 grains per rod per day's exposure should be obtained with fully labelled thymine-requiring organisms. (Note that ^{3}H-thymidine may need purifying to remove traces of labelled contaminating compounds: Forro, 1965; Gross & Caro, 1966.) The large size of the grains in relation to that of an individual bacterium limits the number of grains that can be counted accurately, though larger numbers are countable if each rod is allowed to form a microcolony on agar before being fixed and covered with emulsion (Forro, 1965). The number of grains is expected to differ from cell to cell, even if these are identical and labelled to the same extent, owing to the random disintegration of the isotope. Since the mean number of disintegrations per rod is small, the distribution of grains should be Poissonian, and appropriate methods for estimating the mean and also the homogeneity of the grain counts are discussed on p. 177.

6

QUANTITATIVE ASPECTS OF MICROBIOLOGICAL EXPERIMENTS

Quantitative experiments in microbiology, as in other fields, are generally of two types: those whose object is to determine the intrinsic nature of a biological mechanism such as disinfection by examining the way in which it acts; and those concerned purely with measurement, like viable and total counts. The first type concerns the testing of hypothetical models, while the second in its more sophisticated forms merges into the theoretical designs of biological assay. The two are not exclusive, for many titrations are based on assumptions sufficiently precise to be regarded as the foundations of a model, although others use a relation between, say, dose and intensity of host response which has convenient mathematical properties but no plausible biological interpretation. Typical examples of the latter are the transformed values of response times discussed later on p. 216. This chapter indicates how hypothesis and practice are related in the commoner techniques and describes the elementary treatment of the data they yield. The reader is assumed to understand the broad significance of simple statistical terms, but proofs of the various equations and details of the appropriate methods of statistical analysis have largely been omitted. If needed, they will be found in monographs and reviews dealing with statistics in its general and its microbiological aspects, and with biological assays. These also discuss analysis of variance and the design of multifactorial experiments which test the influence of several variables simultaneously (for examples, see Mitchison & Spicer, 1949; Long, Miles & Perry, 1954; Fothergill & Hide, 1962). Calculation is often avoided by using the nomograms, graphs, and tables mentioned in the text.

QUANTAL (ALL-OR-NONE) RESPONSES

The first part of this chapter concerns *quantal responses*, which, by definition, can only be one or other of two alternatives. Tubes of broth in a dilution count are scored as clear or turbid; bacteria in a disinfection experiment as dead or alive; and colonies as fermenting or non-fermenting in type. The second part of the chapter discusses *quantitative responses* which can take any one of a series of values. Thus, a tube of broth would not be recorded as just clear or turbid, but as having a certain degree of turbidity.

Many experiments entail making one or more dilutions of a concentrated suspension of organisms from which samples are taken for inoculation of counting chambers, tubes of broth, tissue cultures, and so on. On other occasions, phage particles are added to bacterial suspensions, or virus particles placed on monolayers of susceptible animal cells. The feature common to all such experiments is a distribution of organisms amongst different recipients, and, from the statistical point of view, the experiment often resolves itself into the quantitative study of this distribution. In general, therefore, we often have to consider how discrete *units*, like the cells of a bacterial culture, are distributed amongst different *sites*, like tubes of broth.

In microbiology, the fate of each unit can usually be regarded as randomly and independently determined. Thus, it is generally safe to assume that whether a given bacterium or virus particle will be present in a given sample taken from a suspension is purely a matter of chance. An obvious exception would arise if the organisms were clumped, as individual organisms would then not be independently distributed in suspension.

The behaviour of units which are randomly and independently disposed is predictable from the binomial distribution and from its approximations, the Poisson and the normal distributions, which follow here.

THE BINOMIAL DISTRIBUTION

Many experiments are concerned with the proportion of sites, whether they be bacteria, animals, tubes of broth, etc., that give a predetermined quantal response when the chance of a given site responding is x. The binomial theory states that for N sites the probability of $N, N-1, N-2, \ldots,$ 3, 2, 1, 0 sites responding is given by successive terms of the expansion of the binomial $(x+y)^N$. Since each site can only be in one of two states, y necessarily equals $(1-x)$.

In a coin-tossing experiment, x, the chance of a given coin coming up heads, is assumed to be 1/2 and to be unaffected by whether other coins come up heads or tails. For two coins, $N = 2$, and as

$$(x+y)^2 = x^2+2xy+y^2,$$

the binomial $(0{\cdot}5+0{\cdot}5)^2 = 0{\cdot}25+0{\cdot}5+0{\cdot}25$. That is, the chance of two heads is x^2 ($= 0{\cdot}25$); of one head and one tail is $2xy$ ($= 0{\cdot}5$); and of two tails is y^2 ($= 0{\cdot}25$).

Another case covered by the binomial is the proportion of red colonies on a plate also bearing white colonies. Suppose that the fractions of red colonies on two plates are 3/16 and 6/20 respectively and, if the plates were inoculated from the same suspension, that we wish to know the probable limits within which the true proportion of red colonies lay. The binomial applies because this proportion can also be thought of as the probability, x, that a given colony will be red: y is the probability of it being white. The observed value of x is $(3+6)/(16+20) = 9/36 = 0{\cdot}25$. The standard error (S.E.) of x is $\sqrt{(xy/N)}$, where N is the total number of colonies. In this example, the S.E. is $\sqrt{(0{\cdot}25 \times 0{\cdot}75/36)} = \sqrt{0{\cdot}00521} = 0{\cdot}072$. There is, as usual (p. 202), a probability of 0·95 that the 'true' value of the proportion lies in the range,

$$\begin{aligned} 0{\cdot}25 \pm 1{\cdot}96\ (\text{S.E.}) &= 0{\cdot}25 \pm (1{\cdot}96 \times 0{\cdot}072) \\ &= 0{\cdot}25 \pm 0{\cdot}14 \\ &= 0{\cdot}11\text{–}0{\cdot}39. \end{aligned}$$

The confidence limits can be read directly from Figures provided by Berkson (1929) and by Pearson & Hartley (1954, pp. 204, 205).

In other experiments, it might be suspected that the proportions of red colonies on the two plates were really different; for instance, because they were inoculated from different suspensions. The observed proportions of red colonies are respectively 3/16 (= 0·1875) and 6/20 (= 0·3). The S.E. of the difference between two proportions is $\sqrt{[(x_1y_1/N_1)+(x_2y_2/N_2)]}$. If the purpose of the experiment is to see if the two plates are independent samples of the same culture, the best estimates of x and y would be 0·25 and 0·75 as before. The S.E. of the difference is therefore

$$\sqrt{[(0{\cdot}25 \times 0{\cdot}75/16)+(0{\cdot}25 \times 0{\cdot}75/20)]} = 0{\cdot}145.$$

The difference between the observed proportions is $0{\cdot}3 - 0{\cdot}1875 = 0{\cdot}1125$. As 0·1125 is only 0·78 times the S.E. of 0·145, the observed proportions are unlikely to differ significantly (p. 202).

Infectivity titrations which use only one dilution of culture yield data that are analysed in the same way; an example is the proportions of dead animals in groups of normal and of vaccinated animals, produced by inoculating the same number of bacteria. However, the use of a single dilution is often not satisfactory, for, if the dose is misjudged, either 0 % or 100 % responses are obtained and the experiment is a failure (p. 220).

The binomial is unwieldy in practice because N is often large; if it has a value of 1000, there are 1001 separate terms to be calculated for the expansion of $(x+y)^N$. Two approximations are therefore used: the

Poisson distribution, when the mean number of units per site is fairly small ($m < 50$), and the normal distribution, applicable when $m \geqslant 50$.

THE POISSON DISTRIBUTION

A process described by this distribution is the adsorption of phage to bacteria where the units are individual phage particles and the sites are bacteria. It is evident that individual particles are distributed randomly amongst the bacteria, provided the phage is not clumped. Also, the mean number of phage adsorbed per bacterium (m) is usually < 50; in practice, m is usually between 0·01 and 10·0. It follows from the Poisson series that the proportion of bacteria

$$\left.\begin{aligned} &\text{adsorbing 0 particles} = e^{-m}, \\ &\text{adsorbing 1 particle} = me^{-m}/1!, \\ &\text{adsorbing 2 particles} = m^2 e^{-m}/2!, \\ &\text{adsorbing } y \text{ particles} = m^y e^{-m}/y!, \end{aligned}\right\} \tag{6.1}$$

where $y! = 1 \times 2 \times 3 \times \ldots, y$.

Even if a unit arrives at a site, it may not exert an effect: only 1 of every 10^4 quanta of ultra-violet irradiation absorbed by phage T2 causes inactivation (Zelle & Hollaender, 1954). Thus, it is necessary to introduce p, the probability that a unit at a site will exert a detectable effect, and, in this example, p would be 10^{-4}. The product, pm, is therefore the mean number of *effective* units per site, whereas m is the mean number of *all* units per site. Although the distinction between effective and ineffective units may arise from intrinsic differences between them, as between living and dead bacteria, all the units may be identical in their potential ability to affect a site but a proportion fail to do so because the sites are not fully sensitive. In the above example, quanta are not likely to differ, and p is 10^{-4} because a phage can adsorb *ca.* 10^4 quanta on the average without being inactivated.

The experiments described by the Poisson series can be classified by whether it is possible for phenomena to be produced by 1 unit (p. 177) or whether the co-operative action of 2 or more units is invariably required (p. 198). Phage again provides examples of both classes. A single particle of normal virulent phage is able to infect a bacterium but, after irradiation, several damaged particles have to co-operate in a bacterium before vegetative phage growth begins (Luria & Dulbecco, 1949). The large number of phenomena initiated by only one effective unit, such as initiation of growth in a tube of broth by a single bacterium, accounts for the prominence of the first term of the Poisson series in microbiology. As the first term, e^{-pm}, is the probability per site of receiving 0 effective particles, it is also the mean proportion of sites unaffected. Conversely, $1-e^{-pm}$ is the mean proportion of sites that are affected. The second and higher terms are of less practical importance, as far fewer phenomena are produced by the co-operation of 2 or more effective units. To make this point in another way: we often deal with the results of bacterial or viral growth, and, since these organisms are normally self-replicating, an individual cell or particle usually suffices to produce a response.

The following sections concern cases in which there is only a single value of m. These yield data whose compatibility with a Poisson distribution can be tested in two ways. The usual method is a straightforward χ^2 test (see Fisher, 1950, §15, 16). The alternative is a graphical method often found in the analysis of grain counts in autoradiography (though it could equally well be used to demonstrate the random distribution of cells in a counting chamber). The numbers of grains around individual foci is expected to be Poissonian since isotopes disintegrate at random. Hence, if m is the mean number of grains per focus, it follows from equation (6.1) that y grains should occur with an average frequency of $P(y) = m^y e^{-m}/y!$. Rearranged, this gives $\log_e (P(y)\, y!) = y \log_e m - m$. Thus, if the observed frequency, $P(y)$, at which y grains occur is measured, then the observations should fall on a straight line when $\log_e(P(y)y!)$ is plotted against y if the distribution is Poissonian. This is indeed found for grain counts in autoradiographs of thymine-labelled bacteria (e.g. Hanawalt *et al.* 1961). Such plots provide two estimates of m, here the mean number of grains per focus: first, from the slope of the line; and second, from its intercept with the ordinate at $y = 0$. Moreover, heterogeneity in the observed counts can be detected by inspection. When bacteria possessing DNA fully labelled with ^3H divide, the semi-conservative replication of DNA results in the appearance of half-labelled and of completely unlabelled cells in the first and second generations of growth without ^3H, and the sizes of these fractions are evident from the plots (Lark, 1966, fig. 14).

The multiplicity of phage infection

The 'multiplicity of infection' is defined as the mean number of phage particles adsorbed per bacterium (Ellis & Delbrück, 1939) and is often estimated from the proportion of uninfected bacteria (i.e. from the proportion adsorbing 0 particles). If this proportion is S and the multiplicity is m, $S = e^{-m}$, so that if 61 % of bacteria are uninfected, $S = 0{\cdot}61 = e^{-m}$. Table 6.9 shows $m \simeq 0{\cdot}5$. There is usually no direct check that the bacteria are all equally likely to adsorb phage as the model requires. This is never likely to be strictly true because individual bacteria differ in size. This disturbs the calculations only when m is large enough for a high proportion of the organisms to adsorb more than 1 particle (see appendix to Luria & Dulbecco, 1949) and the results then agree better with prediction if allowance is made for size differences (Cairns & Watson, 1956).

The terminal dilution method for the isolation of clones

This method was devised for the isolation of pure clones by Lister (1878). If the progeny of individual bacteria is to be studied, the culture is diluted and distributed amongst a number of tubes of broth. The degree of dilution is chosen so that m, the mean number of bacteria inoculated per tube, is fairly small, for example $< 0{\cdot}05$. The Poisson series then shows that the probability of a given tube receiving any given number of organisms when $m = 0{\cdot}05$ should be as follows:

$$\text{probability per tube of receiving 0 cells} = e^{-0{\cdot}05} = 0{\cdot}9512,$$
$$\text{probability per tube of receiving 1 cell} = 0{\cdot}05\,e^{-0{\cdot}05} = 0{\cdot}0476,$$
$$\text{probability per tube of receiving 2 cells} = \frac{0{\cdot}05^2\,e^{-0{\cdot}05}}{2!} = 0{\cdot}001189,$$
$$\text{probability per tube of receiving 3 cells} = \frac{0{\cdot}05^3\,e^{-0{\cdot}05}}{3!} = 0{\cdot}00001982.$$

The mean proportion of tubes receiving 1 or more cells $= 1 - 0{\cdot}9512 = 0{\cdot}0488$, and of these $0{\cdot}0476/0{\cdot}0488 = 0{\cdot}975$ probably received only 1 cell. In other words, if only the tubes becoming turbid after incubation are considered, the probability that a given turbid tube contains a clone is 0·975. This probability decreases with increase in m, the mean number of bacteria inoculated per tube (Fig. 6.1, Table 6.9).

This method is frequently used in virology for the isolation of pure clones following inoculation of animals or eggs. It is worth noting that the method must succeed if every organism inoculated invariably causes the host to respond ($p = 1$); but, at present, the validity of the method is more doubtful in the numerous systems in which $p < 1$ (p. 229). Even if each inoculated organism is independent, the physiological effects produced by one organism and the clone it generates might affect the rest of the inoculum. If $p = 10^{-4}$, for example, and animals are inoculated with a dose such that 5 % respond, the concentration of 'effective organisms' is *ca.* 0·05 per animal ($0{\cdot}05 \simeq 1 - e^{-0{\cdot}05}$), while the total number of organisms inoculated is $10^4 \times 0{\cdot}05 = 500$ organisms per animal. The assumption commonly made when applying the method is that in such animals only the single 'effective organism' will multiply to cause the response while the 499 other organisms can be totally neglected. This assumption did not hold in the systems where it was tested by the simultaneous inoculation of mixtures of marked strains, so that its general validity must be doubtful (Liu & Henle, 1953; Meynell & Stocker, 1957). Exactly the same considerations apply

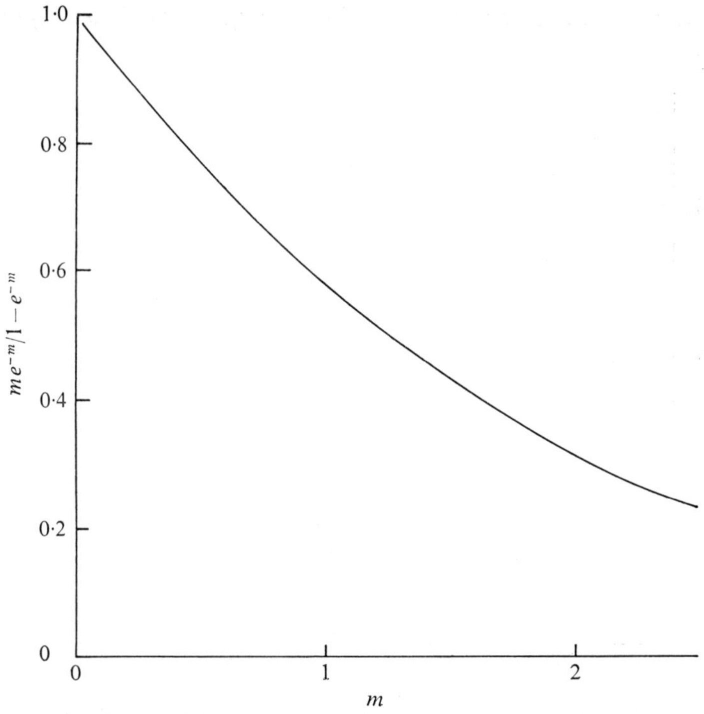

Fig. 6.1. Isolation of clones by the terminal dilution method. Ordinate: probability that a given turbid tube contains only one clone (Table 6.9). Abscissa: m, mean number of bacteria inoculated per tube.

to the isolation of pure clones by subculture of an isolated lesion like a pock on the chorio-allantois, which is assumed to be a pure clone regardless of any 'ineffective' organisms which may have been present in the area occupied by the lesion.

A striking use of the terminal dilution method for the separation of clones was in showing that bacteria killed by an antibacterial agent did not regain viability on incubation with various metabolites, as had been claimed, since the increase in viable count was shown to be due solely to multiplication of cells that had never died (Hurwitz, Rosano & Blattberg, 1957).

Total bacterial counts

Only one dilution of culture is used to fill the chamber, so that the mean number of bacteria per square necessarily has but one value. The usual practice is to count many squares and to calculate the count from the mean number per square. The number of organisms per square is Poissonially dis-

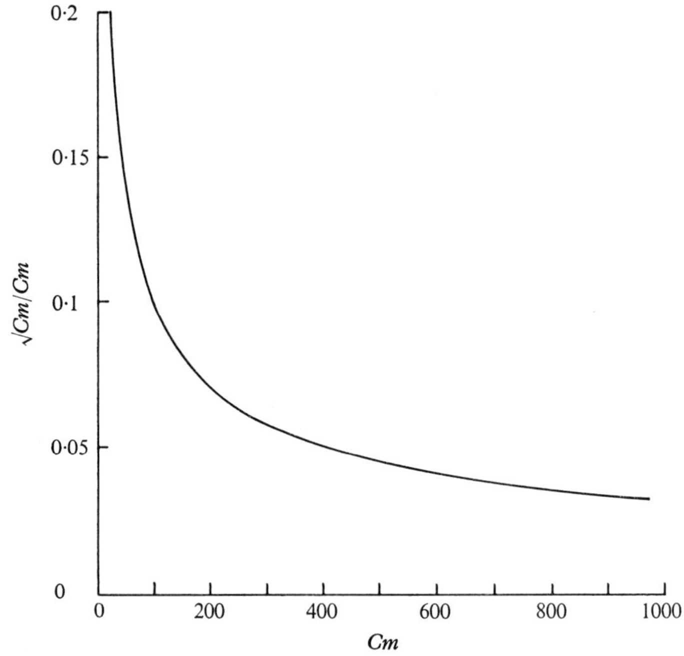

Fig. 6.2. Precision of a total bacterial count. Ordinate: S.E. $(Cm)/Cm$ $(= \sqrt{Cm}/Cm)$. Abscissa: *Cm*, the total number of bacteria counted. The curve shows that the precision increases only slowly after 600–1000 organisms have been counted. S.E. (Cm) is calculated here on the assumption that its size is determined solely by chance. In practice, it is substantially increased by technical errors (p. 22).

tributed under ideal conditions, and can be shown to be so by a χ^2 test (see Fisher, 1950, §§ 15, 16) or by the graphical method described on p. 177.

Suppose C squares are counted, each containing an average of m organisms, to give a total of Cm organisms. The precision of the count evidently depends on the total number of organisms counted, not on the mean; i.e. on Cm, not merely on m. The mean of a Poisson distribution is known to have a standard error (S.E.) equal to its square root. The total,

Cm, must also be Poissonially distributed, for it is irrelevant if the Cm organisms are in C squares or in one square. The S.E. of Cm, like that of m, is therefore equal to its square root so that in 95 % of counts on the same suspension, the observed value of Cm would be expected to lie in the range $Cm \pm 1{\cdot}96$ S.E. $(Cm) = Cm \pm 1{\cdot}96\sqrt{Cm}$, provided chance alone was responsible for differences between repeated counts of this suspension. The 95 % limits for the mean number per square, from which the cell concentration is calculated, are therefore $(Cm + 1{\cdot}96\sqrt{Cm})/C$ to $(Cm - 1{\cdot}96\sqrt{Cm})/C$.

The precision of Cm can be expressed by S.E. $(Cm)/Cm$ $(=\sqrt{Cm}/Cm)$ which is plotted against Cm in Fig. 6.2. This shows that the precision increases (i.e. $\sqrt{Cm}/Cm$ decreases) less and less rapidly, the larger Cm becomes. The maximum practicable precision is obtained by counting 600–1000 organisms, as any appreciable further increase is achieved only by counting many thousands.

Significance of two total counts. A common practical problem is to determine whether or not two total counts (or two colony counts, for the same arguments apply) are likely to differ significantly. Suppose the same volume of a bacterial suspension is counted in two chambers and that the observed total numbers of organisms are 201 and 183 respectively. On the null hypothesis that the counts do not differ significantly, the best estimate of

Table 6.1. *Significance of the difference between two total counts*

Method 1

$$\chi^2 = \frac{\Sigma(O-E)^2}{E},$$

where Σ indicates summation, and O and E are the observed and expected counts respectively

O	E	$(O-E)$	$(O-E)^2$	
201	192	9	81	$\Sigma(O-E)^2 = 162$
183	192	−9	81	$\chi^2 = 162/192 = 0{\cdot}844$

Method 2 If O_1 and O_2 are the two counts,

$$\chi^2 = \frac{(O_1 - O_2)^2}{O_1 + O_2} = \frac{(201-183)^2}{201+183} = 0{\cdot}844$$

the true count is $(201+183)/2 = 192$. The probability that the observed counts differ significantly is then determined by a χ^2 test in the usual way (Table 6.1). A more rapid method of calculation is also given. Both give $\chi^2 = 0{\cdot}844$.

Values of χ^2 are given in statistical tables. The entries for 1 degree of freedom are applicable here and show that 0·844 falls between the values of χ^2 for $P = 0{\cdot}5$ and 0·3. In other words, it is highly probable that the two counts in this example differ solely by chance. If they had differed significantly at the 5 % level ($P = 0{\cdot}05$), $\chi^2 \geqslant 3{\cdot}841$.

Colony counts

A colony on a plate is analogous to a bacterium in the square of a counting chamber, and the calculations for a colony count are therefore the same as those given above for a total count.

The expected frequency of a given number of colonies. Ideally, the number of colonies present on different plates receiving the same inoculum should differ by no more than is accountable to chance. Whether this is true for a particular set of plates can be examined by a χ^2 test (see Eisenhart & Wilson, 1943, for a worked example). Part of the test requires the frequency that a plate will have 1, 2, 3, ... colonies when the mean number per plate is m. When m is less than *ca.* 15, the probabilities can be either calculated from the individual terms of the Poisson series or taken from the tables of Pearson & Hartley (1954) or Molina (1942). However, the tabulated values do not extend to large values of m, but calculation is avoided by remembering that a Poisson distribution with large mean approximates to a normal distribution whose standard deviation is equal to the square root of its mean. Thus, in a colony count, if the observed mean number of colonies per plate is 250, its standard error $= \sqrt{250} = 15{\cdot}8$. A table of values of the standardized normal distribution (p. 202) shows that the proportion of plates bearing 250 colonies is expected to be 0·399. Similarly, the probability of a plate having 220 colonies $= 0{\cdot}066$, since 220 is $(250-220)/15{\cdot}8 = 1{\cdot}9$ times the standard deviation from the mean of 250.

Analysis of colony and local lesion counts. The mean number of bacterial colonies produced by inoculating plates of nutrient agar with successive dilutions of a culture is almost always proportional to the degree of dilution (Fig. 6.3). This is also true of the number of lesions formed by *Pfeifferella whitmori* in the lungs of inoculated animals (Dannenberg & Scott, 1956) and by many viruses on susceptible surfaces like plant leaves, tissue culture monolayers, or the chorio-allantoic membranes of chick embryos. Similarly, the number of mutants induced in a cell population by radiation is often directly proportional to the dose (see Muller, 1954), just as the number of bacteria killed by colicin is proportional to the initial colicin concentration when an excess of bacteria is present (p. 283).

The feature common to these systems is that each response is produced by a single organism or unit of radiation or colicin acting independently and distributed according to the Poisson series in conditions where pm, the mean number of effective units per site, is extremely small. The proof follows as it is generally omitted.

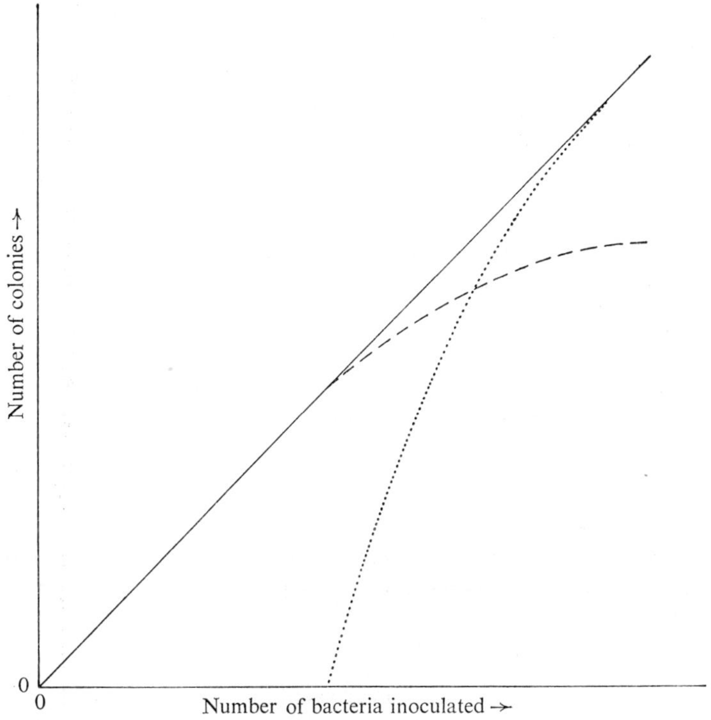

Fig. 6.3. Relation between count and size of inoculum in bacterial colony counts. Ordinate: mean number of colonies. Abscissa: mean number of bacteria inoculated. The count is usually proportional to the inoculum, as shown by the continuous line. If inoculated organisms interact synergistically, the count increases disproportionately faster on increasing the size of the inoculum (dotted line). Antagonism is reflected in the count failing to increase in proportion to the size of the inoculum (dashed line).

Consider a plate in a colony count which receives an inoculum of b bacteria. If there are N possible sites where a colony might form, the mean number of inoculated bacteria per site $= b/N = m$. If y colonies form after incubation, the proportion of sites bearing colonies,

$$y/N = 1-e^{-pm} = 1-e^{-pb/N}. \tag{6.2}$$

$$e^{-pm} = 1-pm+\frac{(pm)^2}{2}-\frac{(pm)^3}{3}+\ldots,$$

but when pm is very small, as is the case here owing to the magnitude of N,

$$e^{-pm} \simeq 1-pm.$$

Substituting in equation (6.5),

$$\begin{aligned} y/N &= 1-(1-pm) \\ &= pm \\ &= pb/N, \\ \therefore \quad y &= pb \quad \text{or} \quad y \propto b. \end{aligned}$$

In other words, y, the number of colonies per plate is proportional to b, the mean number of organisms inoculated per plate.

The term p is often called the *efficiency of plating* in this context (Ellis & Delbrück, 1939) and is usually determined by plating the same suspension of organisms in two ways, for example on two different media, the E.O.P. being taken as the smaller count divided by the larger. A figure that allows the confidence limits of the E.O.P. to be read directly from the two counts is given by Williams (1968).

Synergy. If similar calculations are made on the assumption that a colony is formed only by the co-operation (i.e. the synergistic action) of either 2, 3, 4 ... or x inoculated organisms, the curve has the general form shown on Fig. 6.6 by the dotted line (see Dulbecco & Vogt, 1954). The reason for its convexity is that organisms would be more and more likely to be close enough to co-operate after inoculation, the larger the mean number inoculated. Curves suggesting synergy between a limited number of inoculated plant virus particles are not uncommon (Fulton, 1962). They have not been reported for bacteria, although much more marked synergy is often noticeable, as is seen in recombination experiments, when mating occurs on the plate; on inhibitory media (p. 55); or when sulphonamide-sensitive cells are plated on sulphonamide-containing medium, as shown by, say, confluent growth at the 10^0 dilution and complete absence of colonies at the 10^{-1} dilution. In this case, growth presumably results when the inoculated cells alone can produce sufficient sulphonamide antagonist to neutralize the sulphonamide.

Antagonism. The contrary phenomenon of antagonism is occasionally seen in plate counts, where the number of colonies does not decrease in proportion to the degree of dilution of the culture, for example in genetic experiments where recombinant and parental clones compete (Grigg, 1958).

Variance of the mean. On the Poissonian model, the variance of the mean number of lesions at each dilution is expected to equal the mean itself. This is usually observed in colony counts (p. 31). However, in some virus infections the variance is often larger, which indicates that the precision of the counting method in question is less than that attainable under ideal conditions. In titrations of variola virus on the chorio-allantoic membrane, the variance was thirteen times the mean (McCarthy, Downie & Armitage, 1958) although, with vaccinia virus, the variance and the mean may be nearly equal (Mai & Bonitz, 1963). In seeking to improve an imprecise technique, it is useful to see if the precision depends on the mean number of lesions by plotting the observed variances against the respective means. Since they are equal if the Poissonian model applies, the points should fall on a straight line with slope of unity under ideal conditions. Even when the model does not apply, log variance and log mean are more or less linearly related but the slopes are greater than unity, for example 1·1–1·7 for various animal viruses reviewed by Armitage (1957). Another test is to plot the mean number of lesions against dilution, but in analysing the results, it should be remembered that a regression line cannot be fitted to the points immediately because the size of the variance is correlated with that of the mean (see Kleczkowski, 1949).

Dilution counts

The classic example of this method is a viable bacterial count, but the same procedure is used for measuring the concentration of virus suspensions by inoculating tissue cultures or animals.

In a bacterial count, the suspension is serially diluted and a known volume of each dilution inoculated into 1 or more tubes of broth. After incubation, the number of turbid and clear tubes at each dilution is noted. The tubes remaining clear are each usually assumed not to have received even 1 organism ($p = 1$), though this is often not justified. In practice, the results for all the tubes are considered as a whole to obtain the viable count of the undiluted suspension. The count thus obtained is generally referred to as the 'most probable number' (M.P.N.) of organisms, because it is derived by calculating the number most likely to have given the particular combination of turbid and clear tubes observed in practice (see Eisenhart & Wilson, 1943; Finney, 1964). Calculation is often avoided by planning the experiment so that it conforms to an arrangement for which the most likely results have been tabulated. If every viable organism is assumed capable of producing turbidity ($p = 1$), the proportion, S, of tubes remaining clear after inoculation with a mean of m organisms per tube equals the proportion receiving 0 organisms. As on p. 176,

$$S = e^{-m}. \tag{6.3}$$

If P is the proportion of tubes becoming turbid, $P = 1-S$, or

$$P = 1-e^{-m}. \tag{6.4}$$

The results of dilution counts are often graphed by plotting P against $\log m$. As it is impractical to make more than about 10 inoculations from each dilution, the values of P must be 0, 0·1, 0·2, ..., 0·8, 0·9, 1; while, at the same time, the most dilute inoculum is often only a small fraction of the most concentrated, for example $1/10^9$. Equation (6.4) gives the curve shown

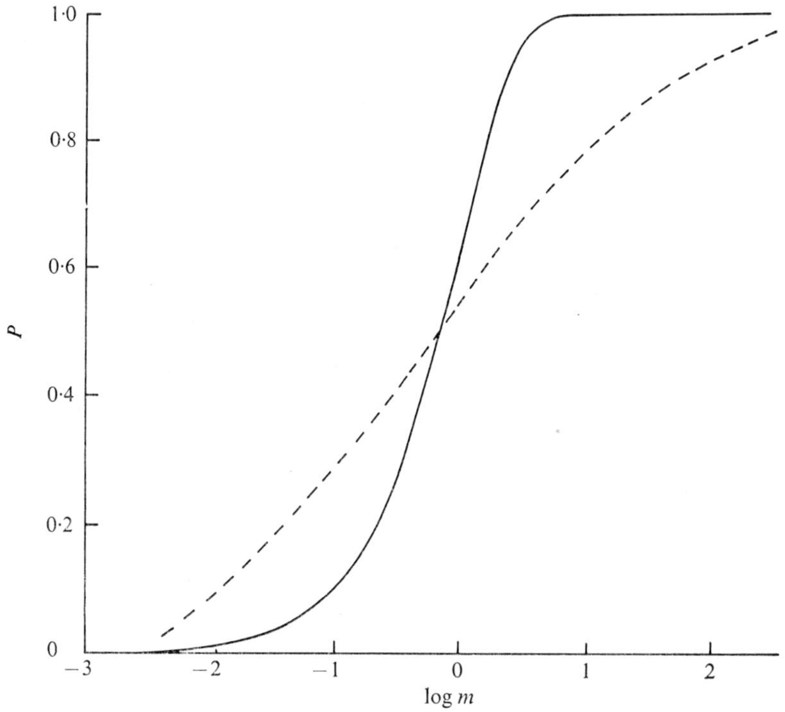

Fig. 6.4. Dose-response curve for the dilution count derived from the equation, $P = 1-e^{-m}$. Ordinate: P, the proportion of inoculated tubes becoming turbid. Abscissa: logarithm of the mean number of organisms inoculated per tube (m). If the tubes of culture medium differ in suitability for bacterial growth, the curve is flatter, as shown by the dashed line.

in Fig. 6.4. If the value of p is less than 1 and some organisms are 'ineffective', the curve shifts to the right on the abscissa but its shape is unchanged. If the tubes differ in responsiveness, the curve becomes flatter (dashed curve in Fig. 6.4: for proofs, see Fazekas de St Groth, 1955). Curves flatter than the Poissonian are occasionally found in bacterial counts (Savage &

Halvorson, 1941; Eisenhart & Wilson, 1943) but are very common in infectivity titrations (see Meynell, 1957, and p. 228). Flattening could also be produced by systematic technical error, as may occur in making dilutions (p. 31). Although Fig. 6.4 is commonly found in the literature, there are, however, several situations in which a log-log plot is preferable, since it gives a linear relation between P and m that allows departures from a Poisson model to be detected more readily (p. 194).

Interpretation. The method is only valid when the results are compatible with equation (6.4). Incompatibility is nearly always due to differences between tubes, as shown in Fig. 6.4, and the count cannot then be estimated for it must depend on whether favourable or unfavourable tubes are considered.

As the dilution count estimates the concentration of organisms by observing the effects of their growth, it necessarily measures the concentration of 'effective' organisms (p. 176). It does not follow, however, that all the organisms are effective ($p = 1$) though this is usually assumed in bacterial counts. In virus titrations particularly, the cells or animals may be partly resistant to viral growth. The value of p can only be estimated if another form of count is done. A total count gives the same figure as the M.P.N. if $p = 1$, but is greater if $p < 1$. When the total count is unknown, as is often the case with viruses, a second infectivity titration can be done in another host which may give a higher titre than the dilution method and so show that $p < 1$ and provide a relative estimate of its value (see Isaacs, 1957). With phage, the dilution method may give higher titres than plaque counts, presumably because some phage fails to infect bacteria in the agar overlay soon enough to form a plaque (Ellis & Delbrück, 1939).

If the observed curve is flatter than the Poissonian, p must be less than 1 for at least some hosts, because flattening indicates heterogeneity in host resistance; i.e. some hosts must then be more resistant than others, so that for them, $p < 1$. For the more susceptible hosts, p may, of course, be unity; i.e. these hosts may be completely susceptible.

Planning. The planning of a dilution count depends on its purpose and on how precisely the concentration of organisms is known beforehand (see Cochran, 1950). In routine water analysis, the bacterial count may only vary within narrow limits over long periods and it may be less important to determine the exact count than to show that it is below an accepted maximum figure (see Swaroop, 1956). In experimental work, the count itself is usually required and its value more or less unknown. Three points have then to be decided:

(1) The maximum and minimum volumes of culture to be inoculated;

(2) the dilution factor; and

(3) the number of tubes to be used at each dilution.

The range of volumes inoculated must be sufficiently wide for one or more dilutions to give some turbid and some clear tubes, since the method inevitably fails if all the tubes inoculated in the experiment are either turbid or clear. Bearing in mind the likely limits of the count, the range should therefore be wide enough to be certain of having turbid and clear tubes at one dilution at least, which implies a mean inoculum of *ca.* 1 organism per tube. Thus, if the count is believed to lie between 10 and 10^3 organisms/ml., the maximum volume of culture inoculated should be 1/10 ml. = 0·1 ml. of undiluted culture, and the minimum should be $1/10^3$ ml. = 10^{-3} ml. or 0·1 ml. of a 10^{-2} dilution, and so on.

The dilution factor and number of tubes used per dilution depend on the precision required and on the amount of effort to be put into the count. Fortunately, it turns out that for any given total number of tubes (i.e. number of dilutions × number of tubes per dilution), the precision is very little affected by the size of the dilution factor, provided it is less than 10 (Fig. 6.5). The periodicity of the curve for 10-fold dilutions in Fig. 6.5 arises because the greatest precision (i.e. the greatest information in the statistical sense) is obtained with a mean inoculum of *ca.* 1·6 organisms per tube (Fisher, 1950). This mean is more likely to be obtained with a small than with a large dilution factor, and the precision therefore fluctuates less with the smaller dilution factor. In general, therefore, although the use of a small dilution factor, such as 2, gives greater precision, the loss in using a factor as large as 10 is not as great as might be supposed. The final choice also depends on which dilution factors are given in tables published for estimating the M.P.N. from the results (classified by Swaroop, 1956, Table III).

As the total number of tubes increases, the precision of the count also increases, although it is subject to a law of diminishing returns. This is evident from Table 6.5, which shows that for a dilution factor between 2 and 10, the precision increases rapidly on increasing the number of tubes per dilution from 1 to 5, but the gain in using 10 tubes per dilution is relatively far less.

Analysis. There are two steps in estimating the concentration of effective organisms:

First, to see if the results are compatible with equation (6.4). If this turns out to be so, as is usual in viable counts, the most probable estimate of the count can often be obtained from published tables (p. 231). On the other hand, if tests show that the results are unlikely to conform to equation (6.4), the viable count cannot be estimated by this method (p. 187). Compatibility of the observations and the expected curve can be tested by a χ^2 test

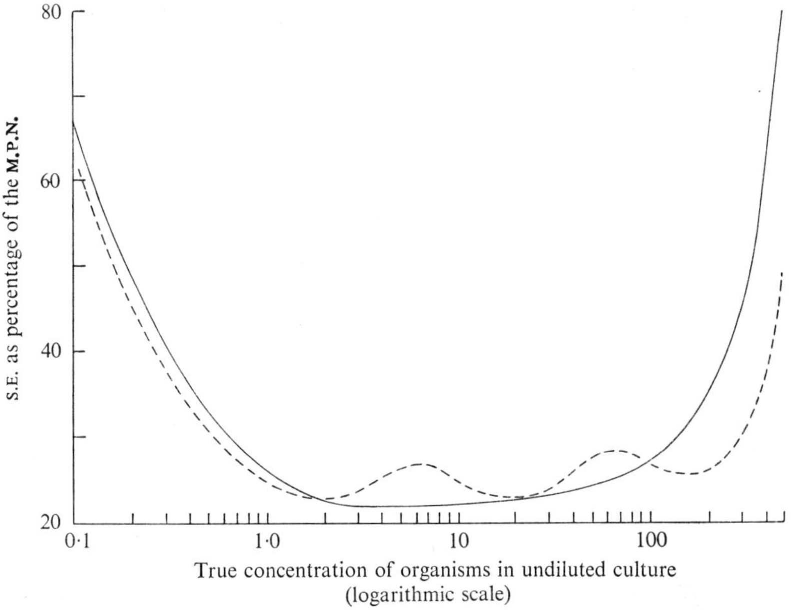

Fig. 6.5. Effect of the dilution factor on the standard error (S.E.) of the most probable number (M.P.N.) in a hypothetical dilution count (Cochran, 1950). Ordinate: S.E. expressed as a percentage of the observed M.P.N. Abscissa: true concentration of organisms in the undiluted culture. The curves are for dilution factors of 2 (continuous line) and 10 (broken line). The respective inocula were assumed to range from 1·28 to 0·01 ml. and 1·00 to 0·01 ml. undiluted culture with a total of 72 tubes in each case.

The S.E. rises at each side, where the concentration is very small or very large, because all the tubes are then increasingly likely to be either clear or turbid. In the extreme cases where 100% clear or turbid tubes are obtained, the count necessarily fails as it depends on observing both turbid and clear tubes at one dilution at least. The S.E. is least (i.e. the precision of the count is greatest) for an M.P.N. of intermediate size which gives approximately equal numbers of clear and turbid tubes. The S.E. for a dilution factor of 10 fluctuates more than that for a factor of 2 because the optimal mean inoculum (*ca.* 1·6 organisms per tube) is more likely to be missed by chance (see text).

(see Haldane, 1939), but simple and more powerful methods have been devised by Moran (1954*a*, *b*, 1958) and by Stevens (1958). Moran's test is only applicable when the extreme dilutions give approximately 0% and 100% turbid tubes respectively. Comparisons by eye of the observed points and the theoretical curve of Fig. 6.4 are unreliable for the dilution method, since the points do not lie unambiguously on one curve owing to the small number of tubes used at each dilution.

Example. Suppose a dilution count gave the result shown in Table 6.2. Moran's test is performed by first calculating $T = \Sigma f_m(N-f_m)$, where f_m is the number of turbid tubes at each dilution, N is the total number of tubes per dilution, and Σ indicates summation (Moran, 1954*a*). In this example, the observed value of

$$\begin{aligned} T &= 10(10-10)+10(10-10)+9(10-9)+2(10-2)+1(10-1)+0(10-0) \\ &= 0+0+9+16+9+0 \\ &= 34. \end{aligned}$$

Table 6.2. *A hypothetical dilution count*

Dilution of culture	10^0	10^{-1}	10^{-2}	10^{-3}	10^{-4}	10^{-5}
No. of tubes inoculated with each dilution	10	10	10	10	10	10
No. of tubes becoming turbid	10	10	9	2	1	0

It will be noted that the steeper the dose-response curve, the smaller will be the observed value of T, because each dilution tends to give 0 % or 100 % turbid tubes and the product, $f_m(N-f_m)$, is therefore 0. Conversely, the flatter the curve, the larger T becomes. The degree of flattening of the observed dose-response curve can therefore be estimated by comparing the observed value of T with the value expected from the Poissonian curve (Fig. 6.4). The expected value of T is here 27·09 from Table 6.3 which also gives the standard error as 9·34. The probability that the observations conform to equation (6.4) is obtained by calculating M, the difference between the observed and expected values of T expressed as a multiple of the S.E. of its expected value:

$$\begin{aligned} M &= \frac{T-E(T)}{\text{S.E. } (T)} \\ &= \frac{34-27{\cdot}09}{9{\cdot}34} \\ &= 0{\cdot}74. \end{aligned}$$

If the observations conform to equation (6.4), values of $M \geqslant 1{\cdot}645$ or $\geqslant 2{\cdot}326$ occur by chance with probabilities $\leqslant 0{\cdot}05$ and $\leqslant 0{\cdot}01$, respectively (Moran, 1954*b*). As M is only 0·74 in this example, the difference between

Table 6.3. *Values of E(T) and* S.E. *(T) for Moran's test*

	E(T)				S.E. *(T)*			
N	2*	4*	√10*	10*	2*	4*	√10*	10*
5	20	10	12·04	6·02	5·69	4·02	4·42	3·12
6	30	15	18·06	9·03	7·63	5·40	5·92	4·19
7	42	21	25·29	12·64	9·75	6·89	7·57	5·35
8	56	28	33·72	16·86	12·02	8·50	9·33	6·60
9	72	36	43·35	21·67	14·46	10·22	11·22	7·93
10	90	45	54·19	27·09	17·03	12·04	13·21	9·34
11	110	55	66·23	33·11	19·74	13·96	15·32	10·83
12	132	66	79·47	39·74	22·58	15·97	17·52	12·39
13	156	78	93·92	46·96	25·55	18·07	19·82	14·02
14	182	91	109·58	54·79	28·63	20·24	22·21	15·71
15	210	105	126·43	63·22	31·83	22·51	24·70	17·74
16	240	120	144·50	72·25	35·14	24·85	27·27	19·28
17	272	136	163·76	81·88	38·56	27·27	29·92	21·16
18	306	153	184·23	92·12	42·07	29·75	32·64	23·08
19	342	171	205·91	102·95	45·70	32·31	35·46	25·08
20	380	190	228·79	114·39	49·41	34·94	38·34	27·11
30	870	435	523·80	261·90	91·53	64·72	71·02	50·22
40	1560	780	939·23	469·61	141·49	100·05	109·79	77·63

N: number of tubes inoculated from each dilution.
* Dilution factor.

the observed numbers of turbid tubes and those predicted from equation (6.4) could very well have arisen by chance.

The test is designed to detect curves that are flatter, not steeper, than the predicted curve. If the observed curve had chanced to be steeper than the Poisson curve, M could be negative. Thus, if the numbers of turbid tubes in Table 6.3 had been 10, 10, 9, 1, 0, 0, then $T = 18$, which is less than 27·09, the value of $E(T)$.

This form of Moran's test has a subjective element in that $E(T)$ is the same whatever the order of turbid tubes, e.g.

0–1–2–9–10–10
0–10–1–9–2–10
9–2–10–10–0–1

These alternatives are obviously incompatible with the Poissonian curve, but this is markedly less easy to judge by eye when the dilution factor is less than 10 and chance inversions in the sequence are more frequent. Moran (1958) has therefore extended the test and gives tables for a dilution factor of 2.

Estimation of the bacterial concentration and its confidence limits. Assuming the observations are compatible with equation (6.4), the M.P.N. is easily obtained if the dilution factor and number of tubes inoculated from each dilution coincide with the entries in one of the published tables. With both factors constant, the only tables giving the M.P.N. directly appear to be those shown in Table 6.4, of which those by Halvorson & Ziegler (1933), Norman & Kempe (1960) and Taylor (1962) are reproduced here on p. 231. Other tables classified by Swaroop (1956) give the results for unevenly spaced inocula, for example the Report (1956) which considers x ml. of a given dilution distributed to each of 5 tubes, and $5x$ ml. of the same dilution to 1 tube. Some tables include sequences of results that are highly unlikely to be compatible with equation (6.4). The confidence limits of the M.P.N. can be obtained from Table 6.5 reproduced from Cochran (1950).

Table 6.4. *Summary of tables giving the most probable number of organisms in dilution counts*

Number of tubes per dilution	Number of dilutions	Dilution factor	Author
5	3	10	McCrady (1918),[1] Taylor (1962)
5	4	10	McCrady (1918)[1]
10	3	10	Halvorson & Ziegler (1933)
3, 5 or 10	3	10	Hoskins (1934)
8	3	10	Norman & Kempe (1960)

[1] Reproduced by Prescott, Winslow & McCrady (1946).

If the dilution factor and number of tubes per dilution do not fit the tables, the M.P.N. can nevertheless be estimated approximately, as shown below, or more precisely by the tables of Fisher & Yates (1963) and Seligman & Mickey (1964), or by the rather more complicated method given with tables by Finney (1951, 1964, §21.5). The confidence limits of the M.P.N. are also obtainable by the last two methods.

In the hypothetical experiment of Table 6.2, the M.P.N. is estimated by Fisher & Yates' table as 270 per inoculum for the 10^0 dilution with 95 % confidence limits of 124–589. On turning to Table 6.12, it is evident that the observations can be entered in four ways: 10–10–9, 10–9–2, 9–2–1, and 2–1–0, each of which gives a different M.P.N. One course would be to consider only the triplet whose middle term corresponded to an inoculum of 1·6 organisms, i.e. 80 % turbid tubes, since this is the result carrying greatest statistical weight ($P = 1-e^{-pm}$; $0{\cdot}8 = 1-e^{-1{\cdot}6}$). In this example, the triplet in question is 10–9–2, with M.P.N. = 228. This procedure therefore discards the results from the highest and lowest dilutions. An alternative might therefore be to average the logarithms of the M.P.N. of all possible triplets, but this is impossible when one or more triplets contain only 100 % or 0 % responses because such triplets do not yield a M.P.N. The third possibility is to discard all the latter triplets and to average the remainder. The average M.P.N. in Table 6.2 is then 252. However, this procedure amounts to giving the same statistical weight to each triplet, although some are known to be more accurate than others. Some authors have therefore devised rules for picking the most informative dilutions or triplets before calculating an average. Taylor (1962) suggests: if some dilutions give 100 % turbid tubes, include only the most dilute; if some give 0 % turbid tubes, include only the most concentrated; and if only one triplet then remains, take the M.P.N. from the tables. If two possible triplets remain, the M.P.N. for each is taken from the tables and their logarithms are averaged to obtain the log average M.P.N. Taylor shows empirically that this procedure gives almost the same M.P.N. as Finney's method when using 5 tubes per dilution and a 10-fold dilution factor, but it is not clear how good the agreement would be in different circumstances. In Table 6.2, for example, there are three triplets admissible under this rule: 10–9–2, 9–2–1, and 2–1–0, with average M.P.N. = 259. It should be noted that all these approximations affect the estimated value of the M.P.N. but, for informative triplets, its precision (v.i.) does not depend on the M.P.N. itself but solely on the number of tubes per dilution and on the dilution factor.

The S.E. is obtained from Table 6.5 by finding the entry corresponding to the number of tubes inoculated from each dilution and to the dilution factor. The left-hand part of Table 6.5 gives the S.E. of the log M.P.N. The right-hand part gives the factor for obtaining the upper and lower 95 % confidence limits directly by respectively multiplying and dividing the M.P.N. itself. The S.E. (log M.P.N.) equals log factor/2.

The hypothetical experiment of Table 6.2 used 10 tubes at each dilution and a dilution factor of 10. The corresponding factor in the right-hand part of Table 6.5 is 2·32. If the M.P.N. is taken as 270, the figure obtained by Fisher & Yates' method, the 95 % confidence limits are (270/2·32) to (270 × 2·32) = 116–626.

Two applications of the dilution method have unusual features. In certain titrations, a virus suspension is smeared on susceptible leaves which subsequently develop discrete lesions that can be counted. The observed number of lesions is often expressed as a proportion of the total number of susceptible sites per leaf. However, this total cannot be measured directly for technical reasons and is therefore estimated from the dose-response curve. Ideally, it should equal the value to which the number of lesions

Table 6.5. *The standard error and 95 % confidence limits of the* M.P.N. *in dilution counts* (*Cochran, 1950*)

	S.E. ($\log_{10}$ M.P.N.)				Factor for 95 % confidence limits			
N	2*	4*	5*	10*	2*	4*	5*	10*
1	0·301	0·427	0·460	0·580	4·00	7·14	8·32	14·45
2	0·213	0·302	0·325	0·410	2·67	4·00	4·47	6·61
3	0·174	0·246	0·265	0·335	2·23	3·10	3·39	4·68
4	0·150	0·214	0·230	0·290	2·00	2·68	2·88	3·80
5	0·135	0·191	0·206	0·259	1·86	2·41	2·58	3·30
6	0·123	0·174	0·188	0·237	1·76	2·23	2·38	2·98
7	0·114	0·161	0·174	0·219	1·69	2·10	2·23	2·74
8	0·107	0·151	0·163	0·205	1·64	2·00	2·12	2·57
9	0·100	0·142	0·153	0·193	1·58	1·92	2·02	2·43
10	0·095	0·135	0·145	0·183	1·55	1·86	1·95	2·32

N: number of tubes inoculated from each dilution.
* Dilution factor.
The S.E. is obtained directly from the left-hand part of the table. The factors in the right-hand part give the 95 % confidence limits which equal M.P.N./factor to M.P.N. × factor. These are valid only for informative triplets.

asymptotes as the concentration of virus inoculated is increased (Fig. 6.4). The difficulty is that no asymptote is sometimes reached, as the number of lesions continues to rise and does not reach a maximum at the level expected from the number produced by smaller inocula. This may be due to changes in the wetting properties of the inoculum as the virus concentration increases (Kleczkowski, 1950).

The second example occurs in titrations where an animal receives only one injection of each dilution. The only possible type of result for each animal is $++-+--$. The usual course is to sum the results for several animals to produce a series like that shown in Table 6.7; but Armitage (1959*a*, *b*) and Armitage & Bartsch (1960) have shown how the results for individual animals can be tested, in order to see how one animal differs in resistance from another.

Log-log (*Weibull*) *plots*. Departures from an exponential dose-response curve are sometimes more easily detected by re-arranging the dose-response equation so as to give a linear plot. From p. 185

$$\log_e S = -m.$$

Hence
$$\log_e(-\log_e S) = \log_e m.$$

Values of $\log_e m$ are now seen to be linearly related to the transformed values of S with slope of 1 (Fig. 6.6). If the data should not be exponential, however, the plot is non-linear, its exact form being determined by the real distribution (see Gart & Weiss, 1967, for graphical tests of heterogeneity).

Plots of this kind have been used in two kinds of experiment. The first is some form of dilution count, notably an infectivity titration (Armitage & Spicer, 1956). A good example of a bimodal distribution of host responses revealed by this plot comes from tomato plants infected by *Corynebacterium michiganense*, where the individual plants probably fell into two groups of differing resistance due to cross fertilization (Ercolani, 1967*a*).

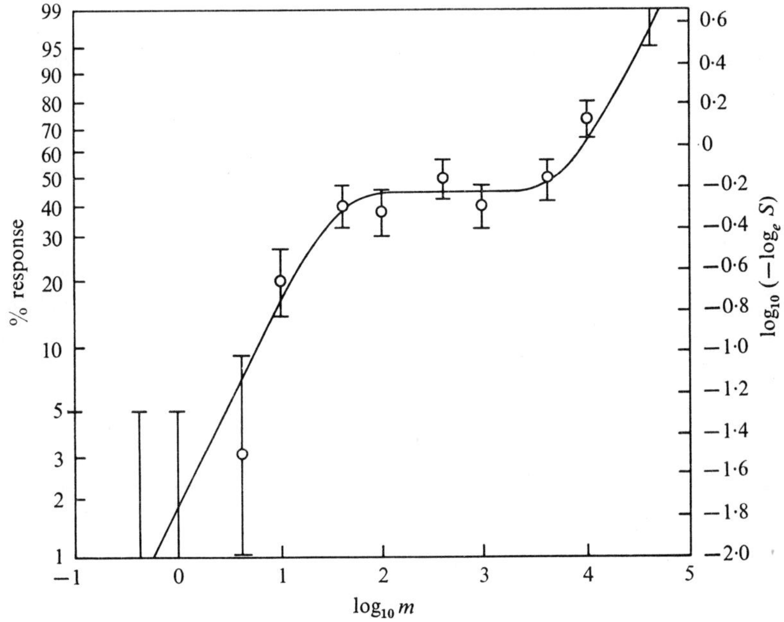

Fig. 6.6. Log-log (Weibull) plot, showing the heterogeneous response of tomato plants to infection by *C. michiganense*. From Ercolani (1967*a*). An exponential dose-response curve derived from a homogeneous system would give a linear plot of slope 1.

In general, however, infectivity titrations yield such imprecise data that only clear cut results should be accepted as evidence for discrete heterogeneity in the host population (see Armitage, Meynell & Williams, 1965). The second application arises in sterilization tests. When sterilization is virtually complete, the random distribution of surviving organisms amongst replicate samples results in some being positive and others negative (p. 95). Suppose a number of samples are taken at each time, then, if killing is exponential, the proportion of sterile samples, S, is related to the time of exposure, m, by equation (6.3): $S = e^{-m}$. Plotting $\log_{10}(-\log_e S)$

against $\log_{10} m$, as in Fig. 6.6, allows a graphical determination of the 'mean single survivor time' corresponding to $m = 1$ (i.e. $\log_{10} = 0$). This is therefore the exposure at which $\log_{10}(-\log_e S) = 0$.

Exponential survival curves

Survival curves are termed 'exponential' when a straight line is obtained by plotting the logarithm of the proportion of survivors against the dose of lethal agent. The dose-response equation is

$$\log_e S = -pm,$$

where S is the proportion of survivors and pm is the mean dose per cell (p. 177). This is another form of equation (6.3), but $\log S$ is usually plotted against m when a large number of organisms is at risk and extremely small survivals, for example 10^{-8}, can be measured (Fig. 6.7*a*). If the data are plotted using logarithms to any base (e, 10, 2), it is usually immediately apparent if the points can be fitted by a straight line. Its slope is $-p$ when $\log_e$ is used. If the observed points are scattered, the most probable value of p can be calculated numerically using the tables given by Spicer (1956).

A typical example of an exponential curve is obtained by exposing a bacterial suspension to radiation whose effects can be thought of as caused by discrete particles, each capable of scoring a lethal 'hit' on a sensitive target in a bacterium (see Fowler, 1964). Since one hit is here assumed sufficient to kill, exponential curves are often referred to as 'one-hit' curves.

Exponential curves are observed repeatedly in many experiments other than those involving irradiation: when bacteria are killed by heat or chemicals (p. 84); and when phage is inactivated by antiserum or is adsorbing to bacteria (see Adams, 1959).

A factor which commonly disturbs the 'one-hit' model is the presence of differences in p, the susceptibilities of individual cells. Suppose the cells are of two distinct kinds, one more resistant than the other so that each has its own value of p. The survival curve is then a composite of the two exponential curves for the two groups (curves A and B in Fig. 6.7*b*). The more susceptible cells would be killed first until virtually all the survivors belonged to the more resistant group and the slope of the survival curve therefore decreased. The proportion of the resistant cells initially present in the mixed population equals X in Fig. 6.7*b*, the value of the ordinate where it is intersected by the extrapolation of curve B. If the sites do not fall sharply into two groups but vary in resistance in a continuous fashion, the curve is correspondingly smoother, as shown by the dotted curve in Fig. 6.7*b*.

Many survival curves are of this type, because microbial populations often differ from non-genetic causes, whether inactivation is due to heat (Lark & Adams, 1953), irradiation (see Alper, 1961), chemical disinfectants, antibiotics, or chilling (Meynell, 1958). Sometimes, heterogeneity appears to develop during exposure to the lethal agent, as when starved cells release material that protects the survivors (Postgate & Hunter, 1963), but, more

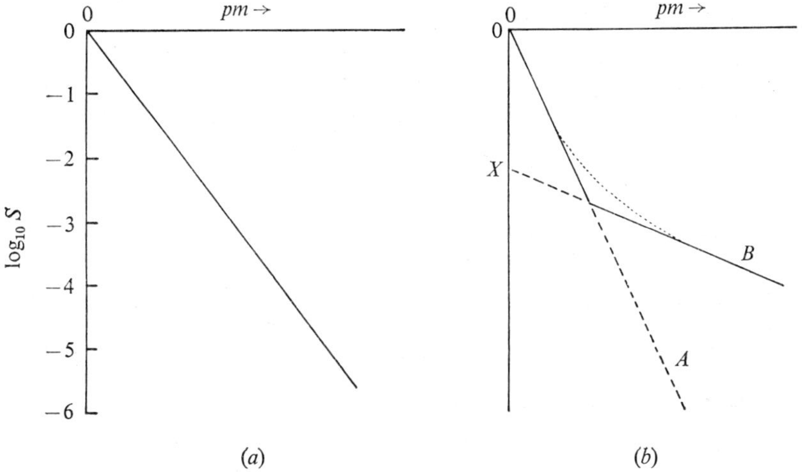

Fig. 6.7. Survival curves for exponentially declining populations.

(*a*) Exponential survival curve of a single population. Ordinate: log *S*, logarithm of the proportion of viable organisms. Abscissa: *pm*, the mean number of effective units per organism (i.e. dose of lethal agent).

(*b*) Composite survival curve for a mixture of two populations, *A*, *B*, of differing susceptibility (continuous line). If the right-hand part of the curve is extrapolated to the ordinate, the point of intersection, *X*, is the proportion of the mixed population initially belonging to the more resistant type, *B*.

If susceptibility is distributed smoothly in the initial population, the survival curve is correspondingly smoother (dotted line).

commonly, the cell population is inherently non-uniform. Thus, the resistance of spores of *Bacillus cereus* to ethylene oxide is correlated with their lipid content and electrophoretic mobility, and the mixed population can be separated into two discrete fractions by electrophoresis (Church *et al.* 1956). Yeast cultures are heterogeneous in resistance to X-rays because they contain budding cells of relatively greater resistance (Beam, 1955). Bacterial cultures often appear heterogeneous when the inactivating agent is more lethal to dividing than to stationary phase cells, for, on passing from the exponential to the stationary phase, the resistance of the culture does not increase uniformly because a completely resistant fraction appears whose size steadily increases. Indeed, an exponentially growing culture may cease to be uniform when the viable count reaches 10^8/ml. (see Meynell, 1958). In a few cases, the resistant organisms differ genetically from the sensitive fraction of the culture, as shown by the isolation of antibiotic-resistant organisms after exposure to antibiotics, or the isolation of u.v.-resistant bacterial mutants from sensitive cultures (Witkin, 1947) and of non-lysogenic cells from lysogenic cultures after exposure to large doses of u.v. irradiation (Lederberg, 1951).

Inactivation dose. The term 'hit' has already been used in connection with radiation. It was therefore proposed that radiation dosage should be expressed in multiples of that dose, termed the *inactivation dose*, which reduced survival to 37 %, since each cell would then have received an average of 1 'hit', i.e. 1 effective unit (from equation (6.1), $0{\cdot}37 = e^{-1}$).

Inactivation rate constant. Heat and antibody often cause exponential inactivation, but as their action is never thought of in terms of discrete units, it is not expressed in terms of 'hits'. Equation (6.3) is therefore usually written

$$S = e^{-kt}, \tag{6.5}$$

where t is the period of exposure and k is a constant, usually named the *inactivation rate constant*, whose magnitude depends on the resistance of the organism to inactivation, the intensity of the inactivating agent, and the time unit. To obtain k from titrations of anti-phage sera, equation (6.5) is often rewritten in $\log_{10}$ (see Adams, 1959):

$$k = \frac{2{\cdot}3D}{t} \log_{10} \frac{P_0}{P},$$

where D is the dilution of serum in the phage–serum mixture (if the serum is diluted 1/1000, $D = 1000$); t is the period of exposure to serum; P_0 is the initial phage concentration in the serum–phage mixture, and P is the concentration of phage still uninactivated at time t. A convenient means for measuring k for an unknown serum is to expose the phage to two-fold dilutions of serum for a fixed time, such as 60 min., and then to note which dilution, D', causes 90 % inactivation: $k = 0{\cdot}0383D'$.

Multi-hit survival curves

The name of these curves arises from the target theory of radiation action, which includes, as well as the one-hit model (p. 196), the possibility that a cell or any other site can only be altered by the joint action of two or more hits (see Fowler, 1964). Two situations can be distinguished:

(1) Where the site is homogeneous but receipt of at least n (> 1) effective units is required to produce an effect. No examples of such a mechanism seem to exist.

(2) Where each site is composite and each of its n parts must be affected by at least one effective unit before a response is produced. An obvious example is killing of a clump of bacteria, each of which has to be 'hit' to prevent colony formation. The dose-response equation is

$$S = 1-(1-e^{-pm})^n. \tag{6.6}$$

All multi-hit curves of the second type have the same general shape; an initial shoulder followed by a final linear portion having the same slope as a one-hit curve with the same value of p (Fig. 6.8). The shoulder is present because each site must first be affected by several units before responding: whereas with $n = 1$, a site will respond to even the first effective unit received so that the curve decreases linearly from time zero. All the curves become linear with the same slope at high pm because virtually all the sites then unaffected will have received $n-1$ effective units and each requires only one more to respond.

These curves have a convenient property. If the linear descending part is extrapolated upwards and to the left, its intersection with the ordinate gives the value of $\log n$ (for proof, see Atwood & Norman, 1949). Curves resembling those shown in Fig. 6.8 are often obtained for responses to agents unlikely to consist of discrete units as the Poisson distribution requires, and Alper, Gillies & Elkind (1960) have therefore proposed the non-committal term, *extrapolation number*, for the value of the ordinate at its intersection with the extrapolated curve. Lellouch & Wambersie (1966) give methods for determining n and its standard error from the data.

If the sites differ in susceptibility, the predicted curves will be altered in shape. Atwood & Norman (1949) give the curves that would be obtained if n differed for individual sites which remained equally susceptible (i.e. if p remained the same throughout). The alternative type of heterogeneity, where n remains constant but each site differs in susceptibility, was considered by Dewey & Cole (1962), who showed that curves for a mixture of two differing populations of sites would be extremely difficult to distinguish from that of a homogeneous population, and that the observed curves may give estimates of the extrapolation number and inactivation dose quite different from those of either component of the mixed population.

By use of the extrapolation number, the mean number of staphylococci in various sizes of air-borne particles has been estimated by irradiation to be 1–6 (Lidwell, Noble & Dolphin, 1959). Death of an irradiated organism appears to require one hit on each chromosome, as shown by observations on a series of yeasts of differing ploidy (Lucke & Sarachek, 1953). A bacterium is haploid, but multi-hit inactivation curves are obtained because

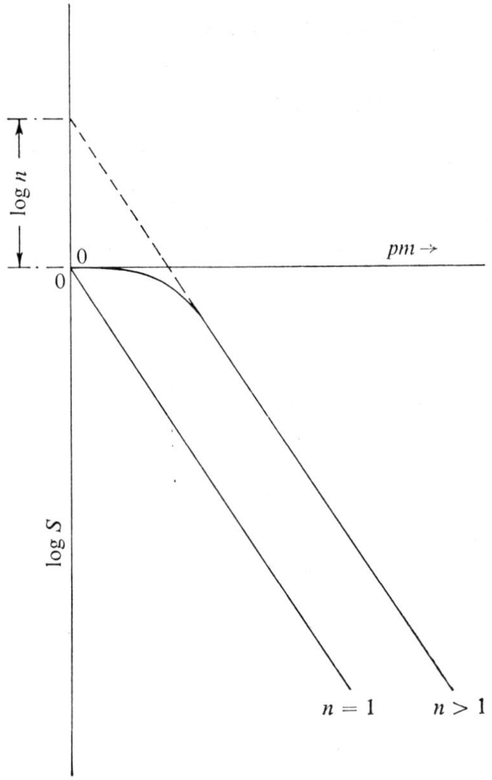

Fig. 6.8. Multi-hit survival curve. Ordinate: log S, logarithm of the proportion of viable organisms. Abscissa: pm, the mean number of hits per cell. n is the extrapolation number.

A curve for $n > 1$ is shown with a 1-hit curve ($n = 1$). The value of n is determined by extrapolation of the linear descending part of each curve to the ordinate: the intersection gives log n.

most of the cells seen microscopically in an asynchronously dividing culture consist of two or more individuals joined together. Hence, the extrapolation number was three in *Escherichia coli*, strain B/r, judging from the inactivation curve of cells dying from decay of ^{32}P incorporated in their DNA (Fuerst & Stent, 1956).

Multi-hit curves can also be used to estimate the number of mature intracellular phage particles present in infected bacteria at different times during the latent period. The method assumes that all the particles matured by the time a bacterium is irradiated have to be inactivated to destroy its plaque-forming ability (Benzer, 1952).

THE NORMAL AND LOG-NORMAL DISTRIBUTIONS

An instance of a normally distributed quantity is the number of colonies on individual plates, each inoculated with an average of several hundred viable organisms (p. 182). The shape of this distribution becomes clear if each

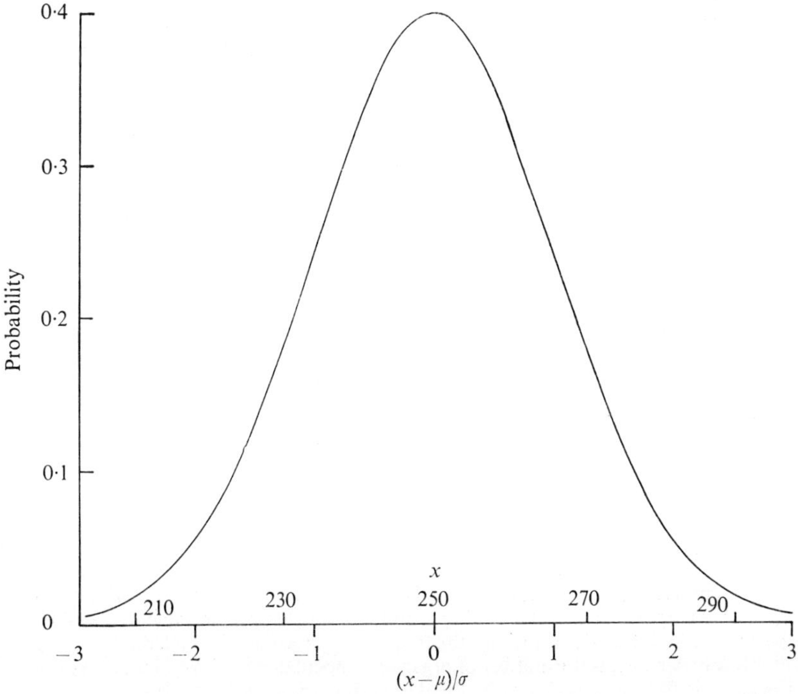

Fig. 6.9. The normal distribution. Ordinate: probability of occurrence of a given value of the variable, x. Abscissa: upper scale, values of x, here the number of colonies per plate (p. 182); lower scale, values of the standardized deviate $(x-\mu)/\sigma$, where μ is the mean and σ the standard deviation of the distribution.

possible number of colonies observed per plate is plotted against the frequency with which it occurs. After counting many plates, the graph approximates to a symmetrical bell-shaped curve (Fig. 6.9). The equation for this curve gives the probability plotted on the ordinate that the variable,

x, will take a particular value plotted on the abscissa. There are two parameters, μ, the mean, and σ^2, the variance; these vary independently in a normal distribution whereas in the Poisson distribution, $\mu = \sigma^2$. The standard deviation of μ is the square root of the variance.

The upper scale on the abscissa of Fig. 6.9 gives values of x, here the number of colonies per plate; but the properties of the distribution become clearer by plotting $(x-\mu)/\sigma$, usually referred to as a *standardized deviate*. The unit on the lower scale of the abscissa is then the difference between a

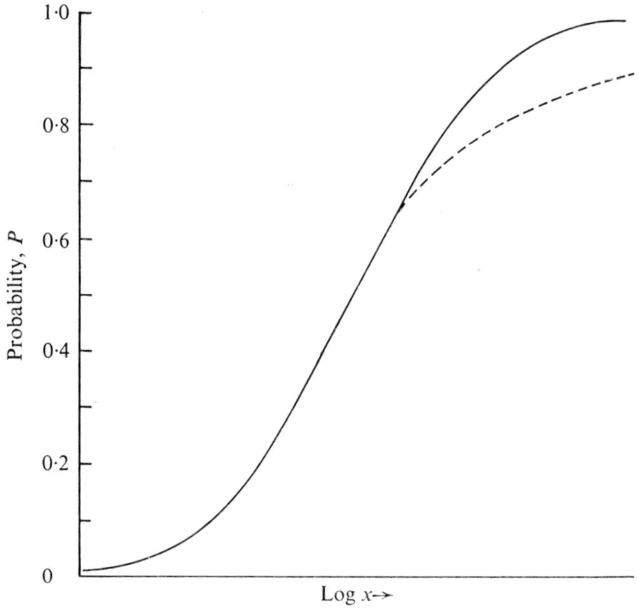

Fig. 6.10. Cumulative (integrated) log-normal distribution. Ordinate: cumulative probability, P, of occurrence of all values of log x up to and including the value in question. Abscissa: log x. In infectivity titrations, the frequency of response is often related to log dose in this way (where x is the number of organisms inoculated per host), but, occasionally, the curve is flattened, as shown by the dashed line (p. 222).

particular value of x and the mean, μ, expressed as a multiple of the standard deviation, σ. It should be appreciated that the probability shown on the ordinate for any value of $(x-\mu)/\sigma$ is the same for all normal distributions whatever the actual values of μ and σ. Two probabilities that recur in discussions of significance are 0·01 and 0·05 which correspond to $\mu \pm 2{\cdot}58\sigma$ and $\mu \pm 1{\cdot}96\sigma$ respectively (see pp. 175, 181, 237).

In a *log-normal distribution*, log x, not x, is normally distributed and a curve resembling Fig. 6.9 is therefore obtained by plotting log x on the abscissa.

These distributions are often used in their cumulative (or integrated) forms which are S-shaped curves obtained by plotting on the ordinate the probability, P, of occurrence of all values of x or log x (Fig. 6.10) up to and including the value in question.

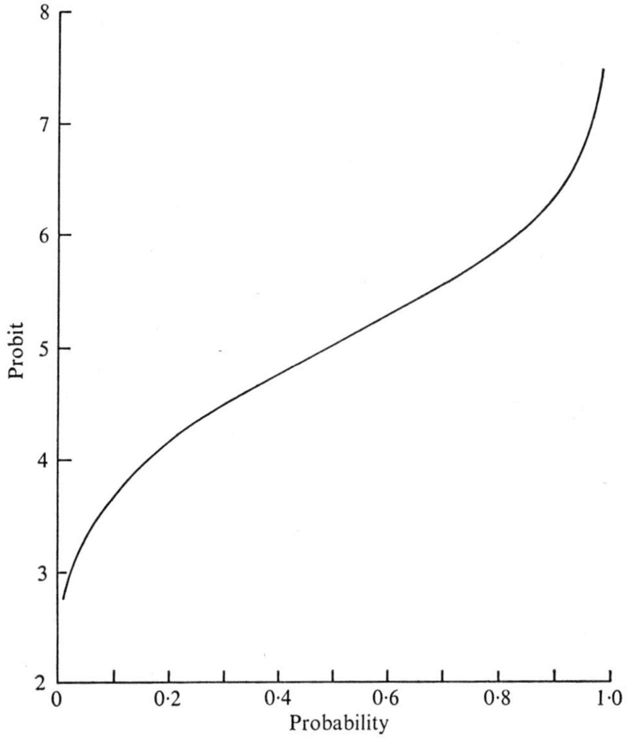

Fig. 6.11. Relation between probability and probit. Ordinate: probit. Abscissa: probability.

The probit transformation

It would be convenient in many ways if the points giving the S-shaped curve in Fig. 6.10 could be plotted to give a straight line. For instance, this might show at once if the dose-response curve was likely to be normal or log-normal, since deviation from a straight line is more easily detected by eye than deviation of one curve from another. Linearity is achieved by plotting on the ordinate, not P, but the function of P named probit P (see Finney, 1952). The relation between the two is shown in Fig. 6.11 and

Table 6.6. *Transformation of probabilities to probits* (*Finney*, *1952b*)

P	0	0·01	0·02	0·03	0·04	0·05	0·06	0·07	0·08	0·09
0	—	2·67	2·95	3·12	3·25	3·36	3·45	3·52	3·59	3·66
0·1	3·72	3·77	3·82	3·87	3·92	3·96	4·01	4·05	4·08	4·12
0·2	4·16	4·19	4·23	4·26	4·29	4·33	4·36	4·39	4·42	4·45
0·3	4·48	4·50	4·53	4·56	4·59	4·61	4·64	4·67	4·69	4·72
0·4	4·75	4·77	4·80	4·82	4·85	4·87	4·90	4·92	4·95	4·97
0·5	5·00	5·03	5·05	5·08	5·10	5·13	5·15	5·18	5·20	5·23
0·6	5·25	5·28	5·31	5·33	5·36	5·39	5·41	5·44	5·47	5·50
0·7	5·52	5·55	5·58	5·61	5·64	5·67	5·71	5·74	5·77	5·81
0·8	5·84	5·88	5·92	5·95	5·99	6·04	6·08	6·13	6·18	6·23
0·9	6·28	6·34	6·41	6·48	6·55	6·64	6·75	6·88	7·05	7·33
0·99	0·0	0·1	0·2	0·3	0·4	0·5	0·6	0·7	0·8	0·9
	7·33	7·37	7·41	7·46	7·51	7·58	7·65	7·75	7·88	8·09

The entries in the body of the table are the probits corresponding to the probabilities, *P*. Thus, the probit of a probability of 0·63 is 5·33.

More comprehensive tables are given by Finney (1952), Fisher & Yates (1963), and Pearson & Hartley (1954), amongst others.

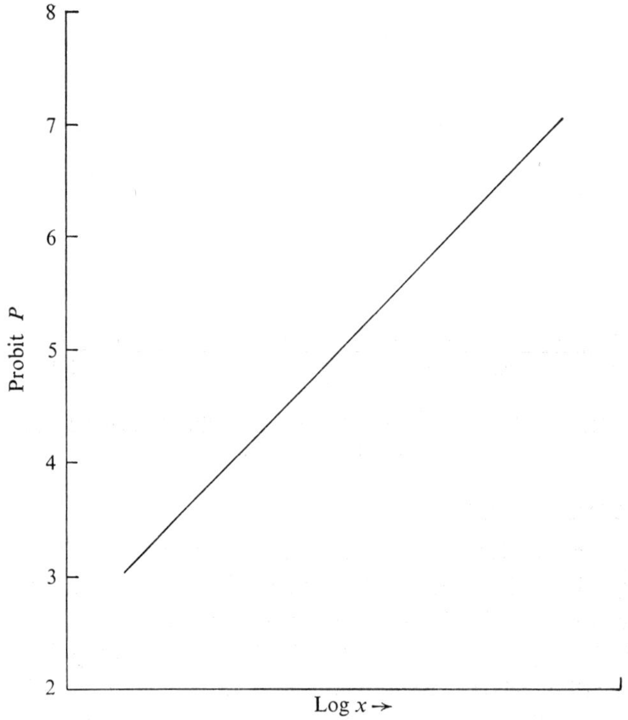

Fig. 6.12. Cumulative log-normal distribution, plotted in probits. Ordinate: probit of the cumulative probability, *P*. Abscissa: log *x*.

Table 6.6. Fig. 6.10 replotted with probit P now gives a straight line (Fig. 6.12). The parameters of this straight line are its mean, μ, corresponding to a probit of 5 ($\equiv$ 50 %), and slope, b, the change in probit P produced by unit change in log x. If probit P decreased from 5·5 to 4·5 as log dosage decreased from 7 to 3, the value of $b = (5{\cdot}5 - 4{\cdot}5)/(7 - 3) = 0{\cdot}25$. The slope is obviously inversely related to the standard deviation, σ, of the distribution and, in fact, $b = 1/\sigma$.

Applications of the log-normal curve

One of the most common arises from titrations of toxins and micro-organisms in living hosts, since P, here the proportion of hosts responding to inoculation, is often related to log dose by an integrated log-normal curve (Fig. 6.10). The results of such titrations are expressed by the dose causing 50 % of hosts to respond, as it corresponds to the steepest part of the curve and is therefore most precisely determined. This dose is generally termed the ED 50 (effective dose 50), but more specific terms having the same significance are also used, for example LD 50 (lethal dose 50) and TCD 50 (tissue culture dose 50). In a few infections, the curve for small doses is log-normal but, at large doses, is unexpectedly flattened, as shown by the dashed curve in Fig. 6.10. The flattening may be caused by an increase in host resistance following inoculation (p. 222; see Bryan, 1959).

Probit plots are also used in sterilization experiments whose data are often fitted equally well by plotting probit S (survival) against log exposure with $b = 2$ ($\sigma = 0{\cdot}5$) as by an exponential survival curve (Fig. 6.7*a*). Systems in which this had been done are described by Withell (1942*a*, *b*), Baten & Stafseth (1956), and Fernelius *et al.* (1958). The last paper shows the advantage in exposition of plotting probit S against log time of exposure, rather than log S against time as in Fig. 6.7*a*, when several curves with widely differing death rates are to be shown on one graph. Using the first method, the plots are parallel, indicating that the distribution of log survival times is the same for each curve, whereas, with the second method, the curves at either high or low death rates are crowded against the axes (Fig. 6.13*a*, *b*).

Estimation of the parameters of a log-normal dose-response curve

Comparison of two dose-response curves shows that the slope as well as the ED 50 has often to be estimated from the observations (p. 223). The numerous methods for obtaining these parameters and their confidence limits are reviewed by Armitage & Allen (1950) and Finney (1952). It would seem from the worked examples given by these authors that the various methods, though differing considerably from the statistical point of view, may give estimates of the ED 50 and slope differing by no more than a few per cent. This may be immaterial in a system which is inherently incapable of giving a very precise result, such as an infectivity titration (p. 228).

Of the various methods, probit analysis is the most efficient (see Finney,

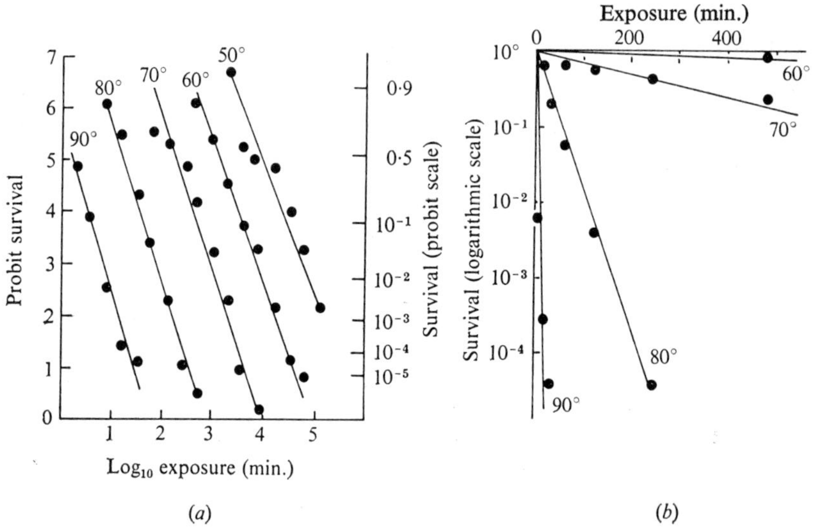

Fig. 6.13. Five survival curves plotted in different ways. Data of Fernelius *et al.* (1958) for anthrax spores heated in aqueous suspension at the temperatures shown on each curve.

(*a*) Ordinate: probit survival. Abscissa: log exposure. Note that, since the curves are approximately equidistant when the exposure is plotted on a logarithmic scale, the exposure producing a given survival is decreased by a constant proportion for each 10° rise in temperature; i.e. the Q_{10} is constant (see p. 93; Fig. 4.7).

(*b*) Ordinate: log survival. Abscissa: exposure.

1952). All the parameters of the dose-response curve and their confidence limits are estimated numerically, even in experiments using different numbers of subjects per dose and a varying dilution factor. The calculations are moderately difficult but the ED 50 and slope can often be estimated sufficiently well by graphical means alone. The proportion of responses for each dose is calculated and the corresponding probit obtained from Table 6.6. These are plotted on ordinary graph paper against the logarithm of the dose. Alternatively, the percentage of responses and the dose in arithmetic units are plotted directly on probability paper. A straight line is then fitted to the points by eye, the best-fitting line being chosen by considering the vertical, not the horizontal, deviations from the points. Points with probits outside the range 2·5–7·5 (0·4–99·6 %) carry little weight. Finally, the compatibility of the fitted line with the data can be checked by a χ^2 test (for an example, see Finney, 1952, § 10). Graphical fitting in this way gives the ED 50 and slope, but not their confidence limits. To obtain these, either further calculation or nomograms like those of Litchfield & Wilcoxon (1949) are required.

A well-known numerical method for obtaining the ED 50 and its confidence limits, but not the slope of the curve, is that of Reed & Muench (1938), which Finney (1964) suggests should be abandoned, partly because it is statistically less efficient than alternatives which are equally simple to compute (Tint & Gillen, 1961, give a nomogram for obtaining the ED 50 by this method). Another method is the Spearman–Kärber method which requires both 0 % and 100 % responses to be produced by doses symmetrically arranged about the ED 50. The *moving average* method described by Thompson (1947) can also be used and is given here with tables that largely eliminate calculation (Table 6.13). This method does not presuppose that the observations conform to the log-normal or any other curve, and obtains the ED 50 by a simple operation involving interpolation, which, unlike probit analysis, does not have to be reiterated.

Table 6.7. *Hypothetical infectivity titration*

Dilution of culture	Log_{10} dilution	Dead/total inoculated	Mortality	Mean proportional mortality ($\bar{m}$)
10^{-5}	−5	0/10	0	—
10^{-4}	−4	1/10	0·1	0·100 (= $\bar{m}_0$)
10^{-3}	−3	2/10	0·2	0·400 (= $\bar{m}_1$)
10^{-2}	−2	9/10	0·9	0·700 (= $\bar{m}_2$)
10^{-1}	−1	10/10	1·0	0·966 (= $\bar{m}_3$)
10^{0}	0	10/10	1·0	—

Estimation of the ED 50 by moving averages. Suppose that the data of Table 6.7 are the results of an infectivity titration in which the response was the death of an inoculated animal. First, the log doses and the mean proportional mortality, $\bar{m}$, are tabulated, taking the dose-groups in threes. Thus, for the first three groups, the mean proportional mortality = (0+0·1+0·2)/3 = 0·1, and for the next three groups = (0·1+0·2+0·9)/3 = 0·4. The two values of $\bar{m}$ immediately less than 0·5 are denoted by $\bar{m}_0$ and $\bar{m}_1$ and the two immediately greater by $\bar{m}_2$ and $\bar{m}_3$.

The ED 50 is estimated from $\bar{m}_1$ and $\bar{m}_2$. The log ED 50 will be a proportionate distance, f, between d_1 and d_2 (the log dilutions corresponding to $\bar{m}_1$ and $\bar{m}_2$: here -3 and -2)

$$\begin{aligned} f &= \frac{0{\cdot}5-\bar{m}_1}{\bar{m}_2-\bar{m}_1} \\ &= \frac{0{\cdot}5-0{\cdot}4}{0{\cdot}7-0{\cdot}4} \\ &= \frac{0{\cdot}1}{0{\cdot}3} \\ &= 0{\cdot}333. \end{aligned}$$

$$\begin{aligned} \text{The log ED 50} &= \log d_0 + \log R(f+1), \quad \text{where } R \text{ is the dilution factor} \\ &= -4+1(0{\cdot}333+1) \\ &= -2{\cdot}667. \end{aligned}$$

If one of the mean proportionate mortalities $= 0{\cdot}5$, the corresponding dose is the estimated ED 50. The data can be analysed by calculating $\bar{m}$ for groups of any number of doses, but groups of 3 are optimal. Taking two doses may give inversions in the sequence of values of $\bar{m}$ so that more than two values appear next to 0·5; on the other hand, most titrations are too small to allow more than one grouping of, say, 5 or 6 doses.

The standard error (S.E.) of the log ED 50 equals

$$\frac{\log R}{a_3-a_0}\sqrt{\frac{(1-f)^2 a_0 b_0 + a_1 b_1 + a_2 b_2 + f^2 a_3 b_3}{n-1}},$$

where f has its previous meaning; a_0 and b_0 are respectively the numbers of fatal and non-fatal infections produced by the dose d_0; and n is the total number of hosts per dose.

In Table 6.7, the S.E. (log ED 50) is therefore

$$\frac{\log 10}{10-1}\sqrt{\frac{\{(1-0{\cdot}333)^2\times 1\times 9\}+(2\times 8)+(9\times 1)+(0{\cdot}333^2\times 10\times 0)}{(10-1)}} = 0{\cdot}1995.$$

The ED 50 and its standard error are rapidly obtained from tables prepared by Weil (1952), reproduced here as Table 6.13 on p. 236. This covers titrations using 2, 3, 4, 5, 6 or 10 subjects per dose; 4 or more doses; grouping 3 doses in calculation of each value of $\bar{m}$; and for any dilution factor, provided it is constant so that the doses form a geometrical progression. The example of Table 6.7 fulfils these conditions. Table 6.13 gives n, the number of subjects inoculated per dose; K, the number of dose-groups used in calculating each value of $\bar{m}$; four values of r, the number of subjects responding at the four doses surrounding the ED 50 as judged by inspection: f has the same meaning as before; and σ_f is used to obtain the S.E. of the

ED 50. The two middle r-values in each entry can be exchanged: thus, the r-values for Table 6.7 are 1, 2, 9, 10, but the same entry is used for a titration giving 1, 2, 9, 10. The table does not include sequences of responses like 1, 0, 1, 1, for which the estimated ED 50 would lie outside the dose-range used.

The ED 50 of the titration shown in Table 6.7 is then obtained from Table 6.13 by looking up the section for $n = 10$ and $K = 3$ and finding the sequence of r-values, 1, 2, 9, 10. The tabulated value of f is again seen to be 0·3333. The log ED 50 $= \log d_0 + \log R(f+1)$, as before, $= -4+1(1{\cdot}3333)$ $= -2{\cdot}667$.

The S.E. of the log ED 50 $= \sigma_f(\log_{10} R)$. In this example it is

$$0{\cdot}19945(\log_{10} 10) = 0{\cdot}19945,$$

as before.

Horn (1956), also using Thompson's method, tabulated the calculated ED 50 and its confidence limits for titrations using 4 or 5 subjects per dose and a dilution factor of $\sqrt{10}$ ($= 3{\cdot}16$) or $\sqrt[3]{10}$ ($= 2{\cdot}15$) where the observations come from 4 successive dilutions.

The estimate of the S.E. is itself subject to considerable sampling error and a rather more precise estimate of its value is obtained by using the mean proportional mortalities in place of the observed mortalities (Armitage, personal communication), where $\overline{w}_0 = n_0 - \overline{m}_0$, etc.

$$\text{S.E. } (\log \text{ED } 50) = \frac{\log R}{(\overline{m}_3 - \overline{m}_0)} \sqrt{\left\{\frac{(1-f)^2 \overline{m}_0 \overline{w}_0}{n_0} + \frac{\overline{m}_1 \overline{w}_1}{n_1} + \frac{\overline{m}_2 \overline{w}_2}{n_2} + \frac{f^2 \overline{m}_3 \overline{w}_3}{n_3}\right\}}.$$

QUANTITATIVE (GRADED) RESPONSES

A quantitative response takes any one of a series of possible values, for example 0, 1, 2, 3, 4 . . ., in contrast to a quantal response which can have only one of two values, like + or −. Many types of quantitative response are used in microbiology, such as response time, which may be the time between challenge and death of an animal; diameters of skin lesions (Long, Miles & Perry, 1954); loss of weight, used in titration of vaccines; degrees of consolidation in pulmonary infections; extent of cytopathic change in virus-infected tissue cultures; rise in temperature produced by injection of bacterial pyrogens; and size of inhibition zone in antibiotic assays (see Cooper, 1963). Some of these responses, like the diameters of skin lesions or zones, can be measured in cm. or any convenient unit of length, but, in other cases, the response is not readily measurable although it is easy enough to place a series of specimens in order of magnitude. Examples are the degree of pulmonary consolidation or of trachoma infection of the eye, which both have to be measured in arbitrary units according to a predetermined scoring system. Some responses previously considered as quantal become quantitative if the site is redefined. For instance, the number of discrete lesions found in the lungs of an infected animal was discussed as a quantal response on p. 182 where the site was taken as the smallest part of the lung capable of supporting formation of a lesion, and the response was the presence or absence of a lesion. But if the whole organ is defined as the site, the response then becomes quantitative, as the number of lesions per lung can be 0, 1, 2, 3, ...

Response times have been singled out for discussion, partly because they are more often used in infectivity titrations than other quantitative responses and more is therefore known about them, and partly because they reflect how micro-organisms behave *in vivo* between the moment of inoculation and the response of the host. As regards the second aspect, some simple models are put forward to account for the observations, which, in our view, are both reasonable on biological grounds and useful in providing a framework of discussion. It is to be expected that these models, like those discussed on p. 225, will need qualification as more data accumulate, but, in the meantime, they serve to show how plausible general hypotheses of microbial infection can be stated. These hypotheses apply as much to viruses as to bacteria, because both are self-replicating and because only non-specific features of their growth are involved, like their rates of division or inactivation. For the same reason, the models apply to other self-replicating parasites, like protozoa or tumour cells, though these are not otherwise referred to here.

MEASUREMENT OF RESPONSE TIME IN MICROBIAL INFECTIONS

The response time is the interval between administration of organisms and appearance of the chosen response, whether it be death, loss of consciousness, rise of temperature, and so on.

The response time provides a good example of the different approaches to quantitative experiment mentioned in the Introduction to this chapter. As it varies with the number of organisms inoculated, it has frequently been proposed for assaying the concentration of microbial suspensions (Gard, 1940; Bryan & Beard, 1940; Ipsen, 1944). It has the advantages, compared to a quantal response like death, of greater precision (p. 218) and of rapidity since a dose of several or many ED 50 is given which causes all the hosts to respond within a relatively short time. However, the response time also reflects the way in which organisms multiply in their hosts. The important point, which becomes clear in the following sections, is that the analysis of the data largely depends on why the experiment is done.

Biological interpretations of response time relationships

In practice, there are two relationships to consider: how individuals respond at different times to inoculation of a given dose, which leads to the estimate of the mean response time (RT 50); and how the RT 50 is related to size of dose.

Distributions of individual response times. These are usually skewed with a long tail to the right corresponding to long response times; which is hardly surprising, since no host can respond before inoculation ($t = 0$) although occasional responses are to be expected after times far longer than average.

The observed distribution has often been treated as log-normal because a straight line can be fitted to the points obtained by plotting the cumulative proportion of responses on a probit scale against the logarithm of time since inoculation (Fig. 6.14, *A*). This line is characterized, like a quantal dose-response curve (Fig. 6.12) by its mean, the RT 50 at which 50 % of hosts respond, and by its slope. Both RT 50 and slope and their standard errors can be estimated either graphically by the nomograms given by Litchfield (1949) or numerically (Bliss, 1937). Despite the convenience of a log-normal relationship, it must be said that most experiments are probably not sufficiently precise to distinguish normal distributions of $\log t$ from those of t or of $1/t$ (Meynell & Williams, 1967; Meynell & Maw, 1968). However, accepting a log-normal distribution as reasonable, two interesting points emerge from the data. The first is that the scatter of $\log t$ generally appears the same for all sizes of dose, which suggests that the behaviour of the organisms *in vivo* is independent of dose (curve *A* in Fig. 6.14: G. G. Meynell, 1963; Meynell & Maw, 1968). The second is that the scatter observed in naturally-occurring (Sartwell, 1950, 1966) and experimental infections (Meynell & Maw, 1968) never exceeds a certain extent, as expressed by Sartwell's dispersion factor Δ (= RT 84/RT 50 = RT 50/RT 16 for a log-normal distribution). Values of Δ greater than 1·5 are uncommon.

Occasional exceptions occur. In some infections, although groups of animals given large doses yield log-normal distributions of t, those

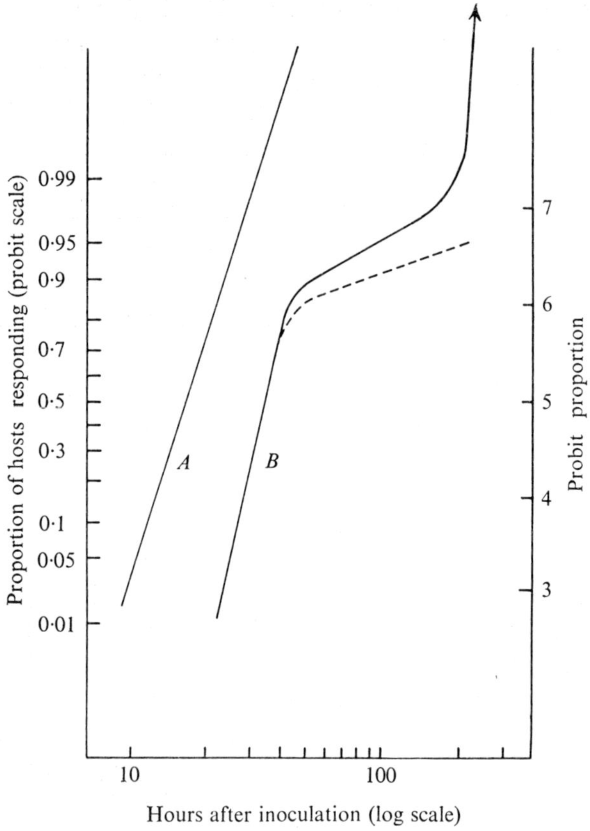

Fig. 6.14. Distributions of death times for mice given pneumococci by intraperitoneal injection (G. G. Meynell, 1963. Data of Cavalli & Magni, 1943). Ordinate: proportion of mice dead at a given time (probit scales). Abscissa: logarithm of number of hours since inoculation.

Curve *A*: 191/191 mice died from a dose of 10^8 cocci. The distribution is log-normal.

Curve *B*: 183/192 mice died from a dose of 10^2 cocci. As less than 100% of mice died, the cumulative proportion of deaths, *P*, is calculated by dividing either by 192, the number inoculated (dashed curve), or by 183, the number fatally infected (continuous curve). The first method gives the type of curve generally described as truncated, since, after the 40th hour, deaths occurred later than expected from the distribution of deaths between 20 and 40 hr. The second method of calculation confirms that the distribution is not log-normal, unlike curve *A*, for, although it appears to be so up to 40 hr., a discontinuity appears which is not an artefact and is probably due to a change in the host–pathogen relationship.

inoculated with smaller doses yield discontinuous distributions in which the earliest responses to small doses occur in the same fashion as those to large doses but the later responses only occur after an unexpectedly long time (curve *B* in Fig. 6.14). Such distributions are described as 'truncated' (Bliss, 1937; Beard *et al.* 1955) and are thought to arise because, at a certain time after inoculation, an immune response causes the organisms to increase less rapidly in those hosts still surviving (Cavalli & Magni, 1943; Beard, Sharp & Eckert, 1955) or the response threshold rises (G. G. Meynell, 1963). The RT 50 has sometimes to be determined from the left-hand part of these distributions; i.e. the RT 50 that would have been obtained if truncation was absent. This is done either graphically by extrapolating the left-hand part of the curve upwards, or numerically (Sampford, 1952). Individual response times can therefore yield valuable information about the course of infection which might otherwise be overlooked.

Truncated distributions occur with mice infected with pneumococci (Cavalli & Magni, 1943) or anthrax bacilli (DeArmon & Lincoln, 1959). No ambiguity is possible when 100 % of hosts respond in the dose group in question but, when some hosts survive, a misleading interpretation is possible which accounted for many of the truncated distributions reported earlier. The question is how to calculate the cumulative proportion of responses. The alternatives are to divide the number of responses that occur by a given time either by the total number of hosts in the dose-group or by the number that have responded when the experiment is ended. If the object is to depict the behaviour of the dose-group as a whole, the first method is correct as it shows the proportion of hosts that fail to respond. On the other hand, if the aim is to study the distribution of response times as an indication of how the organisms produce the response, the second method should be used as non-responders are then irrelevant. The first method may, in fact, be misleading in this situation because it gives plots that superficially suggest truncation and thereby suggest a biological change in the system for which there is otherwise no evidence (G. G. Meynell, 1963).

Relation of RT50 to size of dose. In microbial infections, the RT 50 is usually related to size of dose by the curve shown in Fig. 6.15. (See Meynell & Meynell, 1958. For possible exceptions, see Eckert, Beard & Beard, 1956; Bryan, 1957.) This curve has three parts:

(*a*) A central portion, where RT 50 is inversely related to log dose (the RT 50 is calculated solely for those hosts that respond, for reasons given above). This probably reflects exponential increase of the organisms to a critical concentration, *C*, at which the response occurs (Fig. 6.16: see Williams & Meynell, 1967). If this is so, the slope in Fig. 6.15 is inversely related to the mean rate of increase of the organisms *in vivo* and the curve therefore becomes more closely aligned to the dose axis, the faster the organisms increase. It is evident here that the critical concentration, *C*, and the rate of increase are assumed constant for all doses.

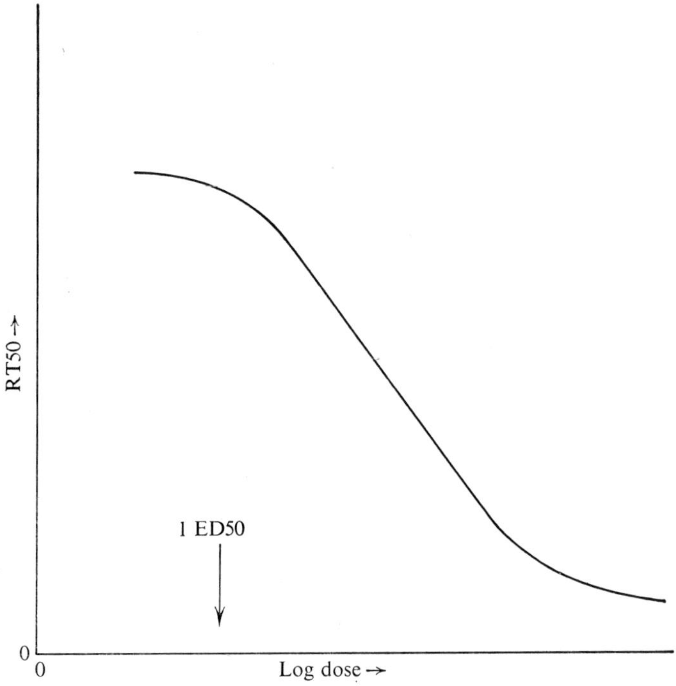

Fig. 6.15. Relation between RT 50, the time at which 50 % hosts respond to each dose, and log dose in microbial infections. Ordinate: RT 50, calculated solely from those hosts that respond (see text). Abscissa: log number of organisms inoculated per host.

(*b*) A portion of lesser slope at high doses, where the curve asymptotes to the dose axis. Obviously, no response will occur instantly, however big the dose, and the responses may here by produced by the toxic effect of the inoculated organisms, not by their multiplication *in vivo*, or, in virus infections, by saturation of the receptors available at the time of inoculation (Golub, 1948).

(*c*) A portion at doses $\leqslant$ 1 ED 50, where the RT 50 may become constant, as found in infection of *Drosophila* by σ virus (Plus, 1954), or of mice by salmonella (Meynell & Meynell, 1958). The reason is thought to be that the fates of organisms in the inocula are determined independently both before and after inoculation. If this is so, and the hosts are of identical susceptibility, the behaviour of the system is described by the Poisson series. This predicts that at a dose of 1 ED 50, the mean number of effective organisms per dose is *ca.* 0·7 (equation (6.3); $S = e^{-pm}$; $0{\cdot}5 \simeq e^{-0{\cdot}7}$). Most of the responses to doses $\leqslant$ 1 ED 50 will then be due to multiplication of only

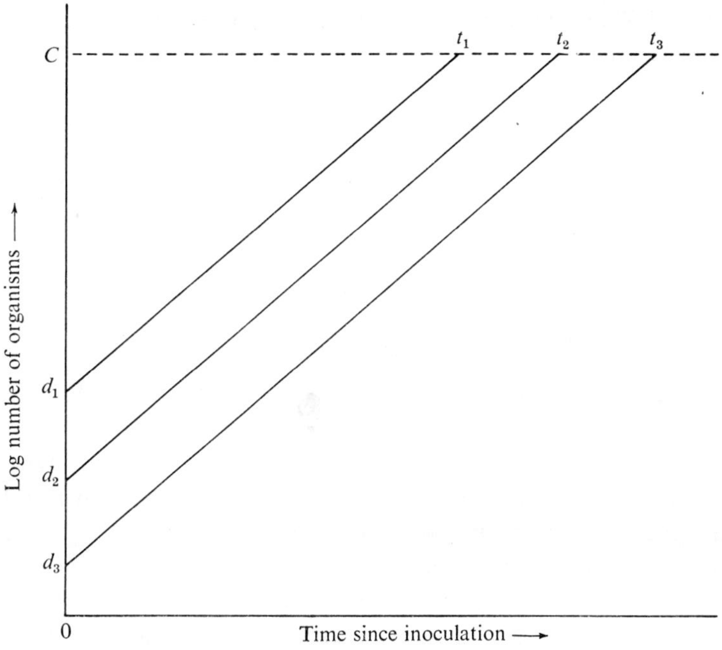

Fig. 6.16. Hypothetical growth curves in infected hosts. Ordinate: logarithm of the number of organisms per host. Abscissa: time since inoculation. The organisms are assumed to increase exponentially to a critical concentration, *C*, at which the host responds. The doses, d_1, ..., cause responses after times t_1, ... It is evident that log *d* is linearly related to *t* as in the central part of the curve shown in Fig. 6.15.

one organism, just as in isolating pure clones by the terminal dilution method (Fig. 6.1), and the RT 50 will therefore tend to constancy. This must be so when a subject invariably responds following receipt of a single organism ($p = 1$); but when $p < 1$, as in most infections, the left-hand part of the curve shown in Fig. 6.15 will hold only if the fates of inoculated organisms are independently determined *in vivo*.

Assays based on response times

The assay is not usually based on the linear relation observed between RT 50 and log dose shown in the central part of Fig. 6.15 owing to the properties of the RT 50. Its confidence limits are symmetrical only when time is expressed in logarithmic units (cf. Fig. 6.14, curve *A*) and become asymmetrical when time is in arithmetic units, as in Fig. 6.15. Moreover, the size of the RT 50 and its variance are positively correlated (Gard, 1940, 1943; Finney, 1964; cf. Figures in Meynell & Meynell, 1958). The RT 50 is therefore determined more precisely for high doses, which produce responses most quickly, than for small doses, a factor which therefore complicates the titration and prevents the immediate fitting of a regression line to the points by the usual methods. This difficulty is avoided by calculating the reciprocal of the harmonic mean response time for each dose group (Gard, 1940).

The reciprocal of the harmonic mean response time. This is:

$$1/\overline{T} = (1/t_1 + 1/t_2 + \ldots + 1/t_N)/N,$$

where t_1, etc., are the individual response times for the N hosts in the dose-group.

When all the hosts respond, the calculation is straightforward. When only a fraction of the hosts respond, these alone could be used in calculation of the harmonic mean. However, the drawback is that no information is derived from the remaining hosts. It has therefore become the practice to assign an arbitrary survival time to each non-responder. Gard (1940) assumed that each of these lived an infinite time, a value which becomes zero when the reciprocals of the individual response times are calculated to obtain the harmonic mean. Other authors assign shorter times like 18 days or 42 days in titrations ended after 16 and 21 days respectively (Finney, 1964; Smith & Westgarth, 1957), although there is no reason to suppose that these subjects would ever have responded if the titration had been continued.

Relation of $1/\overline{T}$ *to log dose.* When $1/\overline{T}$, the reciprocal of the harmonic mean response time, is plotted against log dose, a straight line suitable for an assay is often obtained. As the variance of $1/\overline{T}$ is constant and is not correlated with $1/\overline{T}$ itself, a regression line is then fitted to the points in the usual way (for a worked example, see Smith & Westgarth, 1957).

The precision of a response time assay is therefore determined by (*a*) the precision with which the mean response time is determined for each dose-group, which in turn depends on the variance of the distribution of individual response times (i.e. the slope in Fig. 6.14), and the period between each inspection of the subjects; and (*b*) the slope of the dose-response time curve, which depends on the rate at which the organisms increase *in vivo* (Figs. 6.15, 6.16).

Apart from the points common to all bioassays (p. 218), there are some peculiar to the response time.

The responses should fall in the central portion of the dose-response curve shown in Fig. 6.15 where response time varies with dose.

The smallest dose should, if possible, be large enough for 100 % of hosts to respond. If less than 100 % responses are obtained, the titration is relatively inefficient as little or no information is gained from the non-responders.

These titrations are no exception to the rule that 3 doses of known and unknown suspension respectively are needed (p. 220), although some authors suggest that only one dose of the unknown is necessary (Golub, 1948; Bauer, 1961). This would only be justified if the dose-response time curve remained constant, but this cannot be relied upon in general, and unpredictable changes in slope have been reported in salmonella infection of mice (Meynell & Meynell, 1958) and in psittacosis infection of chick embryos (Dougherty, McClosky & Stewart, 1960).

This discussion of response times in infection therefore illustrates the distinction drawn on p. 173 between the different approaches to quantitative experiment. The first section concerned a model of infection which has features making it unsuitable for titrations, whereas the transformed response times discussed in the second section have no biological significance, yet are appropriate for assays.

GENERAL ASPECTS OF TITRATIONS

All titrations entail certain problems in planning and interpretation, whether they measure sensitivity to irradiation, the bactericidal power of a disinfectant, the concentration of a bacterial growth factor, or the protective power of a vaccine. The type of response has to be decided. If the subjects are expected to differ in susceptibility, they have to be allotted at random to different dose-groups. The number of doses and subjects, and the duration of the titration, have all to be considered. In some experiments, such questions are almost never raised because a certain technique has become established. Thus, in an irradiation experiment the greater precision of colony counts compared to dilution counts is now accepted almost intuitively so that the possibility of measuring survivals by the dilution method hardly comes to mind. In water analysis, on the other hand, colony counts on membrane filters are a relatively recent development and a dilution count is the traditional method. Similarly, it is unusual for the incubation period to be stated explicitly for a colony count of, say, *Escherichia coli* on nutrient agar, although in sterility tests using spores the period of incubation has often to be considered most carefully before a sample can be accepted as sterile (p. 24). Nor is it usual to randomize tubes of broth amongst dose-groups in a bacterial dilution count as the tubes are taken as identical. Questions of planning arise far more forcibly in titrations of living organisms in animals and other susceptible hosts which are usually few in number, so that sampling error is prominent, and, moreover, are usually heterogeneous in susceptibility and may even change during the experiment. These titrations therefore form the main subject of the following sections, though the principles they illustrate apply generally.

PLANNING OF INFECTIVITY TITRATIONS

Choice of response

The response chosen for a titration depends not only on statistical arguments but also on its purpose and on convenience. If the experiment is intended to examine the effect of some treatment like chemotherapy or vaccination on the lethal consequences of an infection of experimental animals, it seems wiser to measure the death rate rather than an indirect host response like antibody titre which is thought to be correlated with mortality. On the other hand, when the infection occurs only in man, indirect responses are often unavoidable: their significance forms a major topic in discussions of immunity in medical textbooks.

Death and other quantal responses may be more convenient to record than response times if hosts respond within a short time of each other, whatever dose they receive, and would have to be counted at, say, 4-hourly intervals to determine the RT 50 at all accurately. For example, with a rapidly growing organism, like *Salmonella typhimurium* in mice, the RT 50 may only fall from 5 to 4 days for a 10^5-fold increase in dose (Meynell & Meynell, 1958).

In general, however, a quantitative response is preferable to a quantal response, since each subject yields more information. With death-time measurements, each animal dies after one of an infinite number of possible times, whereas with a quantal response like death, only one of two results is possible. Moreover, quantal responses are far less precise in infectivity titrations than in, say, toxicity tests, since the slope of the dose-response curve is always comparatively small, possibly due to the intrinsic nature of microbial growth *in vivo* (see p. 228). Hypothesis suggests that, in an infectivity titration, the maximum slope of a probit response–log dose curve will be *ca.* 2 and, in practice, no slopes exceed this value and many are far smaller (see Meynell, 1957).

The relative efficiencies of quantitative and quantal responses, expressed by the number of subjects required to give estimates of equal precision, have been compared in several systems. Use of mean death time reduced the number of subjects to 25 % of that needed for mortality measurements in titrations of *Bacillus anthracis* (Lincoln & DeArmon, 1959). In other systems, only 5 % as many animals were needed for a quantitative response as for a quantal response of the same precision (see Bryan, 1959).

Elimination of systematic differences between dose-groups

It is clearly undesirable in an assay that some dose-groups give a less precise result than others.

One difficulty may be that the precision of each mean response depends on the size of the mean itself, as in local lesion counts (p. 185) and response time assays (p. 216), where the variance increases with the mean. It then follows that the larger the mean, the less precisely is it determined. This difficulty is usually overcome by recalculating the mean response to make it independent of its variance, as in calculating the reciprocal of the harmonic mean response time.

A more general source of bias is for hosts in some dose-groups to differ materially from those in others. The subjects must therefore be allotted at random to different dose-groups, usually by standard tables of random numbers for, if this is not done, the experimenter may unconsciously distribute the subjects non-randomly. Emmens (1948) reported an instance in which the mean weights of groups of mice were positively correlated with the order in which they were removed from the stock box.

The total number of subjects should be divided equally between the known and test groups, which in turn are subdivided into equal numbers of dose-groups, each containing the same number of subjects.

Choice of doses

Unless two doses are used for both the unknown and the test suspensions in a so-called (2+2) assay, the slope and ED 50 of each dose-response curve cannot be determined. It is always preferable, however, to determine three points for each curve in a (3+3) assay in order to confirm its shape (p. 223). Warner (1964) gives nomograms that simplify the calculation of the potency ratio and its confidence limits.

The range of doses required depends on the type of response. If a graded response is used, the doses should exceed that which causes *ca.* 100% responses ($\geqslant$ 5 ED 50). The doses given to the test group should, as far as possible, be chosen to produce the same intensity of response as occurs in the known group; for example if three successive 10-fold dilutions of the known suspension produce 10, 50 and 90% fatal infections, or lesions of 0·1, 0·3 and 0·5 cm. diameter, the concentrations inoculated in the test group should be chosen to give the same mortalities or the same sizes of lesions.

With quantal responses, the doses should theoretically be chosen to lie around that known to be most precise on statistical grounds, i.e. the ED 80 and the ED 50 for exponential and normal (or log-normal) curves respectively (p. 188; Finney, 1952). However, one often has no idea of the potency of the test preparation and, in infectivity titrations at least, the susceptibility of the subjects varies unpredictably from one experiment to the next. A narrow range of doses will then give a useless result if the response turns out to be only 1% of that expected. It is preferable, therefore, to use a series of doses extending from a small fraction to a large multiple of the expected ED 80 or ED 50, for example 1/100–100 ED 50; to inoculate the same number of subjects with each dose; and to use a constant dilution factor. Moreover, the ED 50 is then often obtainable from Tables 6.10–6.12, or 6.13.

Once the range of dilutions is decided, the dilution factor has to be chosen. In general, fewer subjects at closely spaced doses are preferable to more subjects at widely spaced doses, since the smaller the dilution factor, the greater the chance of one dose corresponding to that which carries the greatest statistical weight (see the argument in planning dilution counts, p. 188). When the analysis uses log doses (p. 186, Fig. 6.13*a*), a constant dilution factor should be used so that the doses will be equally spaced on a logarithmic scale. On the other hand, if doses are to be kept in arithmetic units, they should be spaced out at equal distances on an arithmetic scale (Figs. 6.7*a*, 6.13*b*).

Number of subjects

These have been calculated approximately from probit theory for titrations based on a quantal response (DeArmon & Lincoln, 1959). For an ED 50 of given precision, it is evident that the steeper the slope of the dose-response curve, the fewer the subjects are needed: thus, if the slope was ∞, one subject per dose would estimate the ED 50 with maximum precision. Table 6.8, taken from DeArmon & Lincoln (1959), gives the total number of subjects required to estimate the ED 50 with varying degrees of precision, assuming the slope (b on p. 205) to be known in advance and that the centre of the dose-range corresponds roughly to the ED 50. Precision is expressed by the length of the 95 % confidence interval of the log ED 50, that is, by $\log(\text{ED }50 + 1{\cdot}96\sigma) - \log(\text{ED }50 - 1{\cdot}96\sigma)$. Entries are given for slopes of 1, 1·5 and 2. Healy (1950) has made similar calculations for slopes of 1, 2, 4, and 8.

The number of subjects required to estimate the potency ratio with a given precision can also be obtained from Table 6.8.

Table 6.8. *Number of hosts required to estimate the ED 50 with a given precision (DeArmon & Lincoln, 1959)*

Length of 95 % confidence interval[1] in $\log_{10}$	Slope of the dose-response curve (b)[2] 1·0	1·5	2·0
0·21	860	380	220
0·34	330	150	82
0·42	220	96	54
0·77	64	29	16

[1] The 95 % confidence interval here $= \log(\text{ED }50 + 1{\cdot}96\sigma) - \log(\text{ED }50 - 1{\cdot}96\sigma)$. The numbers of hosts required to estimate the potency ratio with a given precision can also be obtained from this table. The first column then becomes the log 95 % confidence interval of the potency ratio. The number of hosts required for each of the two titrations is twice the appropriate entry given above.

[2] See p. 205.

Fewer hosts will be needed if they can be made to respond more uniformly and the slope of the dose-response curve thereby increased. The F_1 hybrid animals obtained by crossing two inbred lines may be more uniform than their parents (see Russell & Burch, 1959; Brown & Dinsley, 1962). In virus titrations on the chorio-allantoic membrane of chick embryos, precision is increased by taking pieces of membrane from young instead of older eggs (White & Fazekas de St Groth, 1959) as well as by other changes in technique (McCarthy, Downie & Armitage, 1958).

Duration of titrations

A point to consider is the length of time for which an infectivity titration is allowed to run before making a final count of the numbers of subjects responding to each dose. The usual practice is to wait until a relatively long time after the last response, and the duration of the titration therefore depends on the slope of the distribution of individual response times (Fig. 6.14). There is always the possibility that another response would have occurred if the titration had been continued, but this point is usually neglected without underestimating the ED 50 since a long gap after the last response implies that an insignificant proportion of subjects is likely to respond subsequently.

This procedure is tantamount, if the observed ED 50 is believed to have any real significance, to postulating that the subjects become of two types following inoculation—those that do and do not respond. In many infections, this assumption is supported by viable counts on the infected subjects which increase progressively if a response eventually occurs, but, otherwise, sooner or later decline. Thus, in a dose-group in which only a few of the hosts respond, viable counts soon become bi-modally distributed amongst the hosts which can therefore be validly regarded as falling into two classes (see G. G. Meynell, 1963). The proportions of responders and non-responders in a dose-group might therefore be forecast from the proportions of hosts giving high or low counts with a considerable saving in time. Schneider & Zinder (1956) found, with mice inoculated with *Salmonella typhimurium*, that the mortalities recorded at the end of 21 days agreed with the proportions obtained from viable counts made on the internal organs only 2 days after challenge.

In some systems, titrations are cut short by the nature of the hosts. Titrations in chick embryos have to be ended by the 21st day after laying because the chicks start to hatch. Titrations in animals may be restricted by an increase in host resistance induced by the infection, manifested by the regression of established lesions (see Bryan, 1959). Host changes occurring during the titration may, however, be detectable by changes in dose-quantal response curves (p. 202) and in the distribution of individual response times (p. 212). Occasionally, the subdivision of a dose-group suggested by the response times is correlated with differing post-mortem appearances. For example, the death times of a group of mice each injected with the same dose of tubercle bacilli were distributed as in curve *B* of Fig. 6.14, which suggested two causes of death. This was confirmed by finding that individual mice fell into the same groups when classified as 'acute' or 'chronic' by the appearance of their internal organs (Hoyt, Moore & Thompson, 1962).

COMPARISON OF DOSE-RESPONSE CURVES

The purpose of many titrations is the comparison of two groups of subjects, for example vaccinated and non-vaccinated, or a group inoculated with a microbial suspension of known concentration with a second group given a suspension of unknown concentration. Each of the two groups yields its own dose-response curve, whether the response is quantal or quantitative, and these curves provide the means for comparing the two groups. The statistical treatment of the data lies within the province of bio-assay and only some general points are mentioned here.

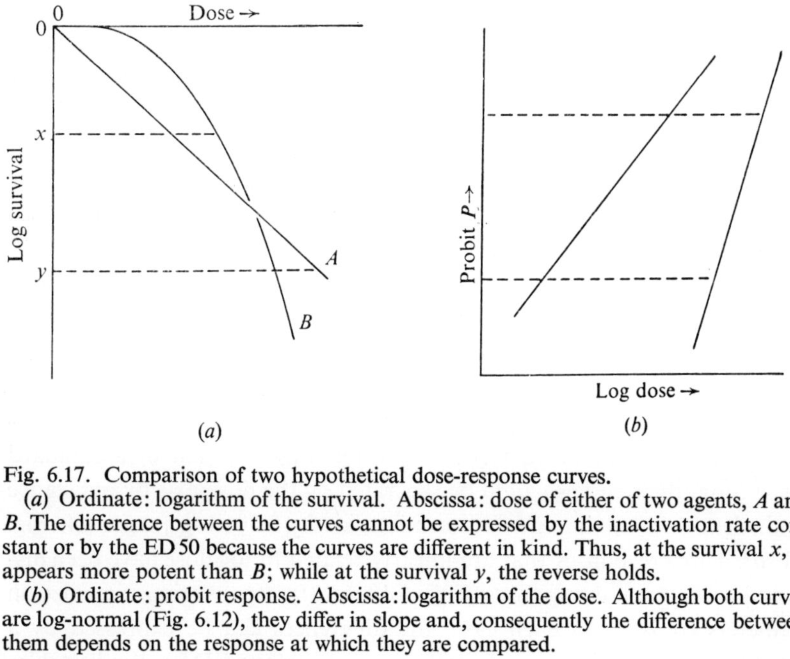

Fig. 6.17. Comparison of two hypothetical dose-response curves.
(*a*) Ordinate: logarithm of the survival. Abscissa: dose of either of two agents, *A* and *B*. The difference between the curves cannot be expressed by the inactivation rate constant or by the ED 50 because the curves are different in kind. Thus, at the survival x, *A* appears more potent than *B*; while at the survival y, the reverse holds.
(*b*) Ordinate: probit response. Abscissa: logarithm of the dose. Although both curves are log-normal (Fig. 6.12), they differ in slope and, consequently the difference between them depends on the response at which they are compared.

Whatever index is used, comparisons are only meaningful if both curves are of the same kind; that is, if both are exponential, or, when they are normal or log-normal, if both have the same slope. Otherwise the difference between the two curves also depends on the survival at which the comparison is made (Fig. 6.17).

Given that the curves are similar in kind, the basis of comparison depends on the form of the dose-response curve. With exponential curves the rate constants are compared and, since these are also the slopes of the curves when log survival is plotted against dose, the comparison is often referred to as being based on a *slope ratio*. Another form of slope-ratio assay arises in the microbiological assay of growth factors like vitamins, whose concentration, C, is often linearly related to N, the yield of organisms obtained in the assay (p. 5): $CK = N$, where K is a constant. If a series of dilutions of a standard preparation of concentration C are tested, a straight line of slope K is obtained by plotting the degree of dilution on an arithmetic scale against the corresponding values of N. The same dilutions of an unknown preparation of differing concentration, C', will also give a straight line but of different slope, K'. The unknown concentration, $C' = C(K'/K)$. Other relationships also occur, for example $\log C = K \log N$ (see Barton-Wright, 1952).

With normal or log-normal curves, the ED 50's are compared by calculating their ratio, usually referred to as the *potency ratio*, either numerically (e.g. by probit analysis) or by nomograms, like those of Litchfield & Wilcoxon (1949) which depend on a preliminary fitting of a dose-response curve to the points by eye. A potency ratio can be similarly calculated for assays using response times or lesion diameters by the ratio of the log doses producing the same response.

In a slope-ratio assay, the difference in slope between two curves arises because the dose is expressed in arithmetic units. On the other hand, if dosage is expressed in logarithmic units, two parallel lines are obtained when the dose-response relationships are basically the same, because one curve then differs only proportionately from the other. Thus, if the concentration of an unknown virus suspension was half that of a standard suspension, their curves would be separated on the dose-axis by a distance equivalent to a halving in concentration, i.e. by 0·301 $\log_{10}$ units (cf. Fig. 6.13*a*). Three responses that are linearly related to log dose have been discussed here: quantal responses (Figs. 6.10, 6.12); response times; and lesion diameters (p. 209). In each case, the differences can be expressed by a potency ratio: $\text{ED }50_{\text{standard}}/\text{ED }50_{\text{unknown}}$. With the ED 50, Weil's tables can be used to obtain the confidence limits of the potency ratio, always provided the dose-response curves are similar.

The log potency ratio $= \log \text{ED }50_{\text{standard}} - \log \text{ED }50_{\text{unknown}}$. The standard error of the log potency ratio is given by:

$$\text{S.E. (log potency ratio)} = \sqrt{\{(\text{S.E. log ED}\,50_{\text{standard}})^2 + (\text{S.E. log ED}\,50_{\text{unknown}})^2\}}.$$

The 95 % confidence limits for the log potency ratio are:

$$\text{Estimated log potency ratio} \pm 1{\cdot}96\ \text{S.E. (log potency ratio).}$$

GENERAL MODELS

On going through the microbiological literature, it is striking how the same models appear repeatedly in interpretations of the data. One postulates a deterministic process, in which a response inevitably follows from a preceding event; the others, a stochastic process whose outcome is governed by chance. Examples of both have already been discussed on p. 84 in the context of chemical disinfection, one of the first microbiological phenomena to which they were applied. Identical arguments have subsequently appeared in many fields, including, amongst others, radiation biology (Lea, Haines & Coulson, 1936), oncogenesis (Armitage, Court Brown & Doll, 1957), microbial infection (Meynell & Stocker, 1957; Armitage, Meynell & Williams, 1965) and colicins (Nomura, 1967). The models are therefore still worth stating in detail, particularly as the stochastic model has been considerably extended in recent years by the development of 'birth-death' processes.

Consider any of those numerous systems in which a single unit is unlikely to produce a response; that is, where many particles are required on the average, as with irradiated phage (p. 176). The problem is why this situation arises. The deterministic solution is that all the units co-operate after administration and that when their number equals or exceeds some threshold value characteristic of the system, the response inevitably follows. The stochastic solution is, on the contrary, that no co-operation occurs and that each unit is independent but has only a small chance, p, of causing a response, so that many units are required on the average for at least one of them to chance to be successful.

A deterministic process presumably occurs when animals are challenged with killed organisms or toxic chemicals, since it is difficult to see how a single molecule or dead organism could affect the whole recipient. For the same reason, lysis-from-without of bacteria adsorbing inactivated phage is presumably another example. For each site (animal or bacterium, as the case may be), there is therefore postulated an Individual Effective Dose (I.E.D.) of units such that receipt of the I.E.D. invariably produces a response, which will not occur, however, on receipt of even 1 unit fewer. The dose-response curve is derived as follows for experiments performed like a dilution count, in which serial dilutions of a suspension of units are prepared, from which samples are taken for the test. Individual units are distributed randomly in suspension. The probability of a given sample containing a given number of units when the mean is μ can therefore be calculated from the corresponding term of the Poisson series (p. 176) for $\mu < 50$ and from

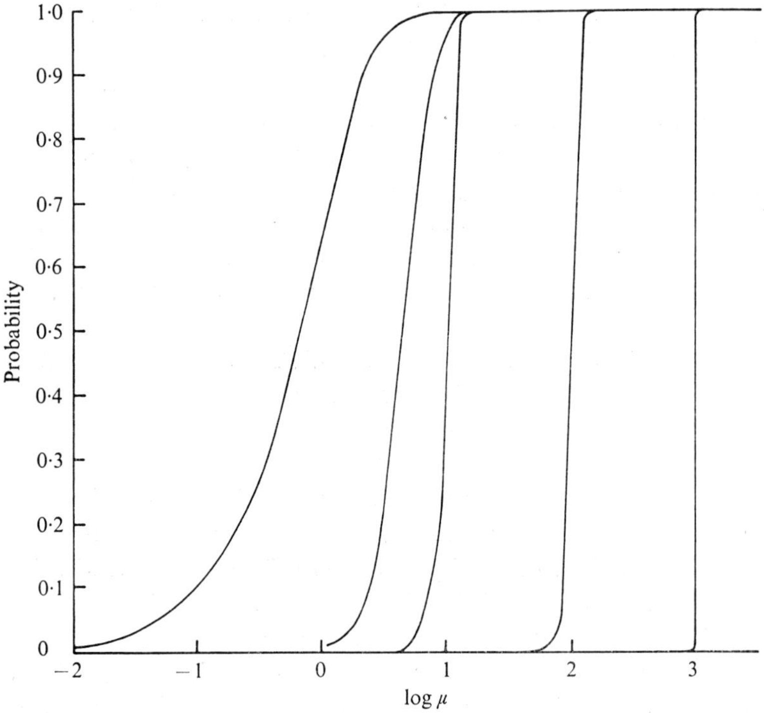

Fig. 6.18. Dose-response curves predicted by the hypothesis of the Individual Effective Dose for hosts of identical susceptibility. Ordinate: probability that a given dose contains x or more toxic molecules. Abscissa: logarithm of μ, the mean number of toxic molecules per dose. The curves are, from left to right, for $x = 1, 5, 10, 100$ and 1000.

the normal distribution for $\mu \geqslant 50$ (as in obtaining the expected frequency of colonies per plate on p. 182). These individual probabilities can then be summed to find the probability of a dose containing x or more molecules for different means. The results are shown in Fig. 6.18 for $x = 1$ (a 1-hit curve, as in Fig. 6.4) and for $x = 5, 10, 100$ and 1000. These curves are therefore the dose-response curves calculated on the assumption that the sites are identical in susceptibility. The random distribution of units in suspension gives the curves a moderate slope for $x < 10$ but this increases until, for $x > 1000$, the slope is almost infinite and the curve appears perpendicular to the dose axis. Such a curve is, of course, never found in practice. On the contrary, observed curves are often compatible with an integrated log-normal distribution which is taken on the co-operative hypothesis to reflect a log-normal distribution of I.E.D.s amongst the sites (Trevan, 1927). Technical errors, like differences in the volumes of reagent injected into different animals or in the amount of inoculum escaping after injection, also contribute to the slope but are often ignored (see, however, Harper & Hood, 1962).

The stochastic hypothesis, on the other hand, predicts an exponential (1-hit) curve which has quite a small slope, even when all the sites are equally susceptible (Fig. 6.4; $x = 1$ in Fig. 6.18). In contrast to the co-operation between units assumed by the deterministic model, all simple stochastic models assume that each unit acts independently and that each has a small chance of causing the response. What this small chance signifies in biological terms depends on the system. It might represent the probability of a single event. Thus, p could be the chance of a unit encountering a susceptible part of the site, as is likely with quanta of energy killing an irradiated bacterium. Or it could reflect heterogeneity in the population of units which might contain a minority which alone were able to cause the response. The selection of mutants by animal passage provides an example. The alternative stochastic mechanism is that p represents not the probability of a single event but the outcome of a *succession* of events, as embodied in a 'birth-death' process. A typical example is a population of cells which is both dividing and being killed. In each short interval of time dt, each cell is assumed to have a probability $\lambda\,dt$ of dividing, $\mu\,dt$ of dying, and $1-(\lambda+\mu)\,dt$ of neither dividing nor dying. That is, the fate of the cell population as a whole is governed by the relative values of λ and μ. If λ exceeds μ, clearly the viable count will almost inevitably increase; if λ is smaller than μ, it will almost inevitably decrease. The nature of the model becomes more apparent on considering a system where divisions and deaths occur with equal probability ($\lambda = \mu$). Consider a large number of replicate cultures, each containing 2 organisms initially. In a deterministic system, divisions would exactly balance deaths, and each culture would contain exactly 2 viable organisms indefinitely. However, in a stochastic system, both cells of a given culture may die, just as they may chance to divide, so that after 4 generations, many cultures will have 0 viable cells, others will have 1, 2, 3 ..., while a few will have the maximum of 16 viable cells. In other words, although the mean viable count per culture remains at 2, the actual numbers become more and more widely distributed at successive generations, with a minority of cultures containing numbers of cells far above the average. One can imagine that in a microbial infection, the chance, p, per inoculated organism of causing a fatal infection really reflects the chance of its being lucky enough to form a clone sufficiently large to kill its host (Williams, 1965*a*, *b*). The value of p is evidently related to those of λ and μ, and, in fact, turns out to be $1-(\mu/\lambda)$ for λ greater than μ, and 0 for $\lambda \leqslant \mu$. Analogous arguments can be applied to any system giving an exponential dose-response curve which can be interpreted in terms of a unit having a chance of causing the response (p. 85). Thus, in the killing of bacteria by colicins, λ and μ would be the probabilities of a lethal and of a restorative change in a cell with an adsorbed colicin molecule.

The stochastic element in a birth-death process becomes less and less obvious as the number of viable units increases and tends to mask random fluctuations, until, with a population of infinite size, the predictions become the same as those of a deterministic process. Thus, with exponentially-growing cells, the usual growth rate constant (p. 4) can be thought of as the probability of the viable count doubling in each interval of time dt. However, if a series of small replicate cultures is followed when they are subject to a birth-death process, again with $\lambda = \mu$, some inocula will die out rapidly, others will oscillate, while others will chance to increase. As the latter grow progressively larger, their behaviour approximates more and more closely to the usual deterministic growth equation, $N_t = N_0 e^{-ct}$ where $\lambda - \mu = c$. This stochastic growth of clones can be mimicked by computer (Monte Carlo method) and examples are given by Till, McCulloch & Siminovitch (1964).

It is apparent that these models must predict different dose-response curves for titrations performed by the dilution method. The deterministic curves for sites of identical susceptibility were shown in Fig. 6.18 and may be almost vertical to the dose axis; whereas the stochastic curve is exponential and always has appreciable slope, even with uniform sites (Fig. 6.4). In this type of titration, it is noteworthy that the slopes observed in toxicity titrations are usually far greater ($b \simeq 20$: p. 205) than those from infectivity titrations ($b \leqslant 2$), suggesting that drug molecules co-operate *in vivo*, whereas inoculated living micro-organisms are independent (see Meynell, 1957). It follows that, in infectivity titrations, there may be little to gain from trying to improve precision by animal breeding schemes for in many cases the observed slope approaches the predicted maximum. A relevant point is that host variability has relatively little effect in reducing the slope of the 1-hit curve (Armitage & Spicer, 1956).

The dilution method of titration, which assumes independence, estimates the concentration of *effective* units (*pm* on pp. 176, 187); but on a deterministic hypothesis, the results of this type of experiment are usually expressed by the ED 50, the dose observed to produce a response in 50 % of hosts, which is interpreted as the mean I.E.D. of the hosts.

The observations from these titrations are most precise from the statistical point of view when the proportion of sites responding is 0·80, assuming independence (p. 182) and 0·5 if the dose-response curve is normal or log normal (p. 202).

It follows from the stochastic models that at doses less than 1 ED 50, most responses are caused by only 1 unit (Fig. 6.1). Consequently, quantitative responses observed in infectivity titrations, like response time or lesion diameter, should tend to constancy at these doses (Fig. 6.15). Furthermore, the isolation of pure clones by the terminal dilution method evidently rests on the assumption of independence, and the technique and calculations expressed by Fig. 6.1 cannot be rationally applied if co-operation is believed to occur. This is not to say that pure clones may not be isolated if infection is co-operative, for example following asynchronous multiplication of inoculated organisms; merely, that predictions derived from independence cannot be used consistently at the same time.

Finally, a stochastic model predicts that the proportion of sites responding to a total of *m* units will be the same if these are given in one dose or are subdivided amongst different doses, provided the sites do not change in susceptibility (equation (6.3)). The deterministic hypothesis makes no prediction, without further assumptions as to the persistence of each inoculum, as to the outcome of this experiment which has been performed relatively infrequently with microbial infections with results favouring a stochastic mechanism (Goldberg *et al.* 1954).

A birth-death model is of considerably greater predictive power than either the simple deterministic or stochastic models mentioned here, and for this reason, it has been easier to demonstrate its limitations experimentally. In the field of microbial infection, it predicts not only the relation of λ and μ to p (hence to the ED 50: p. 215) but also the distribution of viable counts at intervals after infection and the distribution of individual response times to a given dose (Williams, 1965*a*, *b*). None of the observations agree exactly with prediction, very possibly because λ and μ are not constant as the basic form of the model presupposes (Williams & Meynell, 1967). However, the fact that biological phenomena show no greater tendency than physical systems to be ideal is evidently no reason for despairing of models as such. It is of obvious practical importance to realize, for example, that an exponential survival curve need not signify any sort of single critical event but the outcome of successive events involving damage and repair. The value of models often lies not so much in their validity but in their providing coherent sets of ideas by which to test the data.

Table 6.9. *Values of* e^{-m}. *Values of* $me^{-m}/1-e^{-m}$, *the probability in the terminal dilution method that a given response is produced by a single clone when the mean number of organisms inoculated is m*

m	e^{-m}	$me^{-m}/1-e^{-m}$
0·01	0·9900	0·9801
0·1	0·9048	0·9508
0·2	0·8187	0·9034
0·3	0·7408	0·8575
0·4	0·6703	0·8133
0·5	0·6065	0·7708
0·6	0·5488	0·7298
0·7	0·4966	0·6905
0·8	0·4493	0·6528
0·9	0·4066	0·6166
1·0	0·3679	0·582
1·1	0·3329	0·5489
1·2	0·3012	0·5172
1·3	0·2725	0·4870
1·4	0·2466	0·4582
1·5	0·2231	0·4308
2·0	0·1353	0·313
2·5	0·0821	0·2236
3·0	0·0498	0·1572

$me^{-m}/1-e^{-m}$ = probability of a given tube receiving 1 organism/1 − probability of a given tube receiving 0 organisms (see p. 178).

Tables 6.10–6.12. *The most probable number* (M.P.N.) *of organisms in viable counts by the dilution method*

Each of these tables has been computed for a dilution factor of 10, and for either 5, 8, or 10 tubes inoculated from each dilution of culture (Tables 6.10, 6.11, 6.12 respectively).

The entries in each table are the numbers of turbid tubes observed after inoculation of the same volume from any three successive 10-fold dilutions. The stated M.P.N. is the estimated mean viable count per inoculum taken from the most concentrated suspension.

Example 1. Suppose 1 ml. was inoculated from the 10^0, 10^{-1} and 10^{-2} dilutions of culture into 5 tubes for each dilution, and that the numbers of turbid tubes observed after incubation were 4–2–1. Table 6.10 shows that the M.P.N. = 2·6 per inoculum taken from the 10^0 dilution.

Example 2. Suppose the same results were obtained with tubes inoculated from 10^{-3}, 10^{-4} and 10^{-5} dilutions. The M.P.N. is then 2·6 per inoculum taken from the 10^{-3} dilution, or $2{\cdot}6 \times 10^3$ per inoculum of undiluted culture.

The tables include only those combinations of results likely to be compatible with equation (6.4): the probability that a given result would be obtained from the stated M.P.N. is provided in the original papers.

The published form of these tables differs slightly from that used here as they were originally computed for a different form of the dilution count. Instead of making serial dilutions of culture and taking the same volume from each dilution for inoculation, as described here, the culture was not diluted but different volumes of inoculum were taken from it, for example 10, 1, and 0·1 ml., the resulting differences in the concentration of culture medium being ignored. The published tables therefore give the M.P.N./ml. or per 100 ml. culture.

Table 6.10. *Values of the* M.P.N. *for 5 tubes inoculated from each of three successive 10-fold dilutions* (*Taylor, 1962*)

Numbers of turbid tubes observed at three successive dilutions			M.P.N. (per inoculum of the first dilution)
0	1	0	0·18
1	0	0	0·20
1	1	0	0·40
2	0	0	0·45
2	0	1	0·68
2	1	0	0·68
2	2	0	0·93
3	0	0	0·78
3	0	1	1·1
3	1	0	1·1
3	2	0	1·4
4	0	0	1·3
4	0	1	1·7
4	1	0	1·7
4	1	1	2·1
4	2	0	2·2
4	2	1	2·6
4	3	0	2·7

Numbers of turbid tubes observed at three successive dilutions			M.P.N. (per inoculum of the first dilution)
5	0	0	2·3
5	0	1	3·1
5	1	0	3·3
5	1	1	4·6
5	2	0	4·9
5	2	1	7·0
5	2	2	9·5
5	3	0	7·9
5	3	1	11·0
5	3	2	14·0
5	4	0	13·0
5	4	1	17·0
5	4	2	22·0
5	4	3	28·0
5	5	0	24·0
5	5	1	35·0
5	5	2	54·0
5	5	3	92·0
5	5	4	160·0

Table 6.11. *Values of the* M.P.N. *for 8 tubes inoculated from each of three successive 10-fold dilutions* (*Norman & Kempe, 1960*)

Numbers of turbid tubes observed at three successive dilutions			M.P.N. (per inoculum of the first dilution)	Numbers of turbid tubes observed at three successive dilutions			M.P.N. (per inoculum of the first dilution)
8	8	7	208	8	3	2	7·18
8	8	6	139	8	3	1	5·82
8	8	5	98·2	8	3	0	4·67
8	8	4	70·2	8	2	4	8·07
8	8	3	51·0	8	2	3	6·72
8	8	2	38·5	8	2	2	5·50
8	8	1	30·1	8	2	1	4·45
8	8	0	24·0	8	2	0	3·62
8	7	8	59·6	8	1	3	5·22
8	7	7	50·8	8	1	2	4·27
8	7	6	43·3	8	1	1	3·50
8	7	5	36·9	8	1	0	2·87
8	7	4	31·4	8	0	2	3·38
8	7	3	26·7	8	0	1	2·80
8	7	2	22·6	8	0	0	2·31
8	7	1	19·1	7	7	1	5·47
8	7	0	15·9	7	7	0	4·84
8	6	6	28·4	7	6	2	5·30
8	6	5	25·0	7	6	1	4·71
8	6	4	21·8	7	6	0	4·15
8	6	3	18·9	7	5	2	4·58
8	6	2	16·3	7	5	1	4·04
8	6	1	13·8	7	5	0	3·55
8	6	0	11·5	7	4	3	4·46
8	5	6	21·3	7	4	2	3·95
8	5	5	18·9	7	4	1	3·47
8	5	4	16·6	7	4	0	3·04
8	5	3	14·4	7	3	3	3·86
8	5	2	12·3	7	3	2	3·40
8	5	1	10·30	7	3	1	2·98
8	5	0	8·42	7	3	0	2·59
8	4	5	14·8	7	2	3	3·33
8	4	4	13·0	7	2	2	2·92
8	4	3	11·1	7	2	1	2·55
8	4	2	9·40	7	2	0	2·20
8	4	1	7·74	7	1	3	2·87
8	4	0	6·22	7	1	2	2·51
8	3	5	11·8	7	1	1	2·17
8	3	4	10·2	7	1	0	1·86
8	3	3	8·67	7	0	2	2·14

Table 6.11 (*cont.*)

Numbers of turbid tubes observed at three successive dilutions			M.P.N. (per inoculum of the first dilution)	Numbers of turbid tubes observed at three successive dilutions			M.P.N. (per inoculum of the first dilution)
7	0	1	1·83	4	2	1	1·09
7	0	0	1·55	4	2	0	0·93
6	6	1	3·08	4	1	2	1·08
6	6	0	2·77	4	1	1	0·92
6	5	1	2·73	4	1	0	0·76
6	5	0	2·44	4	0	2	0·91
6	4	2	2·69	4	0	1	0·75
6	4	1	2·41	4	0	0	0·60
6	4	0	2·14	3	4	0	1·01
6	3	2	2·38	3	3	1	1·00
6	3	1	2·11	3	3	0	0·85
6	3	0	1·86	3	2	1	0·85
6	2	2	2·09	3	2	0	0·70
6	2	1	1·84	3	1	2	0·84
6	2	0	1·60	3	1	1	0·70
6	1	2	1·82	3	1	0	0·56
6	1	1	1·58	3	0	2	0·69
6	1	0	1·35	3	0	1	0·55
6	0	2	1·56	3	0	0	0·41
6	0	1	1·34	2	4	0	0·79
6	0	0	1·13	2	3	1	0·79
5	5	1	2·07	2	3	0	0·66
5	5	0	1·85	2	2	1	0·65
5	4	1	1·84	2	2	0	0·52
5	4	0	1·63	2	1	1	0·52
5	3	2	1·82	2	1	0	0·39
5	3	1	1·61	2	0	2	0·51
5	3	0	1·41	2	0	1	0·38
5	2	2	1·60	2	0	0	0·26
5	2	1	1·40	1	3	0	0·49
5	2	0	1·21	1	2	1	0·49
5	1	2	1·39	1	2	0	0·36
5	1	1	1·20	1	1	1	0·36
5	1	0	1·01	1	1	0	0·24
5	0	2	1·19	1	0	2	0·36
5	0	1	1·01	1	0	1	0·24
5	0	0	0·83	1	0	0	0·12
4	4	0	1·28	0	2	0	0·23
4	3	1	1·27	0	1	1	0·23
4	3	0	1·10	0	1	0	0·11
				0	0	1	0·11

Table 6.12. *Values of the* M.P.N. *for 10 tubes inoculated from each of three successive 10-fold dilutions* (*Halvorson & Ziegler, 1933*)

Numbers of turbid tubes observed at three successive dilutions			M.P.N. (per inoculum of the first dilution)	Numbers of turbid tubes observed at three successive dilutions			M.P.N. (per inoculum of the first dilution)
10	10	9	230	10	6	1	9·33
10	10	8	162	10	6	0	7·92
10	10	7	120	10	5	5	13·0
10	10	6	91·8	10	5	4	11·5
10	10	5	70·2	10	5	3	10·2
10	10	4	54·2	10	5	2	8·72
10	10	3	42·8	10	5	1	7·42
10	10	2	34·9	10	5	0	6·22
10	10	1	27·5	10	4	5	10·73
10	10	0	24·0	10	4	4	9·43
10	9	9	60·7	10	4	3	8·18
10	9	8	52·6	10	4	2	7·0
10	9	7	45·8	10	4	1	5·89
10	9	6	39·8	10	4	0	4·93
10	9	5	34·6	10	3	4	7·73
10	9	4	29·8	10	3	3	6·62
10	9	3	26·3	10	3	2	5·61
10	9	2	22·8	10	3	1	4·74
10	9	1	19·7	10	3	0	3·99
10	9	0	17·0	10	2	4	6·31
10	8	7	31·0	10	2	3	5·34
10	8	6	27·8	10	2	2	4·56
10	8	5	24·9	10	2	1	3·88
10	8	4	22·1	10	2	0	3·29
10	8	3	19·6	10	1	3	4·42
10	8	2	17·1	10	1	2	3·76
10	8	1	15·0	10	1	1	3·17
10	8	0	13·0	10	1	0	2·75
10	7	6	21·9	10	0	2	3·14
10	7	5	19·5	10	0	1	2·68
10	7	4	17·4	10	0	0	2·31
10	7	3	15·3	9	8	0	4·99
10	7	2	13·3	9	7	1	4·88
10	7	1	11·6	9	7	0	4·35
10	7	0	10·1	9	6	2	4·74
10	6	6	17·5	9	6	1	4·25
10	6	5	15·7	9	6	0	3·81
10	6	4	14·1	9	5	2	4·16
10	6	3	12·5	9	5	1	3·72
10	6	2	10·9	9	5	0	3·34

Table 6.12 (*cont.*)

Numbers of turbid tubes observed at three successive dilutions			M.P.N. (per inoculum of the first dilution)
9	4	3	4·08
9	4	2	3·65
9	4	1	3·24
9	4	0	2·90
9	3	3	3·62
9	3	2	3·24
9	3	1	2·88
9	3	0	2·55
9	2	3	3·16
9	2	2	2·84
9	2	1	2·53
9	2	0	2·23
9	1	2	2·49
9	1	1	2·21
9	1	0	1·93
9	0	2	2·17
9	0	1	1·91
9	0	0	1·64
8	6	0	2·7
8	5	1	2·67
8	5	0	2·42
8	4	2	2·66
8	4	1	2·4
8	4	0	2·17
8	3	2	2·39
8	3	1	2·14
8	3	0	1·93
8	2	3	2·33
8	2	2	2·1
8	2	1	1·88
8	2	0	1·69
8	1	2	1·87
8	1	1	1·66
8	1	0	1·47
8	0	2	1·66
8	0	1	1·46
8	0	0	1·28
7	6	0	2·12
7	5	1	2·09
7	5	0	1·91
7	4	2	2·08
7	4	1	1·88
7	4	0	1·71
7	3	2	1·88
7	3	1	1·69
7	3	0	1·52
7	2	2	1·67
7	2	1	1·5
7	2	0	1·33
7	1	2	1·49
7	1	1	1·32
7	1	0	1·16
7	0	2	1·32
7	0	1	1·16
7	0	0	1·01
6	5	0	1·55
6	4	2	1·71
6	4	1	1·55
6	4	0	1·39
6	3	2	1·53
6	3	1	1·38
6	3	0	1·23
6	2	2	1·37
6	2	1	1·22
6	2	0	1·07
6	1	2	1·21
6	1	1	1·06
6	1	0	0·92
6	0	2	1·06
6	0	1	0·92
6	0	0	0·78
5	5	0	1·28
5	4	1	1·27
5	4	0	1·14
5	3	1	1·13
5	3	0	1·00
5	2	2	1·13
5	2	1	0·99
5	2	0	0·86
5	1	2	0·99

Table 6.12 (*cont.*)

Numbers of turbid tubes observed at three successive dilutions			M.P.N. (per inoculum of the first dilution)	Numbers of turbid tubes observed at three successive dilutions			M.P.N. (per inoculum of the first dilution)
5	1	1	0·86	3	1	0	0·43
5	1	0	0·73	3	0	2	0·52
5	0	2	0·85	3	0	1	0·42
5	0	1	0·72	3	0	0	0·32
5	0	0	0·6	2	4	0	0·62
4	5	0	1·06	2	3	0	0·51
4	4	0	0·93	2	2	0	0·41
4	3	1	0·92	2	1	1	0·4
4	3	0	0·8	2	1	0	0·3
4	2	1	0·8	2	0	2	0·4
4	2	0	0·68	2	0	1	0·3
4	1	2	0·8	2	0	0	0·2
4	1	1	0·68	1	3	0	0·38
4	1	0	0·56	1	2	1	0·38
4	0	2	0·67	1	2	0	0·29
4	0	1	0·56	1	1	1	0·28
4	0	0	0·45	1	1	0	0·19
3	4	0	0·76	1	0	1	0·19
3	3	2	0·86	1	0	0	0·11
3	3	1	0·75	0	2	0	0·18
3	3	0	0·64	0	1	1	0·18
3	2	1	0·64	0	1	0	0·09
3	2	0	0·53	0	0	1	0·09
3	1	2	0·64				
3	1	1	0·53				

Table 6.13. *The ED 50 and its standard error* (S.E.) *in titrations using a quantal response. Computed by Weil (1952), using the method of moving averages*

The table has six parts for different numbers of subjects inoculated (n) with each dilution ($n = 2, 3, 4, 5, 6$ or 10) and for $K = 3$ the number of dose-groups used to calculate each moving average (p. 207).

Each part has three columns:

(*a*) r-values, the number of subjects responding to the four successive dilutions used in calculating the ED 50. The middle pair of r-values can be reversed so that the same line of the table is used for 0, 0, 3, 5 as for 0, 3, 0, 5;

(*b*) f, used in calculation of the ED 50; and

(*c*) σ_f, used in calculation of the S.E. of the ED 50.

Example. Suppose that the dilutions, 10^{-2}, 10^{-3}, 10^{-4} and 10^{-5} killed 2/10, 10/10, 1/10 and 7/10 of inoculated subjects respectively. The value of n is 10, the dilution factor (R) is 10, and the r-values are 2, 10, 1, 7. The corresponding part of the Table shows 2, 1, 10, 7, with $f = 0·4$ and $\sigma_f = 0·28378$.

The $$\log_{10} \text{ED}\,50 = \log_{10} d_0 + \log_{10} R(f+1),$$

where d_0 is the highest dilution contributing an r-value.

The present $\log_{10} \text{ED}\,50 = \log_{10} 10^{-5} + \log_{10} 10(0{\cdot}4+1)$

$$= -5+1(1{\cdot}4)$$
$$= -3{\cdot}6.$$

The ED 50 $= 10^{-3\cdot6}$.

The S.E. $(\log_{10} \text{ED}\,50) = \sigma_f\,(\log_{10} R)$

$$= 0{\cdot}28378(\log_{10} 10)$$
$$= 0{\cdot}28378.$$

The 95% confidence limits of the log ED 50 equal log ED 50 $\pm$ 1·96 S.E. (log ED 50). That is,

$$-3{\cdot}6 \pm (1{\cdot}96 \times 0{\cdot}28378) = -3{\cdot}6 \pm 0{\cdot}556$$

or $10^{-3\cdot044}$ to $10^{-4\cdot156}$.

$n = 2, K = 3$

r-values	f	σ_f
0, 0, 1, 2	1·00000	0·50000
0, 0, 2, 2	0·50000	0·00000
0, 1, 1, 2	0·50000	0·70711
0, 1, 2, 2	0·00000	0·50000
1, 0, 1, 2	1·00000	1·00000
1, 0, 2, 2	0·00000	1·00000
1, 1, 1, 2	0·00000	1·73205
0, 0, 2, 1	1·00000	1·00000
0, 1, 1, 1	1·00000	1·73205
0, 1, 2, 1	0·00000	1·00000

$n = 3, K = 3$

r-values	f	σ_f
0, 0, 2, 3	0·83333	0·33333
0, 0, 3, 3	0·50000	0·00000
0, 1, 1, 3	0·83333	0·47140
0, 1, 2, 3	0·50000	0·47140
0, 1, 3, 3	0·16667	0·33333
0, 2, 2, 3	0·16667	0·47140
1, 0, 2, 3	0·75000	0·51539
1, 0, 3, 3	0·25000	0·37500
1, 1, 1, 3	0·75000	0·71807
1, 1, 2, 3	0·25000	0·80039
2, 0, 2, 3	0·50000	1·11803
0, 0, 3, 2	0·75000	0·37500
0, 1, 2, 2	0·75000	0·80039
0, 1, 3, 2	0·25000	0·51539
0, 2, 2, 2	0·25000	0·71807
0, 1, 3, 1	0·50000	1·11803

$n = 4, K = 3$

r-values	f	σ_f
0, 0, 2, 4	1·00000	0·28868
0, 0, 3, 4	0·75000	0·25000
0, 0, 4, 4	0·50000	0·00000
0, 1, 1, 4	1·00000	0·35355
0, 1, 2, 4	0·75000	0·38188
0, 1, 3, 4	0·50000	0·35355
0, 1, 4, 4	0·25000	0·25000
0, 2, 2, 4	0·50000	0·40825
0, 2, 3, 4	0·25000	0·38188
0, 2, 4, 4	0·00000	0·28868
0, 3, 3, 4	0·00000	0·35355
1, 0, 2, 4	1·00000	0·38490
1, 0, 3, 4	0·66667	0·35136
1, 0, 4, 4	0·33333	0·22222
1, 1, 1, 4	1·00000	0·47140
1, 1, 2, 4	0·66667	0·52116
1, 1, 3, 4	0·33333	0·52116
1, 1, 4, 4	0·00000	0·47140
1, 2, 2, 4	0·33333	0·58794
1, 2, 3, 4	0·00000	0·60858
2, 0, 2, 4	1·00000	0·57735
2, 0, 3, 4	0·50000	0·57735
2, 0, 4, 4	0·00000	0·57735
2, 1, 1, 4	1·00000	0·70711
2, 1, 2, 4	0·50000	0·81650
2, 1, 3, 4	0·00000	0·91287
2, 2, 2, 4	0·00000	1·00000
3, 0, 2, 4	1·00000	1·15470
3, 0, 3, 4	0·00000	1·41421
3, 1, 1, 4	1·00000	1·41421
3, 1, 2, 4	0·00000	1·82574
0, 0, 3, 3	1·00000	0·47140
0, 0, 4, 3	0·66667	0·22222
0, 1, 2, 3	1·00000	0·60858
0, 1, 3, 3	0·66667	0·52116
0, 1, 4, 3	0·33333	0·35136
0, 2, 2, 3	0·66667	0·58794
0, 2, 3, 3	0·33333	0·52116
0, 2, 4, 3	0·00000	0·38490
0, 3, 3, 3	0·00000	0·47140

Table 6.13 (*cont.*)

$n = 4, K = 3$

r-values	f	σ_f
1, 0, 3, 3	1·00000	0·70711
1, 0, 4, 3	0·50000	0·35355
1, 1, 2, 3	1·00000	0·91287
1, 1, 3, 3	0·50000	0·79057
1, 1, 4, 3	0·00000	0·70711
1, 2, 2, 3	0·50000	0·88976
1, 2, 3, 3	0·00000	0·91287
2, 0, 3, 3	1·00000	1·41421
2, 0, 4, 3	0·00000	1·15470
2, 1, 2, 3	1·00000	1·82574
2, 1, 3, 3	0·00000	1·82574
2, 2, 2, 3	0·00000	2·00000
0, 0, 4, 2	1·00000	0·57735
0, 1, 3, 2	1·00000	0·91287
0, 1, 4, 2	0·50000	0·57735
0, 2, 2, 2	1·00000	1·00000
0, 2, 3, 2	0·50000	0·81650
0, 2, 4, 2	0·00000	0·57735
0, 3, 3, 2	0·00000	0·70711
1, 0, 4, 2	1·00000	1·15470
1, 1, 3, 2	1·00000	1·82574
1, 1, 4, 2	0·00000	1·41421
1, 2, 2, 2	1·00000	2·00000
1, 2, 3, 2	0·00000	1·82574
0, 2, 3, 1	1·00000	1·82574
0, 2, 4, 1	0·00000	1·15470
0, 3, 3, 1	0·00000	1·41421
0, 1, 4, 1	1·00000	1·41421

$n = 5, K = 3$

r-values	f	σ_f
0, 0, 3, 5	0·90000	0·24495
0, 0, 4, 5	0·70000	0·20000
0, 0, 5, 5	0·50000	0·00000
0, 1, 2, 5	0·90000	0·31623
0, 1, 3, 5	0·70000	0·31623
0, 1, 4, 5	0·50000	0·28284
0, 1, 5, 5	0·30000	0·20000
0, 2, 2, 5	0·70000	0·34641
0, 2, 3, 5	0·50000	0·34641
0, 2, 4, 5	0·30000	0·31623
0, 2, 5, 5	0·10000	0·24495
0, 3, 3, 5	0·30000	0·34641
0, 3, 4, 5	0·10000	0·31623
1, 0, 3, 5	0·87500	0·30778
1, 0, 4, 5	0·62500	0·26700
1, 0, 5, 5	0·37500	0·15625
1, 1, 2, 5	0·87500	0·39652
1, 1, 3, 5	0·62500	0·40625
1, 1, 4, 5	0·37500	0·38654
1, 1, 5, 5	0·12500	0·33219
1, 2, 2, 5	0·62500	0·44304
1, 2, 3, 5	0·37500	0·46034
1, 2, 4, 5	0·12500	0·45178
1, 3, 3, 5	0·12500	0·48513
2, 0, 3, 5	0·83333	0·41388
2, 0, 4, 5	0·50000	0·39087
2, 0, 5, 5	0·16667	0·34021
2, 1, 2, 5	0·83333	0·53142
2, 1, 3, 5	0·50000	0·56519
2, 1, 4, 5	0·16667	0·58134
2, 2, 2, 5	0·50000	0·61237
2, 2, 3, 5	0·16667	0·67013
0, 0, 4, 4	0·87500	0·33219
0, 0, 5, 4	0·62500	0·15625
0, 1, 3, 4	0·87500	0·45178
0, 1, 4, 4	0·62500	0·38654
0, 1, 5, 4	0·37500	0·26700
0, 2, 2, 4	0·87500	0·48513
0, 2, 3, 4	0·62500	0·46034
0, 2, 4, 4	0·37500	0·40625
0, 2, 5, 4	0·12500	0·30778
0, 3, 3, 4	0·37500	0·44304
0, 3, 4, 4	0·12500	0·39652
1, 0, 4, 4	0·83333	0·43744
1, 0, 5, 4	0·50000	0·23570
1, 1, 3, 4	0·83333	0·59835
1, 1, 4, 4	0·50000	0·52705
1, 1, 5, 4	0·16667	0·43744
1, 2, 2, 4	0·83333	0·64310
1, 2, 3, 4	0·50000	0·62361
1, 2, 4, 4	0·16667	0·59835
1, 3, 3, 4	0·16667	0·64310
2, 0, 4, 4	0·75000	0·64348
2, 0, 5, 4	0·25000	0·47598
2, 1, 3, 4	0·75000	0·88829
2, 1, 4, 4	0·24000	0·85239
2, 2, 2, 4	0·75000	0·95607
2, 2, 3, 4	0·25000	0·98821
0, 0, 5, 3	0·83333	0·34021
0, 1, 4, 3	0·83333	0·58134

Table 6.13 (*cont.*)

$n = 5, K = 3$		
r-values	f	σ_f
0, 1, 5, 3	0·50000	0·39087
0, 2, 3, 3	0·83333	0·67013
0, 2, 4, 3	0·50000	0·56519
0, 2, 5, 3	0·16667	0·41388
0, 3, 3, 3	0·50000	0·61237
0, 3, 4, 3	0·16667	0·53142
1, 0, 5, 3	0·75000	0·47598
1, 1, 4, 3	0·75000	0·85239
1, 1, 5, 3	0·25000	0·64348
1, 2, 3, 3	0·75000	0·98821
1, 2, 4, 3	0·25000	0·88829
1, 3, 3, 3	0·25000	0·95607

$n = 6, K = 3$		
r-values	f	σ_f
0, 0, 3, 6	1·00000	0·22361
0, 0, 4, 6	0·83333	0·21082
0, 0, 5, 6	0·66667	0·16667
0, 0, 6, 6	0·50000	0·00000
0, 1, 2, 6	1·00000	0·26874
0, 1, 3, 6	0·83333	0·27889
0, 1, 4, 6	0·66667	0·26874
0, 1, 5, 6	0·50000	0·23570
0, 1, 6, 6	0·33333	0·16667
0, 2, 2, 6	0·83333	0·29814
0, 2, 3, 6	0·66667	0·30732
0, 2, 4, 6	0·50000	0·29814
0, 2, 5, 6	0·33333	0·26874
0, 2, 6, 6	0·16667	0·21082
0, 3, 3, 6	0·50000	0·31623
0, 3, 4, 6	0·33333	0·30732
0, 3, 5, 6	0·16667	0·27889
0, 3, 6, 6	0·00000	0·22361
0, 4, 4, 6	0·16667	0·29814
0, 4, 5, 6	0·00000	0·26874
1, 0, 3, 6	1·00000	0·26833
1, 0, 4, 6	0·80000	0·25612
1, 0, 5, 6	0·60000	0·21541
1, 0, 6, 6	0·40000	0·12000
1, 1, 2, 6	1·00000	0·32249
1, 1, 3, 6	0·80000	0·33704
1, 1, 4, 6	0·60000	0·33226
1, 1, 5, 6	0·40000	0·30724
1, 1, 6, 6	0·20000	0·25612
1, 2, 2, 6	0·80000	0·36000

$n = 6, K = 3$		
r-values	f	σ_f
1, 2, 3, 6	0·60000	0·37736
1, 2, 4, 6	0·40000	0·37736
1, 2, 5, 6	0·20000	0·36000
1, 2, 6, 6	0·00000	0·26833
1, 3, 3, 6	0·40000	0·39799
1, 3, 4, 6	0·20000	0·40200
1, 3, 5, 6	0·00000	0·34641
1, 4, 4, 6	0·00000	0·36878
2, 0, 3, 6	1·00000	0·33541
2, 0, 4, 6	0·75000	0·32596
2, 0, 5, 6	0·50000	0·29580
2, 0, 6, 6	0·25000	0·23717
2, 1, 2, 6	1·00000	0·40311
2, 1, 3, 6	0·75000	0·42573
2, 1, 4, 6	0·50000	0·43301
2, 1, 5, 6	0·25000	0·42573
2, 1, 6, 6	0·00000	0·29580
2, 2, 2, 6	0·75000	0·45415
2, 2, 3, 6	0·50000	0·48734
2, 2, 4, 6	0·25000	0·50621
2, 2, 5, 6	0·00000	0·43301
2, 3, 3, 6	0·25000	0·53033
2, 3, 4, 6	0·00000	0·48734
3, 0, 3, 6	1·00000	0·44721
3, 0, 4, 6	0·66667	0·43885
3, 0, 5, 6	0·33333	0·44721
3, 0, 6, 6	0·00000	0·44721
3, 1, 2, 6	1·00000	0·53748
3, 1, 3, 6	0·66667	0·57090
3, 1, 4, 6	0·33333	0·61464
3, 1, 5, 6	0·00000	0·64979
3, 2, 2, 6	0·66667	0·60858
3, 2, 3, 6	0·33333	0·68313
3, 2, 4, 6	0·00000	0·74536
3, 3, 3, 6	0·00000	0·77460
4, 0, 3, 6	1·00000	0·67082
4, 0, 4, 6	0·50000	0·70711
4, 0, 5, 6	0·00000	0·80622
4, 1, 2, 6	1·00000	0·80622
4, 1, 3, 6	0·50000	0·89443
4, 1, 4, 6	0·00000	1·02470
4, 2, 2, 6	0·50000	0·94868
4, 2, 3, 6	0·00000	1·11803
5, 0, 3, 6	1·00000	1·34164
5, 0, 4, 6	0·00000	1·61245

Table 6.13 (*cont.*)

$n = 6, K = 3$			$n = 6, K = 3$		
r-values	f	σ_f	r-values	f	σ_f
5, 1, 2, 6	1·00000	1·61245	2, 2, 3, 5	0·66667	0·66852
5, 1, 3, 6	0·00000	1·94936	2, 2, 4, 5	0·33333	0·66852
5, 2, 2, 6	0·00000	2·04939	2, 2, 5, 5	0·00000	0·68313
0, 0, 4, 5	1·00000	0·26833	2, 3, 3, 5	0·33333	0·70097
0, 0, 5, 5	0·80000	0·25612	2, 3, 4, 5	0·00000	0·74536
0, 0, 6, 5	0·60000	0·12000	3, 0, 4, 5	1·00000	0·80622
0, 1, 3, 5	1·00000	0·34641	3, 0, 5, 5	0·50000	0·65192
0, 1, 4, 5	0·80000	0·36000	3, 0, 6, 5	0·00000	0·67082
0, 1, 5, 5	0·60000	0·30724	3, 1, 3, 5	1·00000	0·92195
0, 1, 6, 5	0·40000	0·21541	3, 1, 4, 5	0·50000	0·90830
0, 2, 2, 5	1·00000	0·36878	3, 1, 5, 5	0·00000	0·92195
0, 2, 3, 5	0·80000	0·40200	3, 2, 2, 5	1·00000	1·02470
0, 2, 4, 5	0·60000	0·37736	3, 2, 3, 5	0·50000	1·01242
0, 2, 5, 5	0·40000	0·33226	3, 2, 4, 5	0·00000	1·11803
0, 2, 6, 5	0·20000	0·25612	3, 3, 3, 5	0·00000	1·16190
0, 3, 3, 5	0·60000	0·39799	4, 0, 4, 5	1·00000	1·61245
0, 3, 4, 5	0·40000	0·37736	4, 0, 5, 5	0·00000	1·61245
0, 3, 5, 5	0·20000	0·33704	4, 1, 3, 5	1·00000	1·94936
0, 3, 6, 5	0·00000	0·26833	4, 1, 4, 5	0·00000	2·04939
0, 4, 4, 5	0·20000	0·36000	4, 2, 2, 5	1·00000	2·04939
0, 4, 5, 5	0·00000	0·32249	4, 2, 3, 5	0·00000	2·23607
1, 0, 4, 5	1·00000	0·40311	0, 0, 5, 4	1·00000	0·29580
1, 0, 5, 5	0·75000	0·31869	0, 0, 6, 4	0·75000	0·23717
1, 0, 6, 5	0·50000	0·17678	0, 1, 4, 4	1·00000	0·43301
1, 1, 3, 5	1·00000	0·48734	0, 1, 5, 4	0·75000	0·42573
1, 1, 4, 5	0·75000	0·44896	0, 1, 6, 4	0·50000	0·29580
1, 1, 5, 5	0·50000	0·39528	0, 2, 3, 4	1·00000	0·48734
1, 1, 6, 5	0·25000	0·31869	0, 2, 4, 4	0·75000	0·50621
1, 2, 2, 5	1·00000	0·51235	0, 2, 5, 4	0·50000	0·43301
1, 2, 3, 5	0·75000	0·50156	0, 2, 6, 4	0·25000	0·32596
1, 2, 4, 5	0·50000	0·48088	0, 3, 3, 4	0·75000	0·53033
1, 2, 5, 5	0·25000	0·44896	0, 3, 4, 4	0·50000	0·48734
1, 2, 6, 5	0·00000	0·40311	0, 3, 5, 4	0·25000	0·42573
1, 3, 3, 5	0·50000	0·50621	0, 3, 6, 4	0·00000	0·33541
1, 3, 4, 5	0·25000	0·50156	0, 4, 4, 4	0·25000	0·45415
1, 3, 5, 5	0·00000	0·48734	0, 4, 5, 4	0·00000	0·40311
1, 4, 4, 5	0·00000	0·51235	1, 0, 5, 4	1·00000	0·53748
2, 0, 4, 5	1·00000	0·53748	1, 0, 6, 4	0·66667	0·30225
2, 0, 5, 5	0·66667	0·42455	1, 1, 4, 4	1·00000	0·68313
2, 0, 6, 5	0·33333	0·30225	1, 1, 5, 4	0·66667	0·55998
2, 1, 3, 5	1·00000	0·64979	1, 1, 6, 4	0·33333	0·42455
2, 1, 4, 5	0·66667	0·59835	1, 2, 3, 4	1·00000	0·74536
2, 1, 5, 5	0·33333	0·55998	1, 2, 4, 4	0·66667	0·66852
2, 1, 6, 5	0·00000	0·53748	1, 2, 5, 4	0·33333	0·59835
2, 2, 2, 5	1·00000	0·68313	1, 2, 6, 4	0·00000	0·53748

Table 6.13 (*cont.*)

$n = 6, K = 3$		
r-values	f	σ_f
1, 3, 3, 4	0·66667	0·70097
1, 3, 4, 4	0·33333	0·66852
1, 3, 5, 4	0·00000	0·64979
1, 4, 4, 4	0·00000	0·68313
2, 0, 5, 4	1·00000	0·80622
2, 0, 6, 4	0·50000	0·44721
2, 1, 4, 4	1·00000	1·02470
2, 1, 5, 4	0·50000	0·83666
2, 1, 6, 4	0·00000	0·80622
2, 2, 3, 4	1·00000	1·11803
2, 2, 4, 4	0·50000	1·00000
2, 2, 5, 4	0·00000	1·02470
2, 3, 3, 4	0·50000	1·04881
2, 3, 4, 4	0·00000	1·11803
3, 0, 5, 4	1·00000	1·67332
3, 0, 6, 4	0·00000	1·34164
3, 1, 4, 4	1·00000	2·09762
3, 1, 5, 4	0·00000	1·94936
3, 2, 3, 4	1·00000	2·28035
3, 2, 4, 4	0·00000	2·23607
3, 3, 3, 4	0·00000	2·32379
0, 0, 6, 3	1·00000	0·44721
0, 1, 5, 3	1·00000	0·64979
0, 1, 6, 3	0·66667	0·44721
0, 2, 4, 3	1·00000	0·74536
0, 2, 5, 3	0·66667	0·61464
0, 2, 6, 3	0·33333	0·43885
0, 3, 3, 3	1·00000	0·77460
0, 3, 4, 3	0·66667	0·68313
0, 3, 5, 3	0·33333	0·57090
0, 3, 6, 3	0·00000	0·44721
0, 4, 4, 3	0·33333	0·60858
0, 4, 5, 3	0·00000	0·53748
1, 0, 6, 3	1·00000	0·67082
1, 1, 5, 3	1·00000	0·92195
1, 1, 6, 3	0·50000	0·65192
1, 2, 4, 3	1·00000	1·11803
1, 2, 5, 3	0·50000	0·90830
1, 2, 6, 3	0·00000	0·80622
1, 3, 3, 3	1·00000	1·16190
1, 3, 4, 3	0·50000	1·01242
1, 3, 5, 3	0·00000	0·92195
1, 4, 4, 3	0·00000	1·02470
2, 0, 6, 3	1·00000	1·34164
2, 1, 5, 3	1·00000	1·94936

$n = 6, K = 3$		
r-values	f	σ_f
2, 1, 6, 3	0·00000	1·67332
2, 2, 4, 3	1·00000	2·23607
2, 2, 5, 3	0·00000	2·09762
2, 3, 3, 3	1·00000	2·32379
2, 3, 4, 3	0·00000	2·28035
0, 1, 6, 2	1·00000	0·80622
0, 2, 5, 2	1·00000	1·02470
0, 2, 6, 2	0·50000	0·70711
0, 3, 4, 2	1·00000	1·11803
0, 3, 5, 2	0·50000	0·89443
0, 3, 6, 2	0·00000	0·67082
0, 4, 4, 2	0·50000	0·94868
0, 4, 5, 2	0·00000	0·80622
1, 1, 6, 2	1·00000	1·61245
1, 2, 5, 2	1·00000	2·04939
1, 2, 6, 2	0·00000	1·61245
1, 3, 4, 2	1·00000	2·23607
1, 3, 5, 2	0·00000	1·94936
1, 4, 4, 2	0·00000	2·04939
0, 2, 6, 1	1·00000	1·61245
0, 3, 5, 1	1·00000	1·94936
0, 3, 6, 1	0·00000	1·34164
0, 4, 4, 1	1·00000	2·04939
0, 4, 5, 1	0·00000	1·61245

$n = 10, K = 3$		
0, 0, 5, 10	1·0	0·16667
0, 0, 6, 10	0·9	0·16330
0, 0, 7, 10	0·8	0·15275
0, 0, 8, 10	0·7	0·13333
0, 0, 9, 10	0·6	0·10000
0, 0, 10, 10	0·5	0·00000
0, 1, 4, 10	1·0	0·19149
0, 1, 5, 10	0·9	0·19436
0, 1, 6, 10	0·8	0·19149
0, 1, 7, 10	0·7	0·18257
0, 1, 8, 10	0·6	0·16667
0, 1, 9, 10	0·5	0·14142
0, 1, 10, 10	0·4	0·10000
0, 2, 3, 10	1·0	0·20276
0, 2, 4, 10	0·9	0·21082
0, 2, 5, 10	0·8	0·21344
0, 2, 6, 10	0·7	0·21082
0, 2, 7, 10	0·6	0·20276
0, 2, 8, 10	0·5	0·18856
0, 2, 9, 10	0·4	0·16667

Table 6.13 (*cont.*)

$n = 10, K = 3$			$n = 10, K = 3$		
r-values	f	σ_f	r-values	f	σ_f
0, 2, 10, 10	0·3	0·13333	1, 2, 7, 10	0·55556	0·23064
0, 3, 3, 10	0·9	0·21602	1, 2, 8, 10	0·44444	0·21842
0, 3, 4, 10	0·8	0·22361	1, 2, 9, 10	0·33333	0·19945
0, 3, 5, 10	0·7	0·22608	1, 2, 10, 10	0·22222	0·17151
0, 3, 6, 10	0·6	0·22361	1, 3, 3, 10	0·88889	0·24034
0, 3, 7, 10	0·5	0·21602	1, 3, 4, 10	0·77778	0·24968
0, 3, 8, 10	0·4	0·20276	1, 3, 5, 10	0·66667	0·25391
0, 3, 9, 10	0·3	0·18257	1, 3, 6, 10	0·55556	0·25331
0, 3, 10, 10	0·2	0·15275	1, 3, 7, 10	0·44444	0·24784
0, 4, 4, 10	0·7	0·23094	1, 3, 8, 10	0·33333	0·23715
0, 4, 5, 10	0·6	0·23336	1, 3, 9, 10	0·22222	0·22050
0, 4, 6, 10	0·5	0·23094	1, 3, 10, 10	0·11111	0·19637
0, 4, 7, 10	0·4	0·22361	1, 4, 4, 10	0·66667	0·25926
0, 4, 8, 10	0·3	0·21082	1, 4, 5, 10	0·55556	0·26392
0, 4, 9, 10	0·2	0·19149	1, 4, 6, 10	0·44444	0·26392
0, 4, 10, 10	0·1	0·16330	1, 4, 7, 10	0·33333	0·25926
0, 5, 5, 10	0·5	0·23570	1, 4, 8, 10	0·22222	0·24968
0, 5, 6, 10	0·4	0·23336	1, 4, 9, 10	0·11111	0·23457
0, 5, 7, 10	0·3	0·22608	1, 4, 10, 10	0·00000	0·21276
0, 5, 8, 10	0·2	0·21344	1, 5, 5, 10	0·44444	0·26907
0, 5, 9, 10	0·1	0·19436	1, 5, 6, 10	0·33333	0·26963
0, 5, 10, 10	0·0	0·16667	1, 5, 7, 10	0·22222	0·26565
0, 6, 6, 10	0·3	0·23094	1, 5, 8, 10	0·11111	0·25690
0, 6, 7, 10	0·2	0·22361	1, 5, 9, 10	0·00000	0·24287
0, 6, 8, 10	0·1	0·21082	1, 6, 6, 10	0·22222	0·27076
0, 6, 9, 10	0·0	0·19149	1, 6, 7, 10	0·11111	0·26736
0, 7, 7, 10	0·1	0·21602	1, 6, 8, 10	0·00000	0·25926
0, 7, 8, 10	0·0	0·20276	1, 7, 7, 10	0·00000	0·26450
1, 0, 5, 10	1·0	0·18518	2, 0, 5, 10	1·00000	0·20833
1, 0, 6, 10	0·88889	0·18186	2, 0, 6, 10	0·87500	0·20465
1, 0, 7, 10	0·77778	0·17151	2, 0, 7, 10	0·75000	0·19320
1, 0, 8, 10	0·66667	0·15270	2, 0, 8, 10	0·62500	0·17237
1, 0, 9, 10	0·55556	0·12159	2, 0, 9, 10	0·50000	0·10534
1, 0, 10, 10	0·44444	0·06172	2, 0, 10, 10	0·37500	0·07365
1, 1, 4, 10	1·00000	0·21276	2, 1, 4, 10	1·00000	0·23936
1, 1, 5, 10	0·88889	0·21631	2, 1, 5, 10	0·87500	0·22902
1, 1, 6, 10	0·77778	0·21419	2, 1, 6, 10	0·75000	0·24116
1, 1, 7, 10	0·66667	0·20621	2, 1, 7, 10	0·62500	0·23246
1, 1, 8, 10	0·55556	0·19166	2, 1, 8, 10	0·50000	0·21651
1, 1, 9, 10	0·44444	0·16882	2, 1, 9, 10	0·37500	0·19151
1, 1, 10, 10	0·33333	0·13354	2, 1, 10, 10	0·25000	0·17678
1, 2, 3, 10	1·00000	0·22529	2, 2, 3, 10	1·00000	0·25345
1, 2, 4, 10	0·88889	0·23457	2, 2, 4, 10	0·87500	0·26393
1, 2, 5, 10	0·77778	0·23843	2, 2, 5, 10	0·75000	0·26842
1, 2, 6, 10	0·66667	0·23715	2, 2, 6, 10	0·62500	0·26717

Table 6.13 (*cont.*)

n = 10, *K* = 3			*n* = 10, *K* = 3		
r-values	f	σ_f	*r*-values	f	σ_f
2, 2, 7, 10	0·50000	0·26021	3, 3, 3, 10	0·85714	0·31018
2, 2, 8, 10	0·37500	0·24694	3, 3, 4, 10	0·71429	0·32546
2, 2, 9, 10	0·25000	0·24296	3, 3, 5, 10	0·57143	0·33926
2, 2, 10, 10	0·12500	0·22140	3, 3, 6, 10	0·42857	0·34291
2, 3, 3, 10	0·87500	0·27043	3, 3, 7, 10	0·28571	0·34574
2, 3, 4, 10	0·75000	0·28106	3, 3, 8, 10	0·14286	0·34480
2, 3, 5, 10	0·62500	0·28603	3, 3, 9, 10	0·00000	0·34007
2, 3, 6, 10	0·50000	0·28565	3, 4, 4, 10	0·57143	0·34588
2, 3, 7, 10	0·37500	0·27990	3, 4, 5, 10	0·42857	0·35589
2, 3, 8, 10	0·25000	0·28260	3, 4, 6, 10	0·28571	0·36488
2, 3, 9, 10	0·12500	0·27081	3, 4, 7, 10	0·14286	0·37017
2, 3, 10, 10	0·00000	0·25345	3, 4, 8, 10	0·00000	0·37192
2, 4, 4, 10	0·62500	0·29204	3, 5, 5, 10	0·28571	0·37104
2, 4, 5, 10	0·50000	0·29756	3, 5, 6, 10	0·14286	0·38223
2, 4, 6, 10	0·37500	0·29789	3, 5, 7, 10	0·00000	0·38978
2, 4, 7, 10	0·25000	0·30619	3, 6, 6, 10	0·00000	0·39555
2, 4, 8, 10	0·12500	0·30117	4, 0, 5, 10	1·00000	0·27778
2, 4, 9, 10	0·00000	0·29166	4, 0, 6, 10	0·83333	0·27592
2, 5, 5, 10	0·37500	0·30369	4, 0, 7, 10	0·66667	0·27027
2, 5, 6, 10	0·25000	0·31732	4, 0, 8, 10	0·50000	0·26058
2, 5, 7, 10	0·12500	0·31799	4, 0, 9, 10	0·33333	0·24637
2, 5, 8, 10	0·00000	0·31458	4, 0, 10, 10	0·16667	0·22680
2, 6, 6, 10	0·12500	0·32340	4, 1, 4, 10	1·00000	0·31914
2, 6, 7, 10	0·00000	0·32543	4, 1, 5, 10	0·83333	0·32710
3, 0, 5, 10	1·00000	0·23809	4, 1, 6, 10	0·66667	0·33178
3, 0, 6, 10	0·85714	0·23536	4, 1, 7, 10	0·50000	0·33333
3, 0, 7, 10	0·71429	0·22695	4, 1, 8, 10	0·33333	0·33178
3, 0, 8, 10	0·57143	0·21695	4, 1, 9, 10	0·16667	0·32710
3, 0, 9, 10	0·42857	0·18962	4, 1, 10, 10	0·00000	0·31914
3, 0, 10, 10	0·28571	0·15587	4, 2, 3, 10	1·00000	0·33793
3, 1, 4, 10	1·00000	0·27355	4, 2, 4, 10	0·83333	0·35428
3, 1, 5, 10	0·85714	0·27941	4, 2, 5, 10	0·66667	0·36711
3, 1, 6, 10	0·71429	0·28057	4, 2, 6, 10	0·50000	0·37679
3, 1, 7, 10	0·57143	0·28074	4, 2, 7, 10	0·33333	0·38356
3, 1, 8, 10	0·42857	0·26877	4, 2, 8, 10	0·16667	0·38756
3, 1, 9, 10	0·28571	0·25517	4, 2, 9, 10	0·00000	0·38889
3, 1, 10, 10	0·14286	0·23536	4, 3, 3, 10	0·83333	0·36289
3, 2, 3, 10	1·00000	0·28965	4, 3, 4, 10	0·66667	0·38356
3, 2, 4, 10	0·85714	0·30278	4, 3, 5, 10	0·50000	0·40062
3, 2, 5, 10	0·71429	0·31122	4, 3, 6, 10	0·33333	0·41450
3, 2, 6, 10	0·57143	0·31857	4, 3, 7, 10	0·16667	0·42552
3, 2, 7, 10	0·42857	0·31536	4, 3, 8, 10	0·00000	0·43390
3, 2, 8, 10	0·28571	0·31122	4, 4, 4, 10	0·50000	0·48025
3, 2, 9, 10	0·14286	0·30278	4, 4, 5, 10	0·33333	0·42913
3, 2, 10, 10	0·00000	0·28965	4, 4, 6, 10	0·16667	0·44675

Table 6.13 (*cont.*)

$n = 10, K = 3$			$n = 10, K = 3$		
r-values	f	σ_f	r-values	f	σ_f
4, 4, 7, 10	0·00000	0·46148	6, 3, 3, 10	0·75000	0·57735
4, 5, 5, 10	0·16667	0·45361	6, 3, 4, 10	0·50000	0·59512
4, 5, 6, 10	0·00000	0·47466	6, 3, 5, 10	0·25000	0·64280
5, 0, 5, 10	1·00000	0·33333	6, 3, 6, 10	0·00000	0·69222
5, 0, 6, 10	0·80000	0·33333	6, 4, 4, 10	0·25000	0·65352
5, 0, 7, 10	0·60000	0·33333	6, 4, 5, 10	0·00000	0·71200
5, 0, 8, 10	0·40000	0·33333	7, 0, 5, 10	1·00000	0·55556
5, 0, 9, 10	0·20000	0·33333	7, 0, 6, 10	0·66667	0·57013
5, 0, 10, 10	0·00000	0·33333	7, 0, 7, 10	0·33333	0·61195
5, 1, 4, 10	1·00000	0·38297	7, 0, 8, 10	0·00000	0·67586
5, 1, 5, 10	0·80000	0·39440	7, 1, 4, 10	1·00000	0·63828
5, 1, 6, 10	0·60000	0·40552	7, 1, 5, 10	0·66667	0·66975
5, 1, 7, 10	0·40000	0·41633	7, 1, 6, 10	0·33333	0·72293
5, 1, 8, 10	0·20000	0·42687	7, 1, 7, 10	0·00000	0·79349
5, 1, 9, 10	0·00000	0·43716	7, 2, 3, 10	1·00000	0·67586
5, 2, 3, 10	1·00000	0·40552	7, 2, 4, 10	0·66667	0·72293
5, 2, 4, 10	0·80000	0·42688	7, 2, 5, 10	0·33333	0·78829
5, 2, 5, 10	0·60000	0·44721	7, 2, 6, 10	0·00000	0·86780
5, 2, 6, 10	0·40000	0·46667	7, 3, 3, 10	0·66667	0·73981
5, 2, 7, 10	0·20000	0·48534	7, 3, 4, 10	0·33333	0·81901
5, 2, 8, 10	0·00000	0·50332	7, 3, 5, 10	0·00000	0·90948
5, 3, 3, 10	0·80000	0·43716	7, 4, 4, 10	0·00000	0·92296
5, 3, 4, 10	0·60000	0·46667	8, 0, 5, 10	1·00000	0·83333
5, 3, 5, 10	0·40000	0·48990	8, 0, 6, 10	0·50000	0·88192
5, 3, 6, 10	0·20000	0·52068	8, 0, 7, 10	0·00000	1·01379
5, 3, 7, 10	0·00000	0·54569	8, 1, 4, 10	1·00000	0·95743
5, 4, 4, 10	0·40000	0·49889	8, 1, 5, 10	0·50000	1·02740
5, 4, 5, 10	0·20000	0·53748	8, 1, 6, 10	0·00000	0·16667
5, 4, 6, 10	0·00000	0·56960	8, 2, 3, 10	1·00000	1·01379
5, 5, 5, 10	0·00000	0·57735	8, 2, 4, 10	0·50000	1·10554
6, 0, 5, 10	1·00000	0·41667	8, 2, 5, 10	0·00000	1·25830
6, 0, 6, 10	0·75000	0·45644	8, 3, 3, 10	0·50000	1·13039
6, 0, 7, 10	0·50000	0·43301	8, 3, 4, 10	0·00000	1·30171
6, 0, 8, 10	0·25000	0·45262	9, 0, 5, 10	1·00000	1·66667
6, 0, 9, 10	0·00000	0·47871	9, 0, 6, 10	0·00000	1·77951
6, 1, 4, 10	1·00000	0·47871	9, 1, 4, 10	1·00000	1·77951
6, 1, 5, 10	0·75000	0·52705	9, 1, 5, 10	0·00000	2·18581
6, 1, 6, 10	0·50000	0·52042	9, 2, 3, 10	1·00000	2·02759
6, 1, 7, 10	0·25000	0·54962	9, 2, 4, 10	0·00000	2·33333
6, 1, 8, 10	0·00000	0·58333	9, 3, 3, 10	0·00000	2·38048
6, 2, 3, 10	1·00000	0·50690	0, 0, 6, 9	1·00000	0·21276
6, 2, 4, 10	0·75000	0·56519	0, 0, 7, 9	0·88889	0·19637
6, 2, 5, 10	0·50000	0·57130	0, 0, 8, 9	0·77778	0·17151
6, 2, 6, 10	0·25000	0·60953	0, 0, 9, 9	0·66667	0·13354
6, 2, 7, 10	0·00000	0·65085	0, 0, 10, 9	0·55556	0·06172

Table 6.13 (*cont.*)

$n = 10, K = 3$

r-values	f	σ_f
0, 1, 5, 9	1·00000	0·24287
0, 1, 6, 9	0·88889	0·23457
0, 1, 7, 9	0·77778	0·22050
0, 1, 8, 9	0·66667	0·19945
0, 1, 9, 9	0·55555	0·16882
0, 1, 10, 9	0·44444	0·12159
0, 2, 4, 9	1·00000	0·25926
0, 2, 5, 9	0·88889	0·25690
0, 2, 6, 9	0·77778	0·24968
0, 2, 7, 9	0·66667	0·23715
0, 2, 8, 9	0·55556	0·21842
0, 2, 9, 9	0·44444	0·19166
0, 2, 10, 9	0·33333	0·15270
0, 3, 3, 9	1·00000	0·26450
0, 3, 4, 9	0·88889	0·26736
0, 3, 5, 9	0·77778	0·26565
0, 3, 6, 9	0·66667	0·25926
0, 3, 7, 9	0·55556	0·24784
0, 3, 8, 9	0·44444	0·23064
0, 3, 9, 9	0·33333	0·20621
0, 3, 10, 9	0·22222	0·17151
0, 4, 4, 9	0·77778	0·27076
0, 4, 5, 9	0·66667	0·26963
0, 4, 6, 9	0·55556	0·26392
0, 4, 7, 9	0·44444	0·25331
0, 4, 8, 9	0·33333	0·23715
0, 4, 9, 9	0·22222	0·21419
0, 4, 10, 9	0·11111	0·18186
0, 5, 5, 9	0·55556	0·26907
0, 5, 6, 9	0·44444	0·26392
0, 5, 7, 9	0·33333	0·25391
0, 5, 8, 9	0·22222	0·23843
0, 5, 9, 9	0·11111	0·21631
0, 5, 10, 9	0·00000	0·18518
0, 6, 6, 9	0·33333	0·25926
0, 6, 7, 9	0·22222	0·24968
0, 6, 8, 9	0·11111	0·23457
0, 6, 9, 9	0·00000	0·21276
0, 7, 7, 9	0·11111	0·24034
0, 7, 8, 9	0·00000	0·22529
1, 0, 6, 9	1·00000	0·23936
1, 0, 7, 9	0·87500	0·22060
1, 0, 8, 9	0·75000	0·19376
1, 0, 9, 9	0·62500	0·15468
1, 0, 10, 9	0·50000	0·08838

$n = 10, K = 3$

r-values	f	σ_f
1, 1, 5, 9	1·00000	0·27323
1, 1, 6, 9	0·87500	0·26363
1, 1, 7, 9	0·75000	0·24869
1, 1, 8, 9	0·62500	0·22738
1, 1, 9, 9	0·50000	0·19766
1, 1, 10, 9	0·37500	0·15468
1, 2, 4, 9	1·00000	0·29167
1, 2, 5, 9	0·87500	0·28877
1, 2, 6, 9	0·75000	0·28144
1, 2, 7, 9	0·62500	0·26933
1, 2, 8, 9	0·50000	0·25173
1, 2, 9, 9	0·37500	0·22738
1, 2, 10, 9	0·25000	0·19376
1, 3, 3, 9	1·00000	0·29756
1, 3, 4, 9	0·87500	0·30055
1, 3, 5, 9	0·75000	0·29938
1, 3, 6, 9	0·62500	0·29398
1, 3, 7, 9	0·50000	0·28413
1, 3, 8, 9	0·37500	0·26933
1, 3, 9, 9	0·25000	0·24869
1, 3, 10, 9	0·12500	0·22060
1, 4, 4, 9	0·75000	0·30512
1, 4, 5, 9	0·62500	0·30557
1, 4, 6, 9	0·50000	0·30190
1, 4, 7, 9	0·37500	0·29398
1, 4, 8, 9	0·25000	0·28144
1, 4, 9, 9	0·12500	0·26363
1, 4, 10, 9	0·00000	0·23936
1, 5, 5, 9	0·50000	0·30760
1, 5, 6, 9	0·37500	0·30557
1, 5, 7, 9	0·25000	0·29938
1, 5, 8, 9	0·12500	0·28877
1, 5, 9, 9	0·00000	0·27323
1, 6, 6, 9	0·25000	0·30512
1, 6, 7, 9	0·12500	0·30055
1, 6, 8, 9	0·00000	0·29167
1, 7, 7, 9	0·00000	0·29756
2, 0, 6, 9	1·00000	0·27355
2, 0, 7, 9	0·85714	0·25170
2, 0, 8, 9	0·71429	0·22283
2, 0, 9, 9	0·57143	0·18786
2, 0, 10, 9	0·42857	0·12834
2, 1, 5, 9	1·00000	0·31226
2, 1, 6, 9	0·85714	0·30094
2, 1, 7, 9	0·71429	0·28531

Table 6.13 (*cont.*)

$n = 10, K = 3$		
r-values	f	σ_f
2, 1, 8, 9	0·57143	0·26753
2, 1, 9, 9	0·42857	0·23934
2, 1, 10, 9	0·28571	0·20146
2, 2, 4, 9	1·00000	0·33333
2, 2, 5, 9	0·85714	0·32970
2, 2, 6, 9	0·71429	0·32261
2, 2, 7, 9	0·57143	0·31430
2, 2, 8, 9	0·42857	0·29838
2, 2, 9, 9	0·28571	0·27725
2, 2, 10, 9	0·14286	0·25170
2, 3, 3, 9	1·00000	0·34007
2, 3, 4, 9	0·85714	0·34318
2, 3, 5, 9	0·71429	0·34305
2, 3, 6, 9	0·57143	0·34194
2, 3, 7, 9	0·42857	0·33423
2, 3, 8, 9	0·28571	0·32261
2, 3, 9, 9	0·14286	0·30838
2, 3, 10, 9	0·00000	0·28966
2, 4, 4, 9	0·71429	0·34960
2, 4, 5, 9	0·57143	0·35496
2, 4, 6, 9	0·42857	0·35400
2, 4, 7, 9	0·28571	0·34960
2, 4, 8, 9	0·14286	0·34318
2, 4, 9, 9	0·00000	0·33333
2, 5, 5, 9	0·42857	0·36034
2, 5, 6, 9	0·28571	0·36234
2, 5, 7, 9	0·14286	0·36246
2, 5, 8, 9	0·00000	0·35952
2, 6, 6, 9	0·14286	0·36867
2, 6, 7, 9	0·00000	0·37192
3, 0, 6, 9	1·00000	0·31914
3, 0, 7, 9	0·83333	0·29310
3, 0, 8, 9	0·66667	0·26254
3, 0, 9, 9	0·50000	0·22567
3, 0, 10, 9	0·33333	0·18703
3, 1, 5, 9	1·00000	0·36430
3, 1, 6, 9	0·83333	0·35000
3, 1, 7, 9	0·66667	0·33487
3, 1, 8, 9	0·50000	0·31672
3, 1, 9, 9	0·33333	0·30089
3, 1, 10, 9	0·16667	0·26692
3, 2, 4, 9	1·00000	0·38889
3, 2, 5, 9	0·83333	0·38423
3, 2, 6, 9	0·66667	0·37816
3, 2, 7, 9	0·50000	0·37060

$n = 10, K = 3$		
r-values	f	σ_f
3, 2, 8, 9	0·33333	0·36571
3, 2, 9, 9	0·16667	0·34731
3, 2, 10, 9	0·00000	0·33793
3, 3, 3, 9	1·00000	0·39674
3, 3, 4, 9	0·83333	0·39997
3, 3, 5, 9	0·66667	0·40190
3, 3, 6, 9	0·50000	0·40254
3, 3, 7, 9	0·33333	0·40572
3, 3, 8, 9	0·16667	0·39707
3, 3, 9, 9	0·00000	0·35573
3, 4, 4, 9	0·66667	0·40965
3, 4, 5, 9	0·50000	0·41759
3, 4, 6, 9	0·33333	0·42793
3, 4, 7, 9	0·16667	0·42703
3, 4, 8, 9	0·00000	0·39674
3, 5, 5, 9	0·33333	0·43509
3, 5, 6, 9	0·16667	0·44125
3, 5, 7, 9	0·00000	0·41944
3, 6, 6, 9	0·00000	0·42673
4, 0, 6, 9	1·00	0·38297
4, 0, 7, 9	0·80	0·35100
4, 0, 8, 9	0·60	0·32028
4, 0, 9, 9	0·40	0·29120
4, 0, 10, 9	0·20	0·26432
4, 1, 5, 9	1·00	0·43716
4, 1, 6, 9	0·80	0·42016
4, 1, 7, 9	0·60	0·40596
4, 1, 8, 9	0·40	0·39486
4, 1, 9, 9	0·20	0·38713
4, 1, 10, 9	0·00	0·38297
4, 2, 4, 9	1·00	0·46667
4, 2, 5, 9	0·80	0·46053
4, 2, 6, 9	0·60	0·45743
4, 2, 7, 9	0·40	0·45743
4, 2, 8, 9	0·20	0·46053
4, 2, 9, 9	0·00	0·46667
4, 3, 3, 9	1·00	0·47610
4, 3, 4, 9	0·80	0·47944
4, 3, 5, 9	0·60	0·48571
4, 3, 6, 9	0·40	0·49477
4, 3, 7, 9	0·20	0·50649
4, 3, 8, 9	0·00	0·52068
4, 4, 4, 9	0·60	0·49477
4, 4, 5, 9	0·40	0·51242
4, 4, 6, 9	0·20	0·53216

Table 6.13 (*cont.*)

$n = 10, K = 3$			$n = 10, K = 3$		
r-values	f	σ_f	*r*-values	f	σ_f
4, 4, 7, 9	0·00	0·55377	7, 0, 6, 9	1·00000	0·95743
4, 5, 5, 9	0·20	0·54045	7, 0, 7, 9	0·50000	0·88976
4, 5, 6, 9	0·00	0·56960	7, 0, 8, 9	0·00000	1·01379
5, 0, 6, 9	1·00	0·47871	7, 1, 5, 9	1·00	1·09291
5, 0, 7, 9	0·75	0·43800	7, 1, 6, 9	0·50	1·06066
5, 0, 8, 9	0·50	0·41248	7, 1, 7, 9	0·00	1·10924
5, 0, 9, 9	0·25	0·40505	7, 2, 4, 9	1·00	1·16667
5, 0, 10, 9	0·00	0·41667	7, 2, 5, 9	0·50	1·16070
5, 1, 5, 9	1·00	0·54645	7, 2, 6, 9	0·00	1·30171
5, 1, 6, 9	0·75	0·52457	7, 3, 3, 9	1·00	1·19024
5, 1, 7, 9	0·50	0·51707	7, 3, 4, 9	0·50	1·20761
5, 1, 8, 9	0·25000	0·52457	7, 3, 5, 9	0·00	1·36422
5, 1, 9, 9	0·00000	0·54645	7, 4, 4, 9	0·00	1·38444
5, 2, 4, 9	1·00000	0·58333	8, 0, 6, 9	1·00	1·77951
5, 2, 5, 9	0·75000	0·57509	8, 0, 7, 9	0·00	2·02759
5, 2, 6, 9	0·50000	0·58035	8, 1, 5, 9	1·00	2·18581
5, 2, 7, 9	0·25000	0·59875	8, 1, 6, 9	0·00	2·33333
5, 2, 8, 9	0·00000	0·62915	8, 2, 4, 9	1·00	2·33333
5, 3, 3, 9	1·00000	0·59512	8, 2, 5, 9	0·00	2·51661
5, 3, 4, 9	0·75000	0·59875	8, 3, 3, 9	1·00	2·38048
5, 3, 5, 9	0·50000	0·61520	8, 3, 4, 9	0·00	2·60342
5, 3, 6, 9	0·25000	0·64348	0, 0, 7, 8	1·000	0·25345
5, 3, 7, 9	0·00000	0·68211	0, 0, 8, 8	0·875	0·22140
5, 4, 4, 9	0·50000	0·62639	0, 0, 9, 8	0·750	0·17678
5, 4, 5, 9	0·25000	0·66471	0, 0, 10, 8	0·625	0·07365
5, 4, 6, 9	0·00000	0·71200	0, 1, 6, 8	1·000	0·29166
5, 5, 5, 9	0·00000	0·72169	0, 1, 7, 8	0·875	0·27081
6, 0, 6, 9	1·00000	0·63828	0, 1, 8, 8	0·750	0·24296
6, 0, 7, 9	0·66667	0·58443	0, 1, 9, 8	0·625	0·19151
6, 0, 8, 9	0·33333	0·58443	0, 1, 10, 8	0·500	0·10534
6, 0, 9, 9	0·00000	0·63828	0, 2, 5, 8	1·000	0·31458
6, 1, 5, 9	1·00000	0·72860	0, 2, 6, 8	0·875	0·30117
6, 1, 6, 9	0·66667	0·69979	0, 2, 7, 8	0·750	0·28260
6, 1, 7, 9	0·33333	0·71722	0, 2, 8, 8	0·625	0·24694
6, 1, 8, 9	0·00000	0·77778	0, 2, 9, 8	0·500	0·21651
6, 2, 4, 9	1·00000	0·77778	0, 2, 10, 8	0·375	0·17237
6, 2, 5, 9	0·66667	0·76712	0, 3, 4, 8	1·000	0·32543
6, 2, 6, 9	0·33333	0·79866	0, 3, 5, 8	0·875	0·31799
6, 2, 7, 9	0·00000	0·79349	0, 3, 6, 8	0·750	0·30619
6, 3, 3, 9	1·00000	0·79349	0, 3, 7, 8	0·625	0·27990
6, 3, 4, 9	0·66667	0·79866	0, 3, 8, 8	0·500	0·26021
6, 3, 5, 9	0·33333	0·84376	0, 3, 9, 8	0·375	0·23246
6, 3, 6, 9	0·00000	0·85346	0, 3, 10, 8	0·250	0·19320
6, 4, 4, 9	0·33333	0·85827	0, 4, 4, 8	0·875	0·32340
6, 4, 5, 9	0·00000	0·88192	0, 4, 5, 8	0·750	0·31732

Table 6.13 (*cont.*)

n = 10, *K* = 3			*n* = 10, *K* = 3		
r-values	f	σ_f	*r*-values	f	σ_f
0, 4, 6, 8	0·625	0·29789	1, 4, 10, 8	0·00000	0·27355
0, 4, 7, 8	0·500	0·28565	1, 5, 5, 8	0·57143	0·26034
0, 4, 8, 8	0·375	0·26717	1, 5, 6, 8	0·42857	0·35496
0, 4, 9, 8	0·250	0·24116	1, 5, 7, 8	0·28571	0·34305
0, 4, 10, 8	0·125	0·20465	1, 5, 8, 8	0·14286	0·32970
0, 5, 5, 8	0·625	0·30369	1, 5, 9, 8	0·00000	0·31226
0, 5, 6, 8	0·500	0·29756	1, 6, 6, 8	0·28571	0·34960
0, 5, 7, 8	0·375	0·28603	1, 6, 7, 8	0·14286	0·34318
0, 5, 8, 8	0·250	0·26842	1, 6, 8, 8	0·00000	0·33333
0, 5, 9, 8	0·125	0·22902	1, 7, 7, 8	0·00000	0·34007
0, 5, 10, 8	0·000	0·20833	2, 0, 7, 8	1·00000	0·33793
0, 6, 6, 8	0·375	0·29204	2, 0, 8, 8	0·83333	0·29163
0, 6, 7, 8	0·250	0·28106	2, 0, 9, 8	0·66667	0·23497
0, 6, 8, 8	0·125	0·26393	2, 0, 10, 8	0·50000	0·15713
0, 6, 9, 8	0·000	0·23936	2, 1, 6, 8	1·00000	0·38888
0, 7, 7, 8	0·125	0·27043	2, 1, 7, 8	0·83333	0·35813
0, 7, 8, 8	0·000	0·25345	2, 1, 8, 8	0·66667	0·32341
1, 0, 7, 8	1·00000	0·28966	2, 1, 9, 8	0·50000	0·28328
1, 0, 8, 8	0·85714	0·25170	2, 1, 10, 8	0·33333	0·23497
1, 0, 9, 8	0·71429	0·20146	2, 2, 5, 8	1·00000	0·41944
1, 0, 10, 8	0·57143	0·12834	2, 2, 6, 8	0·83333	0·39890
1, 1, 6, 8	1·00000	0·33333	2, 2, 7, 8	0·66667	0·37634
1, 1, 7, 8	0·58714	0·30838	2, 2, 8, 8	0·50000	0·35136
1, 1, 8, 8	0·71429	0·27725	2, 2, 9, 8	0·33333	0·32341
1, 1, 9, 8	0·57143	0·23934	2, 2, 10, 8	0·16667	0·29163
1, 1, 10, 8	0·42857	0·18786	2, 3, 4, 8	1·00000	0·43390
1, 2, 5, 8	1·00000	0·35952	2, 3, 5, 8	0·83333	0·42174
1, 2, 6, 8	0·58714	0·34318	2, 3, 6, 8	0·66667	0·40783
1, 2, 7, 8	0·71429	0·32261	2, 3, 7, 8	0·50000	0·40062
1, 2, 8, 8	0·57143	0·29839	2, 3, 8, 8	0·33333	0·37634
1, 2, 9, 8	0·42857	0·26753	2, 3, 9, 8	0·16667	0·35813
1, 2, 10, 8	0·28571	0·22283	2, 3, 10, 8	0·00000	0·33793
1, 3, 4, 8	1·00000	0·37192	2, 4, 4, 8	0·83333	0·42783
1, 3, 5, 8	0·58714	0·36246	2, 4, 5, 8	0·66667	0·42269
1, 3, 6, 8	0·71429	0·34960	2, 4, 6, 8	0·50000	0·43033
1, 3, 7, 8	0·57143	0·33423	2, 4, 7, 8	0·33333	0·40783
1, 3, 8, 8	0·42857	0·31430	2, 4, 8, 8	0·16667	0·39890
1, 3, 9, 8	0·28571	0·28531	2, 4, 9, 8	0·00000	0·38888
1, 3, 10, 8	0·14286	0·25170	2, 5, 5, 8	0·50000	0·44445
1, 4, 4, 8	0·85714	0·36867	2, 5, 6, 8	0·33333	0·42269
1, 4, 5, 8	0·71429	0·36234	2, 5, 7, 8	0·16667	0·42147
1, 4, 6, 8	0·57143	0·35400	2, 5, 8, 8	0·00000	0·41944
1, 4, 7, 8	0·42857	0·34194	2, 6, 6, 8	0·16667	0·42873
1, 4, 8, 8	0·28571	0·32261	2, 6, 7, 8	0·00000	0·43390
1, 4, 9, 8	0·14286	0·30094	3, 0, 7, 8	1·00000	0·40552

Table 6.13 (*cont.*)

$n = 10, K = 3$		
r-values	f	σ_f
3, 0, 8, 8	0·80000	0·34692
3, 0, 9, 8	0·60000	0·28378
3, 0, 10, 8	0·40000	0·21208
3, 1, 6, 8	1·00000	0·46667
3, 1, 7, 8	0·80000	0·42729
3, 1, 8, 8	0·60000	0·38941
3, 1, 9, 8	0·40000	0·35352
3, 1, 10, 8	0·20000	0·32028
3, 2, 5, 8	1·00000	0·50332
3, 2, 6, 8	0·80000	0·47647
3, 2, 7, 8	0·60000	0·45274
3, 2, 8, 8	0·40000	0·43267
3, 2, 9, 8	0·20000	0·41676
3, 2, 10, 8	0·00000	0·40552
3, 3, 4, 8	1·00000	0·52068
3, 3, 5, 8	0·80000	0·50368
3, 3, 6, 8	0·60000	0·49044
3, 3, 7, 8	0·40000	0·48129
3, 3, 8, 8	0·20000	0·47647
3, 3, 9, 8	0·00000	0·47610
3, 4, 4, 8	0·80000	0·51242
3, 4, 5, 8	0·60000	0·50824
3, 4, 6, 8	0·40000	0·50824
3, 4, 7, 8	0·20000	0·51242
3, 4, 8, 8	0·00000	0·52068
3, 5, 5, 8	0·40000	0·51691
3, 5, 6, 8	0·20000	0·52949
3, 5, 7, 8	0·00000	0·54569
3, 6, 6, 8	0·00000	0·55377
4, 0, 7, 8	1·00000	0·50690
4, 0, 8, 8	0·75000	0·42898
4, 0, 9, 8	0·50000	0·36324
4, 0, 10, 8	0·25000	0·31732
4, 1, 6, 8	1·00000	0·58333
4, 1, 7, 8	0·75000	0·53033
4, 1, 8, 8	0·50000	0·49301
4, 1, 9, 8	0·25000	0·47507
4, 1, 10, 8	0·00000	0·47871
4, 2, 5, 8	1·00000	0·62915
4, 2, 6, 8	0·75000	0·59219
4, 2, 7, 8	0·50000	0·57130
4, 2, 8, 8	0·25000	0·56826
4, 2, 9, 8	0·00000	0·58333
4, 3, 4, 8	1·00000	0·65085
4, 3, 5, 8	0·75000	0·62639
4, 3, 6, 8	0·50000	0·61802
4, 3, 7, 8	0·25000	0·62639
4, 3, 8, 8	0·00000	0·65085
4, 4, 4, 8	0·75000	0·63191
4, 4, 5, 8	0·50000	0·64010
4, 4, 6, 8	0·25000	0·66926
4, 4, 7, 8	0·00000	0·69222
4, 5, 5, 8	0·25000	0·68971
4, 5, 6, 8	0·00000	0·71200
5, 0, 7, 8	1·00000	0·67586
5, 0, 8, 8	0·66667	0·56534
5, 0, 9, 8	0·33333	0·51984
5, 0, 10, 8	0·00000	0·55556
5, 1, 6, 8	1·00000	0·77778
5, 1, 7, 8	0·66667	0·70175
5, 1, 8, 8	0·33333	0·68393
5, 1, 9, 8	0·00000	0·72860
5, 2, 5, 8	1·00000	0·83887
5, 2, 6, 8	0·66667	0·78480
5, 2, 7, 8	0·33333	0·78480
5, 2, 8, 8	0·00000	0·83887
5, 3, 4, 8	1·00000	0·86780
5, 3, 5, 8	0·66667	0·83065
5, 3, 6, 8	0·33333	0·84539
5, 3, 7, 8	0·00000	0·90948
5, 4, 4, 8	0·66667	0·84539
5, 4, 5, 8	0·33333	0·87410
5, 4, 6, 8	0·00000	0·94933
5, 5, 5, 8	0·00000	0·96225
6, 0, 7, 8	1·00000	1·01379
6, 0, 8, 8	0·50000	0·84984
6, 0, 9, 8	0·00000	0·95743
6, 1, 6, 8	1·00000	1·16667
6, 1, 7, 8	0·50000	1·05409
6, 1, 8, 8	0·00000	1·16667
6, 2, 5, 8	1·00000	1·25830
6, 2, 6, 8	0·50000	1·17851
6, 2, 7, 8	0·00000	1·30171
6, 3, 4, 8	1·00000	1·30171
6, 3, 5, 8	0·50000	1·24722
6, 3, 6, 8	0·00000	1·38444
6, 4, 4, 8	0·50000	1·26930
6, 4, 5, 8	0·00000	1·42400
7, 0, 7, 8	1·00000	2·02759
7, 0, 8, 8	0·00000	2·02759

Table 6.13 (*cont.*)

$n = 10, K = 3$			$n = 10, K = 3$		
r-values	f	σ_f	*r*-values	f	σ_f
7, 1, 6, 8	1·00000	2·33333	1, 0, 9, 7	0·83333	0·26692
7, 1, 7, 8	0·00000	0·38048	1, 0, 10, 7	0·66667	0·18793
7, 2, 5, 8	1·00000	2·51661	1, 1, 7, 7	1·00000	0·35573
7, 2, 6, 8	0·00000	2·60342	1, 1, 8, 7	0·83333	0·34731
7, 3, 4, 8	1·00000	2·60342	1, 1, 9, 7	0·66667	0·30089
7, 3, 5, 8	0·00000	2·72845	1, 1, 10, 7	0·50000	0·22567
7, 4, 4, 8	0·00000	2·76887	1, 2, 6, 7	1·00000	0·39674
0, 0, 8, 7	1·00000	0·28965	1, 2, 7, 7	0·83333	0·39707
0, 0, 9, 7	0·85714	0·23536	1, 2, 8, 7	0·66667	0·36571
0, 0, 10, 7	0·71429	0·15587	1, 2, 9, 7	0·50000	0·31672
0, 1, 7, 7	1·00000	0·34007	1, 2, 10, 7	0·33333	0·26254
0, 1, 8, 7	0·85714	0·30278	1, 3, 5, 7	1·00000	0·41944
0, 1, 9, 7	0·71429	0·25517	1, 3, 6, 7	0·83333	0·42703
0, 1, 10, 7	0·57143	0·18962	1, 3, 7, 7	0·66667	0·40572
0, 2, 6, 7	1·00000	0·37192	1, 3, 8, 7	0·50000	0·37060
0, 2, 7, 7	0·85714	0·34480	1, 3, 9, 7	0·33333	0·33487
0, 2, 8, 7	0·71429	0·31122	1, 3, 10, 7	0·16667	0·29310
0, 2, 9, 7	0·57143	0·26877	1, 4, 4, 7	1·00000	0·42673
0, 2, 10, 7	0·42857	0·21695	1, 4, 5, 7	0·83333	0·44125
0, 3, 5, 7	1·00000	0·38978	1, 4, 6, 7	0·66667	0·42793
0, 3, 6, 7	0·85714	0·37017	1, 4, 7, 7	0·50000	0·40254
0, 3, 7, 7	0·71429	0·34574	1, 4, 8, 7	0·33333	0·37816
0, 3, 8, 7	0·57143	0·31536	1, 4, 9, 7	0·16667	0·35000
0, 3, 9, 7	0·42857	0·28074	1, 4, 10, 7	0·00000	0·31914
0, 3, 10, 7	0·28571	0·22695	1, 5, 5, 7	0·66667	0·43509
0, 4, 4, 7	1·00000	0·39555	1, 5, 6, 7	0·50000	0·41759
0, 4, 5, 7	0·85714	0·38223	1, 5, 7, 7	0·33333	0·40190
0, 4, 6, 7	0·71429	0·36488	1, 5, 8, 7	0·16667	0·38423
0, 4, 7, 7	0·57143	0·34291	1, 5, 9, 7	0·00000	0·36430
0, 4, 8, 7	0·42857	0·31857	1, 6, 6, 7	0·33333	0·40965
0, 4, 9, 7	0·28571	0·28057	1, 6, 7, 7	0·16667	0·39997
0, 4, 10, 7	0·14286	0·23536	1, 6, 8, 7	0·00000	0·38889
0, 5, 5, 7	0·71429	0·37104	1, 7, 7, 7	0·00000	0·39674
0, 5, 6, 7	0·57143	0·35589	2, 0, 8, 7	1·00000	0·40552
0, 5, 7, 7	0·42857	0·33927	2, 0, 9, 7	0·80000	0·32028
0, 5, 8, 7	0·28571	0·31122	2, 0, 10, 7	0·60000	0·21208
0, 5, 9, 7	0·14286	0·37941	2, 1, 7, 7	1·00000	0·47610
0, 5, 10, 7	0·00000	0·23809	2, 1, 8, 7	0·80000	0·41676
0, 6, 6, 7	0·42857	0·34588	2, 1, 9, 7	0·60000	0·35352
0, 6, 7, 7	0·28571	0·32546	2, 1, 10, 7	0·40000	0·28378
0, 6, 8, 7	0·14286	0·30278	2, 2, 6, 7	1·00000	0·52068
0, 6, 9, 7	0·00000	0·27355	2, 2, 7, 7	0·80000	0·47647
0, 7, 7, 7	0·14286	0·31018	2, 2, 8, 7	0·60000	0·43267
0, 7, 8, 7	0·00000	0·28965	2, 2, 9, 7	0·40000	0·48941
1, 0, 8, 7	1·00000	0·33793	2, 2, 10, 7	0·20000	0·34692

Table 6.13 (*cont.*)

r-values ($n = 10, K = 3$)	f	σ_f	*r*-values ($n = 10, K = 3$)	f	σ_f
2, 3, 5, 7	1·00000	0·54569	4, 0, 9, 7	0·66667	0·50917
2, 3, 6, 7	0·80000	0·51242	4, 0, 10, 7	0·33333	0·40062
2, 3, 7, 7	0·60000	0·48129	4, 1, 7, 7	1·00000	0·79349
2, 3, 8, 7	0·40000	0·45274	4, 1, 8, 7	0·66667	0·67586
2, 3, 9, 7	0·20000	0·42729	4, 1, 9, 7	0·33333	0·61864
2, 3, 10, 7	0·00000	0·40552	4, 1, 10, 7	0·00000	0·63828
2, 4, 4, 7	1·00000	0·55377	4, 2, 6, 7	1·00000	0·86780
2, 4, 5, 7	0·80000	0·52949	4, 2, 7, 7	0·66667	0·77778
2, 4, 6, 7	0·60000	0·50824	4, 2, 8, 7	0·33333	0·64536
2, 4, 7, 7	0·40000	0·49044	4, 2, 9, 7	0·00000	0·77778
2, 4, 8, 7	0·20000	0·47647	4, 3, 5, 7	1·00000	0·90948
2, 4, 9, 7	0·00000	0·46667	4, 3, 6, 7	0·66667	0·83887
2, 5, 5, 7	0·60000	0·51691	4, 3, 7, 7	0·33333	0·82402
2, 5, 6, 7	0·40000	0·50824	4, 3, 8, 7	0·00000	0·86780
2, 5, 7, 7	0·20000	0·50368	4, 4, 4, 7	1·00000	0·92296
2, 5, 8, 7	0·00000	0·50332	4, 4, 5, 7	0·66667	0·86780
2, 6, 6, 7	0·20000	0·51242	4, 4, 6, 7	0·33333	0·86780
2, 6, 7, 7	0·00000	0·52068	4, 4, 7, 7	0·00000	0·92296
3, 0, 8, 7	1·00000	0·50690	4, 5, 5, 7	0·33333	0·88192
3, 0, 9, 7	0·75000	0·39198	4, 5, 6, 7	0·00000	0·94933
3, 0, 10, 7	0·50000	0·27003	5, 0, 8, 7	1·00000	1·01379
3, 1, 7, 7	1·00000	0·59512	5, 0, 9, 7	0·50000	0·75462
3, 1, 8, 7	0·75000	0·51454	5, 0, 10, 7	0·00000	0·83333
3, 1, 9, 7	0·50000	0·44488	5, 1, 7, 7	1·00000	1·19024
3, 1, 10, 7	0·25000	0·39198	5, 1, 8, 7	0·50000	1·00692
3, 2, 6, 7	1·00000	0·65085	5, 1, 9, 7	0·00000	1·09291
3, 2, 7, 7	0·75000	0·58999	5, 2, 6, 7	1·00000	1·30171
3, 2, 8, 7	0·50000	0·54327	5, 2, 7, 7	0·50000	1·16070
3, 2, 9, 7	0·25000	0·51454	5, 2, 8, 7	0·00000	1·25830
3, 2, 10, 7	0·00000	0·50690	5, 3, 5, 7	1·00000	1·36422
3, 3, 5, 7	1·00000	0·68211	5, 3, 6, 7	0·50000	1·25277
3, 3, 6, 7	0·75000	0·63533	5, 3, 7, 7	0·00000	1·36422
3, 3, 7, 7	0·50000	0·60403	5, 4, 4, 7	1·00000	1·38444
3, 3, 8, 7	0·25000	0·58999	5, 4, 5, 7	0·50000	1·29636
3, 3, 9, 7	0·00000	0·59512	5, 4, 6, 7	0·00000	1·42400
3, 4, 4, 7	1·00000	0·69222	5, 5, 5, 7	0·00000	1·44338
3, 4, 5, 7	0·75000	0·65683	6, 0, 8, 7	1·00000	2·10818
3, 4, 6, 7	0·50000	0·63191	6, 0, 9, 7	0·00000	1·77951
3, 4, 7, 7	0·25000	0·63533	6, 1, 7, 7	1·00000	2·44949
3, 4, 8, 7	0·00000	0·65085	6, 1, 8, 7	0·00000	2·33333
3, 5, 5, 7	0·50000	0·64818	6, 2, 6, 7	1·00000	2·66667
3, 5, 6, 7	0·25000	0·65683	6, 2, 7, 7	0·00000	2·60342
3, 5, 7, 7	0·00000	0·68211	6, 3, 5, 7	1·00000	2·78887
3, 6, 6, 7	0·00000	0·69222	6, 3, 6, 7	0·00000	2·76887
4, 0, 8, 7	1·00000	0·67586	6, 4, 4, 7	1·00000	2·82427

Table 6.13 (*cont.*)

n = 10, K = 3		
r-values	f	σ_f
6, 4, 5, 7	0·00000	2·84800
0, 0, 9, 6	1·00000	0·31914
0, 0, 10, 6	0·83333	0·22680
0, 1, 8, 6	1·00000	0·38889
0, 1, 9, 6	0·83333	0·32710
0, 1, 10, 6	0·66667	0·24637
0, 2, 7, 6	1·00000	0·43390
0, 2, 8, 6	0·83333	0·38756
0, 2, 9, 6	0·66667	0·33178
0, 2, 10, 6	0·50000	0·26058
0, 3, 6, 6	1·00000	0·46148
0, 3, 7, 6	0·83333	0·42552
0, 3, 8, 6	0·66667	0·38356
0, 3, 9, 6	0·50000	0·33333
0, 3, 10, 6	0·33333	0·27027
0, 4, 5, 6	1·00000	0·47466
0, 4, 6, 6	0·83333	0·44675
0, 4, 7, 6	0·66667	0·41450
0, 4, 8, 6	0·50000	0·37679
0, 4, 9, 6	0·33333	0·33178
0, 4, 10, 6	0·16667	0·27592
0, 5, 5, 6	0·83333	0·45361
0, 5, 6, 6	0·66667	0·42913
0, 5, 7, 6	0·50000	0·40062
0, 5, 8, 6	0·33333	0·36711
0, 5, 9, 6	0·16667	0·32710
0, 5, 10, 6	0·00000	0·27778
0, 6, 6, 6	0·50000	0·40825
0, 6, 7, 6	0·33333	0·38356
0, 6, 8, 6	0·16667	0·35428
0, 6, 9, 6	0·00000	0·31914
0, 7, 7, 6	0·16667	0·36289
0, 7, 8, 6	0·00000	0·33793
1, 0, 9, 6	1·00000	0·38297
1, 0, 10, 6	0·80000	0·26432
1, 1, 8, 6	1·00000	0·46667
1, 1, 9, 6	0·80000	0·38713
1, 1, 10, 6	0·60000	0·29120
1, 2, 7, 6	1·00000	0·52068
1, 2, 8, 6	0·80000	0·46053
1, 2, 9, 6	0·60000	0·39486
1, 2, 10, 6	0·40000	0·32028
1, 3, 6, 6	1·00000	0·55377
1, 3, 7, 6	0·80000	0·50649
1, 3, 8, 6	0·60000	0·45743

n = 10, K = 3		
r-values	f	σ_f
1, 3, 9, 6	0·40000	0·40596
1, 3, 10, 6	0·20000	0·35100
1, 4, 5, 6	1·00000	0·56960
1, 4, 6, 6	0·80000	0·53216
1, 4, 7, 6	0·60000	0·49477
1, 4, 8, 6	0·40000	0·45743
1, 4, 9, 6	0·20000	0·42016
1, 4, 10, 6	0·00000	0·38297
1, 5, 5, 6	0·80000	0·54045
1, 5, 6, 6	0·60000	0·51242
1, 5, 7, 6	0·40000	0·48571
1, 5, 8, 6	0·20000	0·46053
1, 5, 9, 6	0·00000	0·43716
1, 6, 6, 6	0·40000	0·49477
1, 6, 7, 6	0·20000	0·47944
1, 6, 8, 6	0·00000	0·46667
1, 7, 7, 6	0·00000	0·47610
2, 0, 9, 6	1·00000	0·47871
2, 0, 10, 6	0·75000	0·31732
2, 1, 8, 6	1·00000	0·58333
2, 1, 9, 6	0·75000	0·47507
2, 1, 10, 6	0·50000	0·36324
2, 2, 7, 6	1·00000	0·65085
2, 2, 8, 6	0·75000	0·56826
2, 2, 9, 6	0·50000	0·49301
2, 2, 10, 6	0·25000	0·42998
2, 3, 6, 6	1·00000	0·69222
2, 3, 7, 6	0·75000	0·62639
2, 3, 8, 6	0·50000	0·57130
2, 3, 9, 6	0·25000	0·53033
2, 3, 10, 6	0·00000	0·50690
2, 4, 5, 6	1·00000	0·71200
2, 4, 6, 6	0·75000	0·66926
2, 4, 7, 6	0·50000	0·61802
2, 4, 8, 6	0·25000	0·59219
2, 4, 9, 6	0·00000	0·58333
2, 5, 5, 6	0·75000	0·68971
2, 5, 6, 6	0·50000	0·64010
2, 5, 7, 6	0·25000	0·62639
2, 5, 8, 6	0·00000	0·62915
2, 6, 6, 6	0·25000	0·63191
2, 6, 7, 6	0·00000	0·65085
3, 0, 9, 6	1·00000	0·63828
3, 0, 10, 6	0·66667	0·40062
3, 1, 8, 6	1·00000	0·77778

Table 6.13 (*cont.*)

$n = 10, K = 3$			$n = 10, K = 3$		
r-values	f	σ_f	r-values	f	σ_f
3, 1, 9, 6	0·66667	0·61864	0, 0, 10, 5	1·0	0·33333
3, 1, 10, 6	0·33333	0·50917	0, 1, 9, 5	1·0	0·43716
3, 2, 7, 6	1·00000	0·86780	0, 1, 10, 5	0·8	0·33333
3, 2, 8, 6	0·66667	0·74536	0, 2, 8, 5	1·0	0·50332
3, 2, 9, 6	0·33333	0·67586	0, 2, 9, 5	0·8	0·42687
3, 2, 10, 6	0·00000	0·67586	0, 2, 10, 5	0·6	0·33333
3, 3, 6, 6	1·00000	0·92296	0, 3, 7, 5	1·0	0·54569
3, 3, 7, 6	0·66667	0·82402	0, 3, 8, 5	0·8	0·48534
3, 3, 8, 6	0·33333	0·77778	0, 3, 9, 5	0·6	0·41633
3, 3, 9, 6	0·00000	0·79349	0, 3, 10, 5	0·4	0·33333
3, 4, 5, 6	1·00000	0·94933	0, 4, 6, 5	1·0	0·56960
3, 4, 6, 6	0·66667	0·86780	0, 4, 7, 5	0·8	0·52068
3, 4, 7, 6	0·33333	0·83887	0, 4, 8, 5	0·6	0·46667
3, 4, 8, 6	0·00000	0·86780	0, 4, 9, 5	0·4	0·40552
3, 5, 5, 6	0·66667	0·88192	0, 4, 10, 5	0·2	0·33333
3, 5, 6, 6	0·33333	0·86780	0, 5, 5, 5	1·0	0·57735
3, 5, 7, 6	0·00000	0·90948	0, 5, 6, 5	0·8	0·53748
3, 6, 6, 6	0·00000	0·92296	0, 5, 7, 5	0·6	0·48990
4, 0, 9, 6	1·00000	0·95743	0, 5, 8, 5	0·40000	0·44721
4, 0, 10, 6	0·50000	0·57735	0, 5, 9, 5	0·20000	0·39440
4, 1, 8, 6	1·00000	0·16667	0, 5, 10, 5	0·00000	0·33333
4, 1, 9, 6	0·50000	0·91287	0, 6, 6, 5	0·60000	0·49889
4, 1, 10, 6	0·00000	0·95743	0, 6, 7, 5	0·40000	0·46667
4, 2, 7, 6	1·00000	1·19024	0, 6, 8, 5	0·20000	0·42688
4, 2, 8, 6	0·50000	1·10554	0, 6, 9, 5	0·00000	0·38297
4, 2, 9, 6	0·00000	1·16667	0, 7, 7, 5	0·20000	0·43716
4, 3, 6, 6	1·0	1·28019	0, 7, 8, 5	0·00000	0·40552
4, 3, 7, 6	0·5	1·22474	1, 0, 10, 5	1·00000	0·41667
4, 3, 8, 6	0·0	1·19024	1, 1, 9, 5	1·00000	0·54645
4, 4, 5, 6	1·0	1·32288	1, 1, 10, 5	0·75000	0·40505
4, 4, 6, 6	0·5	1·29099	1, 2, 8, 5	1·00000	0·62915
4, 4, 7, 6	0·0	1·28019	1, 2, 9, 5	0·75000	0·52457
4, 5, 5, 6	0·5	1·31233	1, 2, 10, 5	0·50000	0·41248
4, 5, 6, 6	0·0	1·32288	1, 3, 7, 5	1·00000	0·68211
5, 0, 9, 6	1·0	2·23607	1, 3, 8, 5	0·75000	0·59875
5, 0, 10, 6	0·0	1·66667	1, 3, 9, 5	0·50000	0·51707
5, 1, 8, 6	1·0	2·60342	1, 3, 10, 5	0·25000	0·43800
5, 1, 9, 6	0·0	2·18581	1, 4, 6, 5	1·00000	0·71200
5, 2, 7, 6	1·0	2·84800	1, 4, 7, 5	0·75000	0·64348
5, 2, 8, 6	0·0	2·51661	1, 4, 8, 5	0·50000	0·58035
5, 3, 6, 6	1·0	3·00000	1, 4, 9, 5	0·25000	0·52457
5, 3, 7, 6	0·0	2·72845	1, 4, 10, 5	0·00000	0·47871
5, 4, 5, 6	1·0	3·07318	1, 5, 5, 5	1·00000	0·72169
5, 4, 6, 6	0·0	2·84800	1, 5, 6, 5	0·75000	0·66471
5, 5, 5, 6	0·0	2·88675	1, 5, 7, 5	0·50000	0·61520

Table 6.13 (*cont.*)

$n = 10, K = 3$		
r-values	f	σ_f
1, 5, 8, 5	0·25000	0·57509
1, 5, 9, 5	0·00000	0·54645
1, 6, 6, 5	0·50000	0·62639
1, 6, 7, 5	0·25000	0·59875
1, 6, 8, 5	0·00000	0·58333
1, 7, 7, 5	0·00000	0·59512
2, 0, 10, 5	1·00000	0·55556
2, 1, 9, 5	1·00000	0·72860
2, 1, 10, 5	0·66667	0·51985
2, 2, 8, 5	1·00000	0·83887
2, 2, 9, 5	0·66667	0·68393
2, 2, 10, 5	0·33333	0·56534
2, 3, 7, 5	1·00000	0·90948
2, 3, 8, 5	0·66667	0·78480
2, 3, 9, 5	0·33333	0·70175
2, 3, 10, 5	0·00000	0·67586
2, 4, 6, 5	1·00000	0·94933
2, 4, 7, 5	0·66667	0·84539
2, 4, 8, 5	0·33333	0·78480
2, 4, 9, 5	0·00000	0·77778
2, 5, 5, 5	1·00000	0·96225
2, 5, 6, 5	0·66667	0·87410
2, 5, 7, 5	0·33333	0·83065
2, 5, 8, 5	0·00000	0·83887
2, 6, 6, 5	0·33333	0·84539
2, 6, 7, 5	0·00000	0·86780
3, 0, 10, 5	1·00000	0·83333
3, 1, 9, 5	1·00000	1·09291
3, 1, 10, 5	0·50000	0·75462
3, 2, 8, 5	1·00000	1·25830
3, 2, 9, 5	0·50000	1·00692
3, 2, 10, 5	0·00000	1·01379
3, 3, 7, 5	1·00000	1·36422
3, 3, 8, 5	0·50000	1·16070
3, 3, 9, 5	0·00000	1·19024
3, 4, 6, 5	1·00000	1·42400
3, 4, 7, 5	0·50000	1·25277
3, 4, 8, 5	0·00000	1·30171
3, 5, 5, 5	1·00000	1·44388
3, 5, 6, 5	0·50000	1·29636
3, 5, 7, 5	0·00000	1·36422
3, 6, 6, 5	0·00000	1·38444
4, 0, 10, 5	1·00000	1·66667
4, 1, 9, 5	1·00000	2·18581
4, 1, 10, 5	0·00000	2·23607

$n = 10, K = 3$		
r-values	f	σ_f
4, 2, 8, 5	1·00000	2·51661
4, 2, 9, 5	0·00000	2·60342
4, 3, 7, 5	1·00000	2·72845
4, 3, 8, 5	0·00000	2·84800
4, 4, 6, 5	1·00000	2·84800
4, 4, 7, 5	0·00000	3·00000
4, 5, 5, 5	1·00000	2·88675
4, 5, 6, 5	0·00000	3·07318
0, 1, 10, 4	1·00000	0·47871
0, 2, 9, 4	1·00000	0·58333
0, 2, 10, 4	0·75000	0·45262
0, 3, 8, 4	1·00000	0·65085
0, 3, 9, 4	0·75000	0·54962
0, 3, 10, 4	0·50000	0·53301
0, 4, 7, 4	1·00000	0·69222
0, 4, 8, 4	0·75000	0·60953
0, 4, 9, 4	0·50000	0·52042
0, 4, 10, 4	0·25000	0·45644
0, 5, 6, 4	1·00000	0·71200
0, 5, 7, 4	0·75000	0·64280
0, 5, 8, 4	0·50000	0·57130
0, 5, 9, 4	0·25000	0·52705
0, 5, 10, 4	0·00000	0·41667
0, 6, 6, 4	0·75000	0·65352
0, 6, 7, 4	0·50000	0·59512
0, 6, 8, 4	0·25000	0·56519
0, 6, 9, 4	0·00000	0·47871
0, 7, 7, 4	0·25000	0·57735
0, 7, 8, 4	0·00000	0·50690
1, 1, 10, 4	1·00000	0·63828
1, 2, 9, 4	1·00000	0·77778
1, 2, 10, 4	0·66667	0·58443
1, 3, 8, 4	1·00000	0·79349
1, 3, 9, 4	0·66667	0·71722
1, 3, 10, 4	0·33333	0·58443
1, 4, 7, 4	1·00000	0·85346
1, 4, 8, 4	0·66667	0·79866
1, 4, 9, 4	0·33333	0·69979
1, 4, 10, 4	0·00000	0·63828
1, 5, 6, 4	1·00000	0·88192
1, 5, 7, 4	0·66667	0·84376
1, 5, 8, 4	0·33333	0·76712
1, 5, 9, 4	0·00000	0·72860
1, 6, 6, 4	0·66667	0·85827
1, 6, 7, 4	0·33333	0·79866

Table 6.13 (*cont.*)

$n = 10, K = 3$		
r-values	f	σ_f
1, 6, 8, 4	0·00000	0·77778
1, 7, 7, 4	0·00000	0·79349
2, 1, 10, 4	1·00000	0·95743
2, 2, 9, 4	1·00000	1·16667
2, 2, 10, 4	0·50000	0·84984
2, 3, 8, 4	1·00000	1·30171
2, 3, 9, 4	0·50000	1·05409
2, 3, 10, 4	0·00000	1·01319
2, 4, 7, 4	1·00000	1·38444
2, 4, 8, 4	0·50000	1·17851
2, 4, 9, 4	0·00000	1·16667
2, 5, 6, 4	1·00000	1·42400
2, 5, 7, 4	0·50000	1·24722
2, 5, 8, 4	0·00000	1·25830
2, 6, 6, 4	0·50000	1·26930
2, 6, 7, 4	0·00000	1·30171
3, 1, 10, 4	1·00000	1·79951
3, 2, 9, 4	1·00000	2·33333
3, 2, 10, 4	0·00000	2·10818
3, 3, 8, 4	1·00000	2·60342
3, 3, 9, 4	0·00000	2·44949
3, 4, 7, 4	1·00000	2·76887
3, 4, 8, 4	0·00000	2·66667
3, 5, 6, 4	1·00000	2·84800
3, 5, 7, 4	0·00000	2·78887
3, 6, 6, 4	0·00000	2·82427
0, 2, 10, 3	1·00000	0·67586
0, 3, 9, 3	1·00000	0·79349
0, 3, 10, 3	0·66667	0·61195
0, 4, 8, 3	1·00000	0·86780
0, 4, 9, 3	0·66667	0·72293
0, 4, 10, 3	0·33333	0·57013
0, 5, 7, 3	1·00000	0·90948
0, 5, 8, 3	0·66667	0·78829
0, 5, 9, 3	0·33333	0·66975
0, 5, 10, 3	0·00000	0·55556
0, 6, 6, 3	1·00000	0·92296
0, 6, 7, 3	0·66667	0·81901
0, 6, 8, 3	0·33333	0·72293
0, 6, 9, 3	0·00000	0·63828
0, 7, 7, 3	0·33333	0·73981
0, 7, 8, 3	0·00000	0·67586
1, 2, 10, 3	1·0	1·01379
1, 3, 9, 3	1·0	1·19024
1, 3, 10, 3	0·5	0·88976

$n = 10, K = 3$		
r-values	f	σ_f
1, 4, 8, 3	1·0	1·30171
1, 4, 9, 3	0·5	1·06066
1, 4, 10, 3	0·0	0·95743
1, 5, 7, 3	1·0	1·36422
1, 5, 8, 3	0·5	1·16070
1, 5, 9, 3	0·0	0·09291
1, 6, 6, 3	1·0	1·38444
1, 6, 7, 3	0·5	1·20761
1, 6, 8, 3	0·0	1·16667
1, 7, 7, 3	0·0	1·19024
2, 2, 10, 3	1·0	2·02759
2, 3, 9, 3	1·0	2·38048
2, 3, 10, 3	0·0	2·02759
2, 4, 8, 3	1·0	2·60342
2, 4, 9, 3	0·0	2·33333
2, 5, 7, 3	1·0	2·72845
2, 5, 8, 3	0·0	2·51661
2, 6, 6, 3	1·0	2·76887
2, 6, 7, 3	0·0	2·60342
0, 3, 10, 2	1·0	1·01379
0, 4, 9, 2	1·0	1·16667
0, 4, 10, 2	0·5	0·88192
0, 5, 8, 2	1·0	1·25830
0, 5, 9, 2	0·5	1·02740
0, 5, 10, 2	0·0	0·83333
0, 6, 7, 2	1·0	1·30171
0, 6, 8, 2	0·5	1·10554
0, 6, 9, 2	0·0	0·95743
0, 7, 7, 2	0·5	1·13039
0, 7, 8, 2	0·0	1·01379
1, 3, 10, 2	1·0	2·02759
1, 4, 9, 2	1·0	2·33333
1, 4, 10, 2	0·0	1·77951
1, 5, 8, 2	1·0	2·51661
1, 5, 9, 2	0·0	2·18581
1, 6, 7, 2	1·0	2·60342
1, 6, 8, 2	0·0	2·33333
1, 7, 7, 2	0·0	2·38048
0, 4, 10, 1	1·0	1·77951
0, 5, 9, 1	1·0	2·18581
0, 5, 10, 1	0·0	1·66667
0, 6, 8, 1	1·0	2·33333
0, 6, 9, 1	0·0	1·77951
0, 7, 7, 1	1·0	2·38048
0, 7, 8, 1	0·0	2·02759

7

GENETIC TECHNIQUE

A useful distinction can be drawn between genetic experiment and genetic technique. In this sense, the measurement of mutation rates falls within the scope of genetics proper whereas mutagenesis is a technique used in many physiological and biochemical experiments. The distinction lies, in other words, between pure genetics and its applications to microbiology as a whole.

Most investigations are concerned essentially with the enzymic makeup of an organism as determined by its environment and genetic composition. Although a great deal can always be learnt from manipulating the environment, it is no less valuable to alter the genotype, by isolating variants which differ in the way they form the enzymes in question. Mutants are the natural example, and here the problem generally lies not so much in accumulating sufficient cells to be sure of having one of the right sort but in devising a method to select it. Alternatively, the genotype may be changed by introducing new genes from another strain. In the usual methods of chromosome transfer, by which chromosomal genes are transferred by conjugation, transduction with phage or transformation with DNA, the usual outcome is a recombinant derived largely from the recipient, part of whose chromosome has been replaced by some of the donor's genes. Such an approach presupposes a more or less extensive genetic analysis of the system (see Clowes & Hayes, 1968), and it is partly for this reason that only two or three bacterial species have been examined in any detail. On the other hand, there are many instances of gene transfer of considerable general importance that nevertheless demand far less background knowledge. These concern bacterial plasmids, which are sets of genes that are independent of the host chromosome. A cell acquiring a plasmid becomes altered genetically, not by the donated genes being substituted for part of its own genome by recombination, but by the addition of a small independent supernumerary chromosome. The best-known plasmids are R factors, responsible for transmissible drug resistance, and colicin factors, which determine the synthesis of antibiotics acting on *E. coli* named 'colicins' and which have incidentally proved useful as epidemiological markers. R factors in particular are of growing clinical importance, quite apart from their genetic properties, but analogous plasmids have no less significance for other fields like biochemistry. If a plasmid includes a segment of bacterial chromosome, the host strain necessarily becomes diploid for those genes. If the duplicated segments carry different alleles, it immediately becomes possible to see how their diffusible products interact. Two well-known phenomena analysed in this way are the repression of genetic function (Jacob & Monod, 1961) and complementation between inactive polypeptide chains to form an active protein (Siddiqi, 1965). More recently, the scope of this approach has been extended by a relatively simple method for incorporating any predetermined region of chromosome in a plasmid which can then be passed from one strain to another by conjugation (Low, 1968). Phages are plasmids no less than R factors or Col factors but their techniques have now been described in such detail that they are only mentioned here in passing (Adams, 1959; Campbell, 1968; Eisenstark, 1968). Mutants and plasmids are now frequently symbolized according to Demerec *et al.* (1966) and, in the cases of *E. coli* K12 and *Salm. typhimurium* LT2, their genetic and biochemical properties can be traced from Ames & Martin (1964), Sanderson (1967), Taylor & Trotter (1967) and Clowes & Hayes (1968).

MUTAGENESIS

In choosing amongst the large number of mutagenic agents, there is a natural tendency to use those giving the highest numbers of mutants. However, although mutagens are not specific in the sense that a predetermined character can be altered at will, their effects on the genome are certainly specific, as a mass of biochemical and genetic evidence shows (see Orgel, 1965; Zamenhof, 1967). Thus, one of the most potent mutagens, nitrosoguanidine, produces multiple linked mutations at the replicating point of the DNA (Cerda-Olmeda, Hanawalt & Guerola, 1968), whereas nitrous acid generally produces isolated mutations.

Whichever mutagen is chosen, there are certain elementary technical points to observe. First, for two or more mutants of the same phenotype to be accepted as independent, they must have arisen in independent clones, e.g. in broth cultures inoculated from different isolated colonies. Second, a mutant organism does not give rise immediately to a pure mutant clone since it contains a non-mutant as well as a mutant nucleus which segregate at cell division. Thus, if a gal^+ (galactose-fermenting) culture is plated on galactose indicator medium to isolate gal^- mutants immediately after exposure to a mutagen, mixed colonies will be seen containing *gal*$^-$ sectors. On the other hand, if the treated culture is allowed to multiply in broth for a few generations to allow segregation of mutant from non-mutant genomes, homogeneously *gal*$^+$ and *gal*$^-$ colonies will be obtained without sectored colonies. Incubation also allows the segregation of a recessive mutant allele from a dominant wild type, which may be obligatory for a resistance marker. Thus, chromosomal streptomycin-resistance (*str-r*) is recessive to *str-s* so that if a mutagenized culture is plated immediately on streptomycin agar, relatively few resistant colonies are formed, although their count rises as much as 1000-fold during the succeeding few hours (Verly *et al.* 1967). Lastly, during exposure to a mutagen, a varying proportion of the cells will be killed. With u.v. irradiation, killing is marked, whereas with a mutagen like ethyl methane sulphonate it is usually insignificant. In general, the optimal combination of mutagen concentration and exposure has to be determined by trial, so as to strike a balance between an increasing proportion of mutants and a progressively declining population.

With all methods, it is probably best if the organisms are suspended in buffer before treatment to avoid non-specific protection by medium constituents. Nevertheless, satisfactory results are often obtained if the final concentration of broth is less than 10 %, v/v.

Ultra-violet radiation

The procedure is essentially that of producing sterilization by u.v. radiation (p. 121). Technically, the main problem is to remove extraneous u.v.-absorbing material like protein that would protect the organisms and in avoiding photo-reversal of the mutagenic effects (for quantitative details, see Muller, 1954; Zelle, Ogg & Hollaender, 1958).

Nitrous acid

The acid is produced by using sodium nitrite in buffer at pH 4·5 and its action can be stopped simply by diluting into buffer at pH 7·5. Mix 1 ml. cell suspension, 1 ml. 0·05 M-$NaNO_2$ and 1 ml. 3 M-acetate buffer, pH 4·5 (= 100 ml. 3 N-acetic acid + 43 ml. 3 N-NaOH). Incubate in a 37° water bath for up to 20 min. and dilute 1/100 in broth at pH 7·5 (Kaudewitz, 1959).

Ethyl methane sulphonate

This gives up to 5 % mutants of a given phenotype with a survival of 50–70 % with most species (see Loveless, 1966). Most experiments use a

0·05–0·4 M solution in phosphate buffer, pH 7·4, which is readily prepared since EMS is liquid at room temperature. Satisfactory results are obtained if 0·02–0·04 ml. EMS is pipetted into 1 ml. buffer where it will form a drop at the bottom of the tube. This dissolves to give a 0·2–0·4 M solution on swirling the tube in a water bath at 37°; shaking is to be avoided as a turbid emulsion may form. The solution is stable for many hours. Once the EMS is completely dissolved, 0·1 ml. of culture is added (the growth phase seems immaterial), left for 20 min. in a 37° bath and diluted 1/100 into broth to allow segregation. Since EMS can be tasted during use and is evidently volatile, it seems sensible to use a fume cupboard and to wash glassware with care.

Nitrosoguanidine

Exponentially-growing cells are washed and suspended at a concentration of *ca.* 5×10^8/ml. in buffer, pH 6·0. 0·1 ml. of a freshly prepared aqueous solution of *N*-methyl-*N'*-nitro-*N*-nitrosoguanidine (1–10 mg./ml.) is added to a final concentration of 100 μg./ml. The mixture is left at 37° for 30 min. and then diluted 1/100 into broth (Adelberg, Mandel & Chen, 1965).

Hydroxylamine

Although predominantly used with phage, hydroxylamine is also mutagenic for bacteria (see Lie, 1964; Phillips & Brown, 1967). Under ordinary conditions, its solutions cause rapid inactivation, probably because of oxidation products whose formation is catalysed by trace metals. Inactivation is therefore avoided without affecting mutagenesis by including a chelating agent like EDTA (Tessman, 1968). A stock solution of 1 M hydroxylamine and 0·001 M-EDTA, adjusted to pH 6 with NaOH, can be kept frozen at −20° and used at a final concentration of 0·4 M, again at pH 6 with 0·001 M-EDTA. The reaction is stopped by diluting 1/100 into broth + 0·001 M-EDTA: without chelation, this would rapidly inactivate the organisms.

ISOLATION OF MUTANTS

Some classes can be isolated directly, as in selecting antibiotic-resistant mutants by plating sensitive organisms on antibiotic agar. *Direct selection* of this kind is simple but *indirect selection,* which shows that mutants can be isolated without ever coming into contact with the selective agent, is equally possible either by replica-plating on solid medium (Lederberg &

Lederberg, 1952) or by progressive enrichment in broth (Cavalli-Sforza & Lederberg, 1956). When the mutant character does not confer a selective advantage over the wild-type, the mutagenized culture must be plated on a non-selective medium and some form of indicator system introduced to distinguish mutant from non-mutant colonies. Conventional indicator media (p. 61) are often used in this way to isolate clones unable to ferment a sugar incorporated in the medium.

Conditional lethal mutants

The value of this type of mutation is that it permits the genetic analysis of functions vital to cell survival. An absolute defect in such a function is necessarily lethal and no mutants survive to be analysed; but conditional lethal mutants are, by definition, expressed only under certain 'non-permissive' conditions. Such mutants can therefore be isolated and maintained in the appropriate 'permissive' environment where function, although often somewhat impaired, is adequate for survival. Mutations of this class occur throughout the genome and comprise *temperature-sensitive* mutations, expressed at temperatures higher than normal, and *nonsense* (amber and ochre) mutations, expressed in those hosts that lack specific 'suppressor' genes.

Temperature-sensitive mutants are readily isolated by plating a mutagenized wild type culture to obtain discrete colonies, incubating at the low permissive temperature (usually 30°), and then replica-plating to complete medium incubated at the higher non-permissive temperature (usually 42°). Mutant colonies are recognized by their failure to grow on the replica. The nature of the temperature-sensitive mutation has then to be determined by further tests as, not surprisingly, a mutational defect in any one of a large number of functions is sufficient to prevent growth.

Since the effects of *nonsense mutations* are more or less nullified in a permissive strain, they are 'suppressible'. The corresponding suppressor loci of a permissive (i.e. suppressing) strain are designated su^{+} and those of a non-permissive strain, *su*, with different loci indicated by additional symbols, like su_{B}^{+} (Brenner & Beckwith, 1965) or *Su-1* (Garen, Garen & Wilhelm, 1965). At the molecular level, a suppressible nonsense mutation is now known to cause loss of function in an *su* host by prematurely interrupting synthesis of the corresponding peptide chain which is therefore incomplete and non-functional. In an su^{+} host, on the other hand, the nonsense codon is not read as chain-terminating but as specifying an amino acid in the same way as other codons. A complete peptide with some degree of function is then synthesized, albeit with a novel amino acid at the point corresponding to the mutation (see Gorini & Beckwith, 1966; Garen, 1968). Any set of mutants prepared in an *su* strain will include some nonsense mutants which can be identified by the restoration of function that follows introduction of the mutated region into an su^+ host. Thus, of 220 phosphatase-negative (pho^-) mutations in the *su* strain, HfrH, 15 behaved as pho^+ in the su^+ strain, W1 (Garen & Siddiqi, 1962).

Growth factors

Auxotophic mutants with a new requirement for an amino acid or nucleic acid base are fairly readily isolated: vitamin-requiring mutants present more difficulty because only trace amounts of most vitamins are needed for growth (p. 42; thiamine is an exception) and it then becomes harder to eliminate the unwanted growth factor from culture media. Selective methods are usually needed. Specific methods are available for thymine-requiring mutants and for mutants constitutive for the synthesis of some amino acids.

Mutagenized cultures often contain as many as 5% of auxotrophic mutants of various kinds and no selective method is then needed, for sufficient mutant clones will be found simply by replica-plating to the appropriate defined medium. Particular phenotypes can be isolated more readily by employing either delayed or marginal enrichment of the culture medium which enables mutant colonies to be identified by their smaller size. Selection becomes necessary when mutants are rare (e.g. those arising spontaneously) or when a particular auxotroph is needed. A general selective method is provided by *penicillin-selection*, which kills non-mutant cells (see Lederberg, 1950), while specific selective methods are available for certain classes of mutants.

Enrichment methods. In *delayed* enrichment, about 200 mutagenized cells are added to an overlay of 2·5 ml. defined medium which is poured on a plate of the same medium, supplemented as necessary. Thus, to isolate mutants requiring a given amino acid, the medium should contain all other amino acids: acid-hydrolysed casein (p. 45) is suitable for tryptophan-requiring mutants. When the first overlay has set, a second overlay of 5 ml. is poured, allowed to set, and the plate incubated. Cells with unwanted phenotypes develop into colonies whereas the required mutants do not. Ultimately, a third overlay containing the required growth factor (e.g. 50 μg. amino acid/ml.) is poured on the plate which is returned to the incubator and examined at intervals for new colonies. In *marginal* enrichment, the mutagenized population is plated for discrete colonies, using medium containing sufficient growth factor for mutants to form microcolonies but too little to allow normal colonies to develop. Mutants should thus be recognizable by their smaller size but the usual difficulty is to determine the optimal concentration of growth factor.

Penicillin selection. Penicillin kills growing but not non-growing cells. Hence, if penicillin is added to a mixture of wild type and mutant cells incubated in medium unsupplemented with growth factors, only the wild type is able to grow and, in so doing, is killed.

The original technique has been improved since its introduction: (1) the culture is starved in glucose-buffer for 2–3 hr. before adding penicillin, to avoid growth of mutants on stored growth factors (Davis, 1950); (2) to encourage killing, the cells are brought into exponential growth in minimal medium before adding penicillin and exposure is limited to 90 min. to lessen the risk of mutants growing on the products of lysed wild-type cells (Gorini & Kaufman, 1959); (3) the whole selection can be done in agar so that each mutant clone is independent (Adelberg & Myers, 1953); (4) repetition of the method gives demonstrably greater selection (Lubin, 1962).

A culture growing exponentially in glucose-salts medium is exposed to that dose of u.v. radiation causing 99·95 % killing. 1 ml. irradiated suspension containing 10^7 viable cells initially is poured in an overlay of 7 ml. unsupplemented minimal agar on a base of 5 ml. of the same agar. When set, the inoculated layer is in turn covered with 7 ml. unsupplemented overlay. The plate is next incubated at 37° for 7 hr. to allow segregation of auxotrophs. A fourth overlay of 3 ml. agar containing 1500 u. penicillin/ml. is poured on the plate which is incubated at 4° for 12–24 hr. to let the penicillin diffuse through the agar (its final concentration is *ca.* 200 u./ml.). Next, the plate is transferred to 37° for 24 hr. to allow non-exacting clones to grow and be killed by the penicillin. Penicillin is then inactivated by adding penicillinase in a fifth overlay and the plate returned to 37° for 48 hr. A number of colonies of non-exacting organisms appear and their position is marked on the back of the plate. Lastly, an overlay of nutrient agar is poured to allow the development of auxotrophic colonies during further incubation at 37°, which are then subcultured for identification. The timing of the various steps considerably affects the efficiency of selection (for details, see Adelberg & Myers, 1953) and one would suspect that the details of the technique might well need checking for new strains.

The principles of penicillin selection are obviously applicable whenever a bactericidal agent is available that specifically kills growing cells. Thus,

8-azaguanine (Wachsman & Mangalo, 1962), ^3H of high specific activity (Lubin, 1959) and glycine (Liu & Takahashi, 1964) have been used with vegetative organisms. Heat successfully selects mutants with requirements for germination from spore suspensions incubated in unsupplemented medium (Iyer, 1960). Auxotrophic mutants are readily derived from strains initially requiring thymine or diamopimelic acid by incubating in medium lacking either of these growth factors since their absence leads to death of cells able to divide (Bauman & Davis, 1957).

Thymine-requiring (*thy*) *mutants* are selected by trimethoprim (2,4-diamine-5-(3′4′5′-trimethoxy)-benzyl pyrimidine), an inhibitor of dihydrofolate reductase. If defined medium containing 50 μg. thymine and 5–10 μg. trimethoprim is inoculated with a *thy*$^+$ strain at an initial concentration of 10^7/ml., it is evident that the turbidity of the culture subsequently increases very slowly. After 2–3 days, up to 100% of the viable cells require more than 50 μg. thymine/ml. for growth and are also resistant to as much as 100 μg. trimethoprim/ml. agar (Stacey & Simson, 1965). These mutants fail to form thymidylate synthetase, and secondary mutants able to grow on 2 μg. thymine/ml. occur readily which in addition are defective in catabolism of deoxynucleosides (Beacham *et al.* 1968). Trimethoprim offers greater selection than aminopterin, the first selective agent to be described (Okada, Homma & Sonahara, 1962), which has nevertheless been used successfully in agar (Caster, 1967).

Mutants *constitutive* for the synthesis of biosynthetic enzymes can often be selected on medium containing an analogue, since many analogue-resistant strains owe their resistance to overproduction of the growth factor in question (Adelberg, 1958). Thus, mutants constitutive for the synthesis of arginine and histidine are obtained by selecting for resistance to canavanine and to 1,2,4-triazole-3-alanine, respectively (Maas, 1961; Roth, Antón & Hartman, 1966).

Sugars

Fermenting mutants are usually easily isolated by plating on indicator medium containing 1–2%, w/v, of sugar. Sectors of fermenting growth are sometimes seen after 18–24 hr. incubation, particularly if plating is done immediately after mutagenesis, but the more usual appearance with non-mutagenized cultures is of hemispherical raised areas ('papillae': see Braun, 1965, p. 197) which appear on the surface of non-fermenting colonies after 2–3 days' incubation when the wild-type has stopped growing. By this time, the indicator dye is often bleached but papillae have an unmistakable shape. With some organisms like *E. coli mutabile* or *Shigella sonnei*, acquisition of fermentative ability depends on the appearance of an active permease but in other cases results from mutation in the structural gene for the hydrolytic enzyme.

Non-fermenting mutants are isolated almost equally readily, by plating mutagenized populations on TTC agar (p. 65) on which the mutants form opaque red colonies or sectors in the midst of translucent white non-fermenting growth. Both α- and β-galactosides use the same permease (Pardee, 1957), so a lactose non-fermenting mutant which simultaneously becomes melibiose non-fermenting, has lost its permease.

Constitutive sugar-fermenting mutants can be identified by the fact that colonies of the inducible wild-type will not contain the hydrolytic enzyme in the absence of inducer. Thus, colonies of the mutagenized wild-type population on weakly-buffered nutrient agar are incubated for 2 days and then are first sprayed with an aqueous solution containing 10 %, w/v, sugar and 0·5 %, w/v, chloramphenicol. The chloramphenicol prevents induction of the wild-type: whereas constitutive clones already possess the enzyme and will hydrolyse the sugar. After 30 min. incubation at 37°, the plates are sprayed with 4 %, w/v, TTC in 1 M-phosphate buffer, pH 7, whereupon fermenting colonies (i.e. constitutive clones) turn red (Lin, Lerner & Jorgensen, 1962). Evidently, the fermenting colonies do not become sufficiently acid to inhibit reduction of the TTC, with the result that the colour changes are the reverse of those seen on TTC-fermentation agar.

For an alternative method of selecting constitutive mutants, see Cohen-Bazire & Jolit (1953).

Other enzymes

Phosphatase. The alkaline phosphatase of *E. coli* is repressed by excess phosphate ($\geqslant 6{\cdot}5 \times 10^{-4}$ M) and becomes induced when the concentration of inorganic phosphate falls to low levels (Echols *et al.* 1961). Phosphatase-negative (*pho*$^-$) clones are therefore distinguished from wild-type *pho*$^+$ by plating on tris-buffered minimal medium (p. 39) supplemented with limiting phosphate (e.g. $6{\cdot}5 \times 10^{-5}$ M-KH_2PO_4), incubating for 2 days at 37° and spraying the clonies with an aqueous solution of *p*-nitrophenyl phosphate (15 mg./ml. of 1 M-tris buffer, pH 8). *Pho*$^+$ colonies turn yellow, whereas *pho*$^-$ colonies remain white and survive the spray long enough to be subcultured.

Constitutive mutants are not repressed by excess phosphate and can therefore be isolated in the following way (Torriani & Rothman, 1961). The culture is plated on tris-buffered minimal agar containing 0·01 M-glycerophosphate as sole carbon source and also excess phosphate ($6{\cdot}6 \times 10^{-4}$ M-KH_2PO_4). The inorganic phosphate represses phosphatase synthesis by the wild-type, which therefore cannot hydrolyse the glycerophosphate and fails to grow for lack of a carbon source. Constitutive clones form phosphatase and grow normally.

Penicillinase. On hydrolysis, penicillin yields penicilloic acid: penicillinase-producing colonies are therefore relatively acid and may be detected with a pH indicator. The details of the method used for staphylococci are determined by the level of penicillinase activity to be detected (Novick & Richmond, 1965) but, in principle, the organisms are first grown for 48 hr. on unbuffered complete agar containing acid-hydrolysed casein and yeast extract, pH 7·6. After drying for 2 hr. at 37° with the lids of the plates removed, the agar is flooded with 1·5 ml. of a solution of the pH indicator, *N*-phenyl-1-naphthylamine-azo-*o*-carboxybenzene (B.D.H., Ltd, 0·25 %, w/v, in *N,N*-dimethyl formamide). When dry, 1·5 ml. of a newly-prepared aqueous solution of unbuffered benzylpenicillin G is added, using 0·5 %–10 %, w/v, solutions according to the activity of the colonies. Penicillinase-producing colonies change the indicator from yellow to purple. Sectors are readily seen due to the insolubility of the indicator at low pH, and viable organisms can be recovered for as long as 6 hr. after testing.

Resistant mutants

With antibacterial agents like antibiotics, resistant mutants can usually be isolated by direct plating. In the case of a few agents, completely indifferent mutants occur, but with many others a single mutation produces only resistance to intermediate concentrations of inhibitor, and maximal resistance is reached only after several successive mutations. Gradient plates then offer a convenient means for exposing a sensitive population to a range of concentrations. Amongst the agents associated with one-step maximal resistance is streptomycin ($\geqslant$ 1000 μg./ml.). Either a large number (e.g. 5×10^{10}) sensitive cells are plated, or enrichment in liquid medium can first be used (e.g. by adding 100 ml. broth containing 2000 μg. streptomycin/ml. to 100 ml. unshaken 18 hr. culture, shaking overnight, and then plating on streptomycin agar).

Inhibitors to which resistance occurs in stepwise fashion include nalidixic acid (25–50 μg./ml.), sodium azide (0·002–0·004 M), chloramphenicol and tetracyclines (Cavalli, 1952; Reeve, 1968). Note that a 'resistant' mutant may in reality be drug-dependent (Brock, 1966; Verly *et al.* 1967).

Colicin-resistant mutants are isolated by inoculating part of a nutrient agar plate with a colicinogenic strain, incubating for 6–24 hr. to allow growth, chloroforming, and overlaying with soft agar containing the sensitive strain. After further incubation, a large area of inhibition is seen in the overlay with scattered colonies of resistant mutants. It is safest to subculture a number of resistant colonies for, just as with phage-resistant mutants, some may be highly mucoid and inconvenient to use.

For *phage-resistant* mutants, the simplest method is to inoculate an agar plate with a broth culture of the sensitive strain, taking care to spread the cells uniformly, and to add a drop of high-titre phage stock. Resistant colonies appearing in the area of lysis are purified by re-streaking. Virulent phage generally presents little difficulty although, in some systems, the colonies obtained may not consist of true resistant mutants but of phenotypically-resistant cells which continually segregate sensitive progeny on subculture (Zinder, 1958; Meynell, 1961).

With temperate phage, the bulk of the phage-resistant growth consists of lysogenized cells; these are avoided by selecting with virulent mutants like the *C* mutants of λ or P22.

Phages attaching to known receptors provide a simple means for obtaining specific classes of bacterial mutants.

Bacterial organelles

Flagella. Motile mutants are selected from non-motile populations by semi-solid medium (p. 53). The bacteria are inoculated on a small area of the plate and those which are motile swarm across the medium while the remainder are left at the site of inoculation. If anti-flagellar serum is incorporated, mutants with new flagellar antigens are obtained. Relatively high concentrations of antibody (40 times the tube agglutination titre: Edwards & Ewing, 1962) yield variants in the alternate antigenic phase or with entirely novel antigens; marginal concentrations of antibody give variants whose antigens differ only slightly from those of the parent strain (Joys & Stocker, 1966).

Non-motile mutants can be selected from many motile strains of salmonella by exposure to phage χ which adsorbs to flagella (Meynell, 1961). If only a small proportion of the cells are motile initially, many clones surviving the phage are as motile as their parent on subsequent testing, and presumably merely chanced to be phenotypically non-motile at the time the phage was added. Of the truly non-motile phage-resistant isolates, some are non-flagellated but others are 'paralysed' with abnormal flagella. These no longer adsorb the phage and may either be morphologically normal or occasionally show the usual spiral form but have only half the normal wavelength. Straight flagella also result in paralysis, but without loss of phage adsorption. In strains undergoing phase variation of their flagellar antigens, abnormal flagella are confined to only one antigenic phase, since the effect of the mutation is confined to the structural gene for that particular flagellin (see Joys, 1968).

Pili. Although these appendages were originally referred to as 'filaments' by Houwink & van Iterson, they have subsequently been named either 'fimbriae' (Duguid, Anderson & Campbell, 1966) or 'pili' (Brinton, 1965).

Two main classes exist. 'Sex pili' determined by plasmids like F and colicin factor I which are involved in bacterial conjugation and gene transfer and which also provide the receptors for donor-specific phages (p. 277). 'Common pili', the second class, form a heterogeneous group determined by chromosomal genes, and confer a variety of properties on their hosts like haemagglutinating ability (Duguid, Anderson & Cambell, 1966).

Sex pilus synthesis is nearly always subject to repression and one class of mutations therefore affects the regulatory system (p. 272). Structural mutations also occur which affect the pilus itself and may be selected by exposing a de-repressed strain to the appropriate donor-specific phage. The commonest type of phage-resistant mutant fails to form sex pili and no longer donates (Cuzin & Jacob, 1967): less often, mutant pili are synthesised which adsorb the phage used for selection but do not allow infection. Thus, the exposure of F^+ bacteria to the isometric RNA phage MS2 allows the isolation of mutants which are still sensitive to filamentous DNA phages but resistant to all RNA phages (Silverman, Mobach & Valentine, 1967).

Common pili are present on most strains of enterobacteria grown under suitable conditions, but may be lost following mutation. Non-piliated clones can be selected by the morphology of their colonies: back mutation is said not to occur (Brinton, 1959; Maccacaro & Hayes, 1961). In addition, the synthesis of common pili is greatly affected by the cultural conditions, being favoured by broth incubated in air and inhibited by growth on agar (see Duguid & Wilkinson, 1961; Brinton, 1965). It seems unclear whether this is a purely phenotypic effect or whether pilus synthesis exhibits phase variation in the same way as antigenic type, and the alternate phases are selected by different environments.

Cell wall. Numerous classes of mutation affect cell wall synthesis, as would be expected from the complex structure of the wall itself (see Guze, 1968). Some abolish the mechanical properties of the wall and these mutations are therefore lethal in ordinary culture media. Other mutations affect the synthesis of superficial wall components, like capsules, somatic antigens or phage receptors, that are not essential for cell survival.

The mechanical strength of the cell wall is dependent on its innermost 'rigid layer' and if one of its components cannot be synthesized, plasmolysis and lysis occur unless the osmotic pressure of the medium is increased, as by adding sucrose. Auxotrophic mutants of this type require, for example, lysine, glucosamine or diaminopimelic acid. To isolate mutants defective in wall synthesis, Lederberg & St Clair (1958) plated u.v.-irradiated *E. coli* in agar containing 0·3 M-sucrose, 0·008 M-Mg^{2+} and an intermediate concentration of penicillin which inhibited both bacterial and *L*-form growth, yet allowed auxotrophs to form colonies. Mutants of *Salm. typhimurium* which lack UDP-galactose-4-epimerase and cannot ferment galactose, are inhibited by galactose and, in its presence, form a defective cell wall and then lyse (see Kalckar, 1965; Lüderitz, Staub & Westphal, 1966).

Capsules and smooth somatic antigens. In general, either of these components causes colonies to appear smoother, or even mucoid, compared to a non-capsulated or 'rough' mutant. Occasionally, a capsule forms and quickly dries after overnight incubation, giving the colony a misleadingly rough appearance. Rough colonies may be noted by chance, particularly on plating out stock cultures, or mutants may first appear as rough outgrowths from the edges of smooth colonies after several days' incubation. Well-known examples occur with pneumococci (see Hayes, 1966), salmonella and other enterobacteria (see Lüderitz, Staub & Westphal, 1966; Nikaido, 1968) and also in the segregation of non-capsulated from capsulated anthrax bacilli (Meynell & Meynell, 1966). With many bacteria, the antigens may also be phage receptors, and mutants with altered structural genes can be selected by their phage resistance. Thus, capsulated, fully virulent, back-mutants of *Bacillus anthracis* were selected from non-capsulated avirulent mutants by phage able to attack non-capsulated but not capsulated cells (E. W. Meynell, 1963). Similarly, Vi-negative strains of *Salmonella typhi* are rapidly obtained with Vi phage: while salmonellas with wild-type and mutant somatic antigens can equally well be selected by exposure to 'rough' or 'smooth' phages (Burnet, 1930; Lüderitz, Staub & Westphal, 1966; Nikaido, 1968).

Phage also selects regulatory mutants. Wild-type anthrax bacilli synthesize capsules only when a critical concentration of HCO_3^- is present and are therefore non-capsulated in air. When cultures growing in air are exposed to phage, constitutive mutants are isolated which are phage-resistant because they form capsules without added HCO_3^- (Meynell & Meynell, 1966).

Smooth mutants can be selected from rough strains by their colonial morphology, with phage or, with virulent species, by animal passage. Note, however, that the success of passage depends on using inocula large enough to be sure of including one mutant in the dose (e.g. Sabin, Hennessen & Winsser, 1954).

Spores. The genetics of sporulation has so far been studied largely in *Bacillus subtilis*, the only sporing species at present open to genetic analysis. Various classes of non-sporing mutant are selected by growth on nitrate, or when citrate or histidine are the sole carbon sources, and form colonies of distinctive morphology (Jacob, Schaeffer & Wollman, 1960; Michel, Cami & Schaeffer, 1968). Temperature-sensitive mutants are also known (see Murrell, 1967). Some mutants have an absolute block: others are leaky and their cultures contain a small proportion of cells that sporulate normally (Ryter, Schaeffer & Ionesco, 1966). Microscopy separates mutants into a relatively small number of classes, corresponding broadly to the seven accepted stages in sporulation (p. 169), although the number of structural genes involved must be considerably greater, judging from the number of new functions that appear during sporulation (see Halvorson, 1965; Murrell, 1967).

Nuclei. The genetic analysis of DNA replication has yielded numerous mutants of *E. coli* and *Bacillus subtilis* with manifest nuclear abnormalities, often associated with abnormalities in cell growth as a whole. Since such mutations are lethal, they have generally been studied in temperature-sensitive mutants (Gross, Karamata & Hempstead, 1968). On incubation at the permissive temperature, growth may be visually normal, but at the higher non-permissive temperature, many of the mutants form filaments whose nuclei show characteristic arrangements. Perhaps the most extreme instances are those strains which segregate cells from which nuclei are entirely absent (Adler *et al.* 1967; Hirota *et al.* 1968).

BACTERIAL PLASMIDS

The typical plasmid can be described in various terms (see Meynell, Meynell & Datta, 1968). Genetically, it is an independent extrachromosomal entity able to ensure its own replication. Physiologically, it often enables its host to act as a genetic donor and to form the surface structures necessary for conjugation and gene transfer which, at the same time, are the sites of adsorption for donor specific phages. Physically, it is detected as a short length of covalently-linked circular DNA whose size is about 1–5 % of the chromosome of *Escherichia coli* (Bazaral & Helinski, 1968; Falkow, Haapala & Silver, 1969).

An unsuspected plasmid may therefore show itself in many ways: by the simultaneous acquisition or loss of a number of unrelated functions like resistance to several antibiotics; by bringing about the transfer of otherwise non-transmissible genes, including those of the host chromosome; or as a new fraction of the total bacterial DNA. However, exceptions to the type are far from uncommon. Some plasmids, notably those of *Staphylococcus aureus* (see Richmond, 1968; Novick, 1969) and some colicin (Col) factors (Smith, Ozeki & Stocker, 1963) and drug-resistance (R) factors (Watanabe, 1963; Anderson, 1968) found in *E. coli* and other enterobacteria, are extrachromosomal yet non-transmissible on their own. Others, like F, may cease to be extrachromosomal by recombining ('integrating') with the host genome which may then become transmissible. Plasmids able to integrate in this way were originally classed as 'episomes' (Jacob, Schaeffer & Wollman, 1960), but integration is now known to be also governed by the host and, for this reason, the wider term 'plasmid' is generally preferred.

Plasmids are usually grouped by their most prominent function, like colicinogeny or drug-resistance, as if this divided them into natural classes. This is no more than a historical accident for, not only may different types of function like streptomycin-resistance and colicinogeny be determined by a single plasmid but it is now known that apparently different plasmids may possess the same set of genes (the 'sex factor') responsible for their transfer. Thus, the same type of sex factor is present in F, ColV, ColB, and *fi*$^{+}$ (F-like) R factors; while another type is present in ColI, ColE1a and *fi*$^{-}$ R factors. The two classes have therefore been designated 'F-like' and 'I-like' after their first members to be identified (see Meynell, Meynell & Datta, 1968). In the sections that follow, the general properties of plasmids have therefore been discussed first, leaving the particular characteristics of Col factors and R factors to the end, as there is evidently little point in seeking fixed categories. Rather, it is more realistic to regard plasmids as highly labile genetic elements, losing and gaining genes by spontaneous segregation and recombination with each other and with the chromosome of their host. The plasmid population of a culture may therefore be highly heterogeneous in some instances, with the plasmid forming a single linkage group in some cells while, in others, it may have segregated to give two or more independent genomes. More and more bacterial characters have been found to be plasmid-determined, and two examples falling outside the better-known categories are somatic antigen K88 (Ørskov & Ørskov, 1966) and enterotoxin formation in *E. coli* (Smith & Halls, 1968). At present, there seems to be no bar to the functions that may be determined in this way.

Transmissibility

The genetic criterion of a plasmid as an extrachromosomal genome is that it behaves independently of the host chromosome in genetic experiments. One of the simplest ways in which this can be demonstrated is by showing that the plasmid is transmitted when the chromosome is not. Even when chromosomal transfer occurs, the independence of a plasmid will be evident. Thus, F, the sex factor originally discovered in *E. coli*, is transfered from an F^+ host about 10^5 times more frequently than any chromosomal marker. If plasmid and chromosomal transfer occur equally frequently, as with an Hfr strain carrying ColI, then transfer of the plasmid is independent of which chromosomal region is first donated (Clowes, 1964). Absence of linkage may indeed provide the first indication that a plasmid is present in the test strain.

Transmission by conjugation is simple to demonstrate in principle although the details depend on the system. The test strain is mixed with a differentially-marked recipient, left for a short time to allow conjugation and gene transfer, and the recipient then isolated by plating on selective medium. Thus, a methionine-requiring strain carrying F^-*lac*$^+$, a plasmid comprising F linked to genes conferring the ability to hydrolyse lactose, could be mixed with an F^-*lac*- tryptophan-requiring recipient for 20 min. at 37° and *lac*$^+$ recipients isolated on salts agar containing lactose and tryptophan but lacking methionine. The most important factor determining the frequency of transfer is the proportion of donor cells capable of conjugation (i.e. on the extent to which the sex factor is de-repressed as discussed later) but, with any pair of strains, non-genetic factors are also relevant. In general, donor ability is greatest with cultures growing exponentially with gentle aeration; the parental cultures must be sufficiently dense ($\geqslant 10^8$ cells/ml.) for cell contacts to occur rapidly after mixing (even so, conjugation may not occur readily or be stable unless the parents are mated on a membrane filter: Matney & Achenbach, 1962; Sanderson, 1967); pairing is encouraged if at least one parent is motile and forms common pili; and to estimate donor ability accurately, an excess of recipient must be used, by mixing donor and recipient in the ratio of 1/10 or less, and plasmid transfer should be measured after no more than 20 min. or some recipient cells will have had time to become donors (this is largely avoided by using starved or stationary phase recipients). Plating immediately after mating is not possible when selecting for transfer of phage-resistance because cells acquiring the resistance genes may take several generations to develop the resistant phenotype ('phenotypic lag': Hayes, 1957). The mating mixture should then be diluted 1/1000 in broth, agitated vigorously to separate mating pairs and incubated at 37° to allow expression of resistance before plating on selective medium. However, with transmissible drug-resistance the lag is short, less than one generation (Watanabe, 1963).

The sex factors of plasmids like F and ColV are not repressed, so that up to 100% of their host cells may donate. However, with all other wild type plasmids, the sex factor is subject to repression, so that although the plasmid is present throughout the population, it is spontaneously de-repressed in only 0·01–1% of cells which are therefore the only ones able to conjugate at a given time. De-repression of conjugating ability occurs in two circumstances. First, immediately after transfer of the plasmid to a new host, until repression supervenes. In the interval, the new host transmits with up to 100% efficiency to other cells which are themselves converted to donors and so transmit in their turn. Transfer of a repressible plasmid to a new culture is therefore characterized by 'high frequency transfer' (an 'HFT' system: Stocker, Smith & Ozeki, 1963), leading to its epidemic spread through the culture. HFT donor cultures are readily prepared by 5×10^6 cells of the plasmid-carrying donor strain with 10^8 cells of an 'intermediate', plasmid-free, strain in 10 ml. broth, incubating without shaking overnight and then for a further 2 hr. after dilution 1/20 in fresh broth. With ColI, 20–80% of the intermediate cells will generally be able to donate their plasmid. The proportion of donors is measured in the same way as chromosomal transfer. The HFT mixture and the recipient are first mixed in the ratio of 1/20 and left for 20 min., which is sufficient to allow transfer of the plasmid, as distinct from the 90–140 min. required for complete chromosomal transfer in *E. coli* (Taylor & Trotter, 1967) or *Salm. typhimurium* (Sanderson, 1967). Mating is ended by diluting the mixture 1/100 and by then agitating vigorously with a Blendor or on a cam to separate pairs of conjugating cells before plating on medium selective for the recipient.

The second method for de-repressing conjugating ability is to isolate plasmids with de-repressed mutant sex factors. Two techniques have been described; the first using replica-plating for plasmids possessing a selective marker like drug-resistance (Meynell & Datta, 1967), and the second based on successive matings in broth for other plasmids (Edwards & Meynell, 1968). In the first method, a mutagenized population of an R^+*lac*$^-$ strain is plated on nutrient agar, incubated until microcolonies form, and then replica-plated to lawns of an R^-*lac*$^+$ recipient spread on minimal agar containing the appropriate drug and lactose as sole carbon source. Provided the colonies on the master plate are not allowed to become large enough to contain some spontaneously de-repressed wild-type sex factors, the only master colonies giving growth on the recipient will be those containing de-repressed mutants. In practice, each master plate should be replicated to two recipient plates and only master colonies represented on both recipients should be considered. When no selective marker is available, although the presence of the plasmid is recognizable in the way a Col factor is detected by colicin production, repeated mating in broth provides a simple means for isolating de-repressed mutants. An established culture will contain a proportion p of wild-type plasmids that are spontaneously de-repressed and transmissible, and also a proportion m of de-repressed mutants. The ratio of mutant to wild-type plasmids transferred to a second host is therefore m/p. This host is then subcultured at low density ($< 5 \times 10^6$ cells/ml.), to prevent cell-to-cell contacts and epidemic spread of the wild-type plasmids it received, which consequently become repressed during the succeeding 14–20 generations of growth. Ultimately, therefore, only a proportion p of the wild-type plasmids is again transmissible, although the mutant factors will all have remained de-repressed. When the second host is then mated in its turn, the ratio of mutant to wild-type plasmids received by the third host is $m/p.p$ or m/p^2. Thus, after each cycle of mating, repression and mating, the proportion of wild-type plasmids in the total plasmid population decreases by p. Plausible values for m and p are, respectively, 10^{-4} after mutagenization, and 10^{-3}, so that in the third host the ratio of mutant to wild-type plasmids should be $10^{-4}/10^{-6}$ or 100/1. In practice, selection is not quite as efficient as this but de-repressed sex factors of colicin factors like I, E1a or B are usually present in about 10% of the Col^+ clones of the third host.

Some sex factors carry no known distinguishing characters save their sex factor ability but their de-repressed mutants could be presumably selected by following the transfer of an otherwise non-transmissible plasmid like ColE2.

The presence of a de-repressed sex factor is confirmed in several ways: (1) the high donor ability of its hosts in straightforward conjugation experiments; (2) the formation of characteristic sex pili on a corresponding high proportion of cells, as seen by electron microscopy; (3) visible lysis by donor-specific phage (p. 277); (4) by colonies being markedly smaller than those of strains carrying the wild-type plasmid; and (5) by increased activity of other functions determined by the plasmid. The last two properties therefore result, with Col factors, in the characteristic appearance of an unusually small colony surrounded by a disproportionately large inhibition zone.

Non-transmissible plasmids fall, genetically, into two classes. The first, and probably the most common, consists of extra-chromosomal genomes which are as stably inherited by their bacterial hosts as a transmissible plasmid, yet do not enable their host to conjugate because they do not contain a functional sex factor. The outstanding group of non-transmissible plasmids is that found in *Staphylococcus aureus* which determines resistance to antibiotics and heavy metals (see Richmond, 1968; Novick, 1969). Conjugation is unknown in cocci and transfer occurs by transduction.

Other non-transmissible plasmids include ColE2 and some R factors, both naturally-occurring and obtained by transduction with phage P22. Although non-transmissible when present alone in the cell, they are often co-transferred when a transmissible plasmid is present, like F or ColI (Smith, Ozeki & Stocker, 1963; Watanabe, 1963, 1969; Meynell, Meynell & Datta, 1968; Anderson, 1968; Meynell & Datta, 1969). The usual method for demonstrating transmission is to set up an HFT system with the transmissible and non-transmissible plasmids (e.g. ColI and ColE2) in the donor and intermediate strains, respectively: the transmissible plasmid enters the intermediate from which both it and the non-transmissible plasmid are passed on to the recipient (Ozeki, 1968).

The second class of non-transmissible plasmid arises when plasmid genes become integrated in the host genome without their sex factor but with the genome of the transducing phage following transduction and lysogenization. Thus, drug-resistance determinants become integrated at the attachment site of phage P22, from which they may be released during vegetative phage growth to yield high-frequency transducing lysates (Dubnau & Stocker, 1964).

Transduction. Both transmissible and non-transmissible plasmids are transferred by transduction. Phage P1 of *E. coli* usually transfers a transmissible plasmid entire, whereas with the salmonella phages P22, ϵ15 and ϵ34, the sex factor often fails to be transferred and a non-transmissible plasmid results, though how frequently this occurs depends on the plasmid in question (Watanabe, Furuse & Sakaizumi, 1968).

In general, a stock of transducing phage is prepared on the plasmid-carrying donor strain and, after any surviving bacteria have been killed or removed, is used to infect the recipient which is then plated on selective medium to isolate transductants. The phage can usually be grown on the donor simply by plating in overlays so that it results from vegetative growth.

The choice of phage is determined by the necessity for generalized transduction. Since phages like P1 for *E. coli* and P22 for *Salm. typhimurium* are temperate, the transductants will usually be lysogenic unless precautions are taken. Either the wild-type phage is used at a multiplicity of infection small enough to ensure that the majority of cells are infected by only one particle each (p. 178): since the transducing particles are defective, non-lysogenic transductants are obtained if unadsorbed phage and phage released after infection is neutralized by adding anti-phage serum to the culture. Alternatively, a high multiplicity of infection can be used, which incidentally gives a higher frequency of transductants, and lysogeny avoided by using non-lysogenizing mutants of the transducing phage. Examples are P1*kc* and P22-L4 which is temperate but does not lysogenize (Smith & Levine, 1967).

Phage infection can be performed under any conditions giving adequate adsorption. P1 requires 0·002 M-Ca^{2+} (which incidentally allows reinfection to be prevented by adding sodium citrate to 1 %, w/v) but P22 adsorbs well in conventional broth. Either stationary phase or growing cells can be used: lysogenic recipients give slightly lower transduction frequencies. Transduced R factors are selected by plating the recipient on agar containing the appropriate antibiotic, while transduced Col factors can be selected by plating in double overlays (p. 281). Transmissible plasmids like F that are not linked to an easily detected marker cannot be selected but, if the recipient is subcultured sufficiently after transduction, they will undergo epidemic spread through the entire population which can then be tested for some property like sensitivity to phage or ability to act as genetic donors. Resistance genes may be integrated in the bacterial chromosome by lysogenization with defective phage P22 (Dubnau & Stocker, 1964) and transducing phage must therefore be prepared by induction of the lysogenic strain or by harvesting phage spontaneously released during growth (see Campbell, 1968).

Barriers to transfer

The most general barrier, which applies equally to conjugation and to transduction, is *restriction* (see Arber & Linn, 1969). The donated DNA becomes genetically inactive and in only a minority of recipient cells are the donated genes expressed. In some systems, the DNA is known to be degraded. The restricting genes may be chromosomal or they may be carried by a plasmid like a prophage or R factor already present in the recipient. It is common, for instance, when testing R^+ strains with DNA donor-specific phage to find that the phage fails to plate with full efficiency (RNA phages are not subject to restriction).

The effects of restriction may be overcome in several ways: (1) using a segregant of the recipient which lacks the restricting plasmid; (2) isolating non-restricting mutants from the minority of recipient cells failing to inactivate the donated genes (Glover *et al.* 1963; Okada & Watanabe, 1968); (3) abolishing the expression of restricting genes non-specifically, as by starvation (Schell & Glover, 1966); or (4) using a different strain as donor which does not 'modify' the transferred DNA and render it susceptible to the restricting system of the recipient.

The role of restriction in limiting genetic recombination between species is not yet clear, but the third and fourth of the methods mentioned above might be worth trying in such crosses. Thus, Drabble & Stocker (1968) failed to transfer R factors from *Shigella* directly to *Salmonella* but succeeded after first introducing them into *Escherichia.*

Transfer may apparently fail, even with a non-restricting recipient, if it already contains a plasmid related to that being transferred. Two related plasmids such as two F factors do not readily co-exist stably and one fails to replicate and segregates away at cell division, the phenomenon of *superinfection immunity*. No means are at present known by which this can be overcome: simultaneous selection for both plasmids yields either unstable strains or an altered or recombinant plasmid (see Meynell & Datta, 1969). Another phenomenon is also encountered in conjugation experiments between parental strains carrying similar plasmids, namely, *exclu-*

sion. Here the proportion of cells acquiring the donor's plasmid is markedly less than with the same strain lacking the plasmid. Exclusion is overcome by starvation (e.g. by shaking in glucose-buffer for 4 hr. at 37°): the recipient then accepts normally and, if it carries the sex factors, F or I, is said to be an 'F^-' or 'I^- phenocopy'.

Donor-specific phages

De-repressed donor cells form specialized filaments ('sex pili') which may provide the channel by which genes pass to the recipient (see Brinton, 1965). These pili also act as receptors for certain phages which therefore specifically attack cells capable of conjugation and gene transfer. Just as the sex pili determined by apparently unrelated plasmids fall into only two major classes typified by those of F and ColI, so there are only two classes of donor-specific phage: F phages, which are either isometric and RNA-containing (Watanabe & August, 1968) or filamentous and DNA-containing (Hoffmann-Berling, Kaerner & Knippers, 1966; Marvin and Hohn, 1969), and I phages, of which only filamentous DNA phages are so far known (Meynell & Lawn, 1968). F and I phages exhibit distinct host ranges inasmuch as F phages attack only cells with F-like sex factors, and I phages attack only cells with I-like sex factors. This is easily shown with F or any other de-repressed sex factor by the visible lysis produced by a drop of phage stock placed on a lawn of the test strain. With ambiguous strains, clearer results are obtained by pouring an overlay inoculated with the test strain, allowing it to set, and placing drops of high titre phage stock on the surface. Failing this, plates evenly inoculated by flooding (not spreading with a rod) can be used. Clearings are usually most distinct after incubating for 4 hr. at 37° and overnight at 22°.

With repressed sex factors, no lysis is seen, whichever method is used. The reason is that lysis requires a high proportion of susceptible cells—in this instance, of cells forming sex pili—and when repression occurs, more than 99 % fail to form pili and are therefore phage-resistant. The only way to detect the minute proportion of spontaneously de-repressed cells is to infect a growing broth culture with donor-specific phage and to look for a subsequent increase in phage titre. This may only be small, so that as much unadsorbed phage as possible must be removed by antibody and washing on a membrane filter (Meynell & Datta, 1966).

Segregation

Since an extrachromosomal plasmid replicates independently of the host genome, it may sometimes fail to be inherited at cell division by one of two daughter cells. The result is that a number of characters are lost simultaneously and irreversibly, often at a rate far higher than could be due to mutation. With F-*lac*, segregation occurs in *E. coli* in perhaps 0·1 % of divisions, whereas in *Proteus*, segregation initially occurs far more frequently, so that a majority of colonies show lac^- sectors on lactose indicator medium. Analogous appearances are seen with penicillinase-forming staphylococci on penicillin indicator plates (Novick & Richmond, 1965). Incomplete segregation is also common. An F′ factor with the linked genes lac^+ ade^+ pro^+ readily yields segregants lacking one or more of its associated determinants until one ends with what appears as an F^+ strain. Similarly, R factors frequently lose one or more determinants, particularly in *Salm. typhimurium* (see Watanabe, 1963). The frequent occurrence of variants is therefore strongly suggestive of a plasmid-determined character.

Segregation can be encouraged by a variety of treatments which depend on replication of the autonomous plasmid being inhibited to a greater degree than that of the bacterial chromosome. For this reason, no segregation occurs when plasmid and chromosome recombine to form one structure, in the way that an Hfr strain is formed by integration of an autonomous F factor. All these methods are presumably best applied to small inocula (e.g. 10^3 cells/ml.), to allow sufficient generations of growth for segregants to accumulate. F, ColV and possibly some R factors are 'cured' by growing their hosts in broth at pH 7·6 containing 20–50 μg. acridine orange/ml. (Hirota, 1960). The related compound, ethidium bromide (see Waring, 1968), is also effective at 25 μg./ml. (Bouanchaud, Scavizzi & Chabbert, 1968). Alternatively, 20 mM Co^{2+} (or Ni^{2+}) will eliminate F after growing F^+ cells in cobalt-broth or by selecting cobalt-resistant clones on cobalt-agar (Hirota, 1956). The mutagens, ethyl methane sulphonate (0·01–0·02 M) and nitrosoguanidine (50–100 μg./ml.) also eliminate F (Willetts, 1967). Enforced segregation of this kind should be distinguished from selection for plasmid-free members of a culture. Thus, R^- segregants of a strain carrying an R factor determining resistance to a bacteriostatic agent like tetracycline may be isolated by penicillin selection in the presence of tetracycline: the latter prevents growth of the R^- cells which therefore survive the penicillin, while the R^+ grow and are killed. Cephalosporin can be used in place of penicillin with R factors determining the synthesis of penicillinase. Another selective method is growth overnight in shaken buffered glucose broth containing 10%, w/v, sodium dodecyl sulphate (Tomoeda *et al.* 1968). F^- segregants and clones with defective F factors are selected from F^+ cultures by passage through semi-solid motility medium (p. 54; Skaar, Richter & Lederberg, 1957).

With all these methods, it is important to check that the supposed segregants have indeed lost the plasmid completely and are not carrying it in a non-functional form. This can be done by testing for (1) loss of phage restriction when this is determined by the plasmid, as with phage T3 and F (Schell *et al.* 1963; the female-specific phages that specifically lyse F^- cultures are phages restricted by F); phage BF23 and ColIb (Strobel & Nomura, 1966); or the sets of phages restricted by R factors described by Anderson (1968) and Bannister & Glover (1968); (2) absence of exclusion and superinfection immunity when the 'cured' strain is tested as recipient for the lost factor; and (3) in the case of transmissible plasmids, failure to propagate the appropriate donor-specific phage.

Curing lysogeny. The new character may be determined, not by a plasmid of the sort considered up to here, but by a typical temperate phage and then appears as an example of 'lysogenic conversion'. A lysogenic strain cannot be cured of its prophage by any of the previous techniques. One method is to expose the strain to an agent, often u.v. light, which induces vegetative growth of the phage. Lysogenic cells are killed more readily than non-lysogenic which are therefore selected. Non-lysogenic clones will be susceptible to the phage released following induction, and consequently their colonies show sectors of lysis and appear 'nibbled'. This is usually more evident with enterobacteria after plating on EMB agar containing 0·1%, w/v, glucose, since the nibbled areas differ in colour (Zinder, 1958). An alternative method for obtaining cured strains is to prevent spontaneous non-lysogenic segregants from being re-infected by free phage by preventing phage adsorption, either by including antiphage serum in the broth or by removing cations required for adsorption with a chelating agent like citrate. Some lysogenic strains are cured as a result of superinfection by a related, but heteroimmune, phage (see Thomas, 1968).

Colicin factors

Many bacteria produce antibiotic substances ('bacteriocins') active against strains of the same species, but, for historical reasons, the colicins formed by many strains of *E. coli* and other members of the *Enterobacteriaceae* have been by far the most intensively studied. Studies of their mode of action show that they comprise a highly heterogeneous group and, indeed, some are tails of defective phages (Nomura, 1967). However, from the point of view of technique, different colicin factors can be handled in very much the same way (see Ozeki, 1968).

The factors now in general use were isolated originally from naturally-occurring enterobacteria and have subsequently been classified by a variety of criteria (see Fredericq, 1965). First, an unknown colicin is tested against a set of standard *indicator strains*, each previously selected for resistance to standard-type colicins just as *str-r* mutants are selected for resistance to streptomycin. This places the unknown colicin in one of several standard classes designated by capital letters like I, V or E. Conversely, a standard universally-sensitive indicator like *E. coli* K12 or *E. coli* ϕ is exposed to the unknown colicin, *resistant mutants* isolated, and their resistance to standard-type colicins determined. Clearly, the two sets of tests should be consistent; and should, ideally, give a clear-cut distinction between the different classes. However, while some mutants are resistant only to one class, others show partial or complete resistance to several. Thus, resistance to colicin I is often accompanied by resistance to colicins V and B. This means that several tests, using different indicator strains, may be required to identify the colicin. Second, the colicin factor is transferred to a standard indicator which thereupon itself produces colicin and becomes *immune* to the colicin concerned. Immunity conferred by colicinogeny is more specific than resistance acquired by mutation, and if the Col^+ strain is tested for sensitivity to other colicins of the same class, further subdivisions are generally apparent. Thus, I colicins are revealed as being of two types, for bacterial mutants selected for resistance to a given colicin I are resistant to all other colicins of the I class, whereas strains carrying a given ColI factor are resistant to only some of them. To avoid confusion, Col factors are now designated by the strain in which they were discovered, e.g. ColIb–P9 from *Shigella sonnei* P9 and ColIa–CA53 from *E. coli* CA53. In practice, the identification of a colicin is often complicated by the presence of multiple Col factors in the strain, and it should be appreciated that other properties of the colicin, like morphology of inhibition zones, heat-stability and susceptibility to inactivation by trypsin, are also of taxonomic value. Electrophoretic separation may be valuable in analysing mixtures of colicins (Maré, Coetzee & de Klerk, 1964).

Colicin production is easily detected. The test strain is grown on agar as a spot or as individual colonies until visible growth is seen (e.g. 6–18 hr. at 37°); the organisms killed by inverting the plate over a watch glass holding a few ml. chloroform for 20–30 min. at room temperature (plastic lids will dissolve); and the chloroform allowed to evaporate by leaving the plate open for a further 30 min. 1 ml. of an overnight broth culture of the colicin-sensitive indicator strain is added to 25 ml. molten soft agar (broth solidified with 0·35 %, w/v, agar) held in a 48° bath and drops of the mixture immediately distributed over the chloroformed plate. When the overlay has set, the plate is incubated at 37°. Inhibition zones are usually visible in 4–6 hr. After overnight incubation, colonies of colicin-resistant mutants will often be seen in the overlay. The proportion of colicinogenic cells in a culture is counted by the same technique. A colony count is made by spreading (p. 26) and the colonies chloroformed as before. When only a small proportion of Col^+ colonies is present, they may be missed unless the indicator culture is fresh and is plated immediately after addition to soft agar. *Double overlays* allow the isolation of Col^+ clones: the colony count is made by the overlay method (p. 28) but, as soon as the overlay is set, it is sealed in by adding a second overlay of 2·5 ml. uninoculated agar followed by chloroform for 1 min. to kill any organisms carried to the surface. After incubation for 18–48 hr., depending on the size of zone produced by the particular colicin, the plates are overlaid with the indicator in soft agar as usual, but without first being chloroformed. The test colonies therefore remain viable, and, after incubation, any that are Col^+ are picked through the overlay for subculture. Salmonellas and shigellas are easily separated from the coliform used as indicator by subculturing to a selective medium like SS agar (e.g. Difco or Oxoid). In these tests, the indicator may be inhibited not by a colicin, but by temperate phage released from a lysogenic test strain. This is not usually a source of confusion: inhibition zones due to colicin have a uniform edge, whereas those due to phage usually show a cloud of small plaques produced by phage particles dislodged from the colonies while applying the indicator. Some colicins are inactivated by trypsin, and zones are then reduced or abolished by including 200 μg. trypsin/ml. of overlay. The most straightforward way to exclude phage is to examine a fragment of the inhibition zone by electron microscopy using a negative-contrast agent, which usually shows not only the phage but also its mode of attachment to the bacteria.

The antibiotic activity of Col^+ cultures is assayed in two ways. Either the number of colicin-producing cells is counted by the 'lacuna' technique of Ozeki, Stocker & de Margerie (1959) or the amount of colicin is assayed.

Lacunae are small clearings seen in a lawn of an indicator strain plated with chloroformed Col^+ cells. Alternatively, a streptomycin-resistant indicator is plated with sensitive Col^+ cells on streptomycin agar. Lacunae are not always easy to count and the method is undoubtedly most successful with those colicins like E2 which produce large zones. With colicin I, which produces small zones, the clearest results appear to be obtained by inoculating 2·5 ml. soft water agar containing streptomycin (0·35 %, w/v, agar in distilled water with 200 μg. streptomycin/ml.) with the streptomycin-sensitive Col^+ cells and 0·1 ml. of an overnight broth culture of a streptomycin-resistant indicator, which is then poured on a plate of nutrient agar. After incubating for 5 hr. at 37°, lacunae are dimly seen on viewing the plate by transmitted daylight. They become much clearer if the overlay is first flooded with broth containing 0·2 %, w/v, glucose and 0·1 %, w/v, TTC, and incubation continued at 37° for 20 min. (Pattee, 1966). The background turns red as the TTC is reduced, leaving the lacunae as colourless areas.

In assaying colicin, higher titres are obtained by lysing the cells, so a distinction should be made between extracellular and intracellular colicin. Chloroforming alone does not produce complete lysis (indeed, it is one method for lacunae) but chloroformed cells from growing cultures are lysed at 37° by 0·2 %, w/v, sodium dodecyl sulphate. Otherwise, cells can be treated with lysozyme (1 mg./ml.) and 0·07 M-EDTA in 0·1 M tris buffer, pH 8, for 10 min. at 37°, followed by 0·5 %, w/v, sodium dodecyl sulphate.

Titres of colicin preparations are assayed in three ways:

(*a*) drop method, where the preparation is serially diluted in two-fold steps and drops of 0·02 ml. placed on an overlay inoculated with the indicator. After incubation at 37° for 5–18 hr., the titre is read as the highest dilution causing inhibition;

(*b*) diffusion assay, as used for antibiotics (see Kavanagh, 1963). This is complicated by the slow diffusion of colicin molecules through agar. Richardson, Emslie-Smith & Senior (1968) describe a plate method in which the colicin is allowed to diffuse from cups on the surface of the agar for 40 hr. at 4° before incubating at 37° to enable the indicator to grow. The colicin concentration of the sample is then obtained from the diameters of the circular inhibition zones around the cups by reference to a standard preparation;

(*c*) concentration of 'lethal units'. Jacob, Siminovitch & Wollman (1952) showed that killing by colicin is a one-hit process (i.e. the colicin molecules act independently; p. 182), since the number of bacteria killed was proportional to the amount of colicin present, provided the bacteria were in excess. Under these conditions, however, only the initial part of each survival curve is exponential because all the colicin is adsorbed before all the cells are affected. The rate of killing therefore soon drops to zero and a fraction of the culture survives. This fraction, F, represents the cells not adsorbing even one 'lethal unit' of colicin each. It follows from p. 177 that $\log_e F = -m$, where m is the initial multiplicity of lethal units per cell. A plot of log F against m (usually represented by the dilution of colicin preparation) should be linear, as is indeed found in practice (see Nomura, 1967). The first step in setting up the assay is to verify this relationship. The colicin preparation is serially diluted and an aliquot of each dilution mixed with a known number of bacteria. As only growing, not stationary, cells may be susceptible (p. 197), growing cultures are used: the viable count can be held constant during the assay while preserving the sensitive phenotype by including chloramphenicol (50 μg./ml.) and 0·001 M-dinitrophenol. Adsorption is usually complete within 10 min. at 37°, provided the cell concentration is 10^8/ml. or more, although this should be checked. Finally, the size of the surviving bacterial fraction in each mixture is measured by a colony count and the absence of free colicin confirmed by centrifuging the mixture and assaying the supernatant. Each dilution should give a consistent estimate of the number of lethal units, which are simple to calculate as a 'lethal unit' is defined as the amount of colicin required to kill one bacterium. Each unit probably corresponds to more than one molecule of colicin; i.e. the probability per molecule of causing a lethal effect is less than 1, as in most biological systems (p. 176).

It is worth remembering that colicin titres are considerably affected by the nature of the bacterial culture medium.

Drug resistance (R) factors

The presence of an R factor is to be suspected when resistance to several unrelated antibacterial agents is present, and is confirmed by showing that the resistance genes behave as a single extrachromosomal entity, judging from their patterns of joint transfer (p. 270) and simultaneous loss by segregation (p. 278).

Transfer is demonstrated by mixing broth cultures of the suspected R^+ strain with an R^- recipient, followed by streaking on a medium that allows only the recipient to grow and that also contains one or more of the agents to which the donor is resistant. If the donor is an auxotroph, a recipient with different nutritional requirements is used (deletion mutants are useful because they do not revert to prototrophy). Thus, a growing culture of a methionine-requiring tetracycline-resistant donor could be mixed with a tryptophan-requiring recipient and, following incubation, the mixture streaked for isolated colonies on glucose-salts medium supplemented with tryptophan and tetracycline. Naturally-occurring strains of *Escherichia* and *Salmonella* are usually non-exacting (Lederberg, 1947) but differ in their ability to use certain carbon sources. *Escherichia* uses lactose but not citrate, while the reverse holds for *Salmonella. Shigella* and *Proteus* usually require nicotinic acid. If the donor and recipient have identical growth requirements, mutants of the recipient completely resistant to streptomycin can be isolated by direct plating. The recipient is then selected from the mating mixture by plating on agar containing streptomycin, 1000 μg./ml., since R factors do not confer this level of resistance on their hosts and the donor is therefore inhibited. Nalidixic acid is a useful alternative (25–100 μg./ml.), at least until R factors come to confer resistance to it, as it allows streptomycin-resistance of the donor to be tested for transmissibility. Transfer is often detectable 20 min. after mixing donor and recipient, but, in doubtful cases, the mixture should be incubated overnight before plating, to allow time for further matings and for epidemic spread of the R factor through the recipient strain.

If no transfer occurs after overnight incubation, the drug-resistant strain evidently does not contain a sex factor conferring the ability to conjugate and to act as a genetic donor. A non-transmissible R factor may nevertheless be present and becomes transmissible when a functional sex factor is also introduced into the strain. A de-repressed sex factor like F is preferable to one that is repressed because conjugation and transfer are correspondingly more frequent. The experiment is set up as for the HFT mixtures described on p. 272. The F^+ and the drug-resistant strains are mixed, diluted to about 10^7 cells/ml., and grown together overnight without shaking. Next day, the mixture is diluted 1/20 in fresh broth and, after 2 hr. incubation, is mated with an equal number of recipient cells. After mating for 1–24 hr., any recipient cells acquiring drug-resistance are isolated on the appropriate selective medium, just as in a cross with an R factor able to bring about its own transmission. Not every sex factor may be able to co-transfer a particular non-transmissible R factor, so, if F gives negative results, it is wise to try an unrelated sex factor like de-repressed colicin factor I.

Another characteristic which reveals the presence of an R factor is *segregation*, the appearance of drug-sensitive clones during subculture of the resistant strain. This often occurs spontaneously, more in *Salm. typhimurium* than in *E. coli* (see Watanabe, 1963), and may involve either the entire R factor, so that all the drug resistances disappear at once, or only one or two resistances may be lost, like sulphonamide- and streptomycin-resistance. The incidence of sensitive segregants in the culture may be increased by the simple procedure of growing the strain overnight in broth containing acridine orange or sodium dodecyl sulphate (p. 278). Whatever the means by which segregants are obtained, the culture is plated on drug-free complete medium to give discrete colonies (say, 50 colonies per plate) whose resistance markers are then scored by replica-plating to medium containing each of the different agents individually.

As a growing number of antibacterial agents are brought into medical and veterinary use, more and more types of resistance are determined by R factors. It is not unusual for a single strain to be resistant to as many as six different agents like sulphonamide, penicillin, streptomycin, tetracycline, chloramphenicol and kanamycin. This is not to say that all these resistances are necessarily carried by a single R factor for a substantial fraction of strains evidently carry two unrelated factors which coexist stably (Romero & Meynell, 1969). The degree of resistance conferred by a given factor may also be considerably influenced by its bacterial host. Thus, the same factor transferred to *Shigella* or *E. coli* and to *Salm. typhimurium* produced resistance to 100 μg. and 10 μg. tetracycline/ml. (see Watanabe, 1963). Furthermore, the level of resistance depends on the culture medium used for determining sensitivity: streptomycin appears less inhibitory on glucose-salts medium than on broth agar (Gunderson, 1965), whereas the opposite is true of chloramphenicol and tetracycline. Sulphonamides form a special case in that sensitive organisms appear completely resistant on broth agar because it contains a potent sulphonamide antagonist whose effect can, however, be removed by adding 5 %, v/v, lysed horse blood (for details, see Harper & Cawston, 1945). The drug concentrations that are usually suitable for distinguishing R^+ from R^- strains are streptomycin, 10–20 μg./ml. (in contrast to the complete indifference produced by chromosomal mutation to streptomycin-resistance); chloramphenicol, 20 μg./ml.; kanamycin, 20 μg./ml.; tetracycline, 10 μg./ml.; ampicillin, 20 μg./ml.; and sulphonamides, 100–200 μg./ml. Most of these agents dissolve readily in water but, as with streptomycin, their solutions may be strongly acid. Chloramphenicol may be dissolved in methanol (e.g. at 20 mg./ml. or 1000 times working strength). Nalidixic acid has to be brought into solution by adding strong NaOH. Stock solutions can generally be treated as stable at

4° with the exception of tetracycline which is best renewed weekly from a concentrated stock solution kept frozen at −10°.

The biochemical basis of the individual resistances is steadily becoming clearer and in several cases has proved to be a plasmid-determined enzyme that inactivates the drug, as opposed to a loss of permeability or an altered metabolic pathway. The known instances include penicillinase, readily detected in colonies (p. 265), chloramphenicol acetylase and streptomycin adenylate synthetase, all of which can be accurately assayed (Citri & Pollack, 1966; Shaw, 1967; Garber, Zohar & Michlin, 1968; Yamada, Tipper & Davies, 1968).

ANCILLARY TECHNIQUES

Centrifugation

Technical aspects of the separation of organisms or macromolecules by centrifugation have now been extensively reviewed by Brakke (1960, 1969), Vinograd (1963), Szybalski (1968) and Mazzone (1969). Table 7.1 gives the weight of CsCl required for solutions of known density, calculated from Vinograd's equation (3 *a*). Three important improvements in isopycnic density gradient centrifugation are the use of angle as opposed to swing-out rotors to increase resolution and capacity (Flamm, Bond & Burr, 1966); the introduction of initially discontinuous gradients, which should greatly lessen the time taken to reach equilibrium (Brunk & Leick, 1969); and the isolation of closed circular DNA by means of ethidium bromide (Bauer & Vinograd, 1968).

Table 7.1. CsCl *solutions of known density*

$\rho^{0,25°}$	g	$\rho^{0,25°}$	g	$\rho^{0,25°}$	g
1·40	0·635	1·54	0·916	1·68	1·236
1·42	0·673	1·56	0·959	1·70	1·285
1·44	0·712	1·58	1·003	1·72	1·336
1·46	0·751	1·60	1·048	1·74	1·387
1·48	0·791	1·62	1·093	1·76	1·44
1·50	0·832	1·64	1·14	1·78	1·493
1·52	0·873	1·66	1·187	1·80	1·548

$\rho^{0,25°}$: density at 25°. g: grams of CsCl to be added to 1 g. water. Calculated from equation 3 *b* in Vinograd (1963).

Gradient plates

Organisms frequently have to be exposed to progressively increasing concentrations of an inhibitor like an antibiotic. Clearly, a series of replicate cultures could be set up, each containing a different concentration, but a gradient plate is both more economical and yields more information. If the maximum concentration to be tested is 100 μg./ml. then 15 ml. of nutrient agar containing that concentration is poured into a plate raised at one edge so that the agar sets in a wedge. The direction of the wedge is marked on the plate which is then placed horizontally and 15 ml. of antibiotic-free nutrient agar poured on the lower layer. After drying, the agar is inoculated evenly by flooding with a fully-grown broth culture or by streaking different clones parallel to the gradient, and incubated. The antibiotic diffuses into the upper layer during growth and the organisms are therefore exposed to a linearly decreasing concentration across the plate. Confluent growth may very well be seen at one side of the plate, partially-resistant mutants at the centre and a few fully resistant mutants at the far end. Drug-dependent clones grow only at higher concentrations.

A more abrupt gradient is obtained with a *ditch plate* in which a strip or quadrant of agar is cut out from a plate of nutrient agar and replaced by agar containing antibiotic.

Although gradient plates are of particular value in selecting mutants possessing partial resistance to antibiotics and other antibacterial agents, the same principle can be used to study the effect of pH (Sacks, 1956; Battley & Bartlett, 1966) or temperature (Fluegal, 1963; Nakae, 1966).

Multipoint inoculators

The scoring of multiple nutritional or fermentation markers of a large number of recombinants is enormously simplified by replicating from a master plate of complete medium to a series of minimal or pH indicator plates. The first of these techniques employed velvet, or more properly velveteen, as the replicating surface (Lederberg & Lederberg, 1952). A square of dry sterile velvet is placed pile upwards on a circular block 8 cm. in diam., and secured by a tight-fitting metal ring. The edge of the master plate is marked at one point and its colonies are pressed against the velvet with the mark opposite a reference mark on the metal ring. The replicas are then made by inverting each of the test plates against the velvet, after first marking its edge and bringing it opposite the reference mark so that corresponding points on the master plate and its replica can later be identified. After incubation, the master plate and its replicas are compared, and the reaction of each of the master colonies noted. Auxotrophic mutants fail to grow on unsupplemented replica plates but their position can be detected by looking on the agar for imprints of master colonies that failed to grow.

Used velvets are best discarded into a beaker of water to avoid drying and matting of the pile. After autoclaving, each piece is gently squeezed, laid flat to dry and brushed to raise the pile. Before use, the pieces are autoclaved in a bag of autoclavable kraft paper or plastic and should be dry enough to use if the autoclave pressure is brought down abruptly at the end of the run (p. 106).

Velvet is excellent for replicating colonies but the pile tends to deteriorate and a modified technique is often used with recombinants. After purification, the recombinant colonies are subcultured in rows on a master plate from which they are replicated to test plates by corresponding rows of metal prongs protruding about 1·5 cm. from a base plate. Metal replicators of this kind are readily sterilized by dipping the prongs into ethanol and igniting in a flame. Many similar devices have been described for phage-typing, colicin-typing, and for biochemical tests in both liquid and solid media (see Gibbs & Skinner, 1966, pp. 117, 125).

A different form of multipoint inoculator are the 'Multodisks' sold by Oxoid Ltd which consist of paper sheets cut in a number of lobes impregnated with antibiotics. A sheet is dropped on a plate which has previously been flooded with the test strain and allowed to dry. Six or eight sensitivities depending on the type of Multodisk can be tested in one step. There are standard combinations of antibiotics or any chosen combination can be prepared to order. The technique could presumably be extended to the identification of growth requirements and of fermentation reactions by impregnating the paper with growth factors or sugars in place of antibiotics.

Stock cultures

It is well recognized that mutation, followed by selective outgrowth of mutants, often occurs in a stock kept on nutrient medium, and that, when a fresh stock is set up, the inoculum has to be chosen carefully to avoid storing a mutant instead of the parental strain. This is usually done by picking a single colony, as mutants are normally fairly rare. However, bacterial variation also arises from non-genetic causes and is then characterized by quasi-stable quantitative differences between clones in characters like fermentative ability, exotoxin production, or colony size (see Wilson & Miles, 1964, p. 384). It follows that the common practice of picking a single colony for subculture, in order to avoid mutants, may then lead to storage of a clone with exceptional characteristics acquired from non-genetic causes. This is also true of strains carrying plasmids which often fail to be inherited at cell division and are absent from many cells in a culture. A satisfactory method of subculture has to minimize the effect of both sources of variation.

The logical course is to plate out the previous stock culture for isolated colonies; to incubate and confirm that no contaminants are present; and finally to avoid selection from either cause by picking off neither one colony nor confluent growth but by taking a sweep of an intermediate number of colonies, say, 50.

Needless to say, this procedure will not prevent mutants accumulating in stocks stored on nutrient media. There is less opportunity of selection with dried cultures, but, even here, suitable inocula have to be chosen for the cultures from which stocks are made.

Bacteriostasis. Storage at *ca.* 4° does not prevent cell division which is only stopped by freezing or by dehydration. A large proportion of the culture may be killed by either treatment or may die when the culture is reconstituted. The cause of death on freezing is not known and, indeed, the literature suggests that different types of cell behave differently. Most theoretical discussions start by considering the culture as an aqueous salt solution which becomes progressively more concentrated during cooling because part of its water separates as crystals of pure ice. The resulting pockets of salts solution formed within the ice freeze only when the temperature has fallen below their eutectic point. Cell death may then reflect denaturation caused by concentrated salts solution or, possibly, structural disorganization produced by intracellular or extracellular crystals (see Muggleton, 1963; Martin, 1964).

Slow drying of bacterial cultures in liquid or on solid medium is lethal to many species, and dehydration is therefore usually restricted to cultures already frozen at temperatures below −20°. Water then passes directly from the solid to the gaseous state (Fig. 4.2). Freezing and drying are often combined, as in freeze-drying, by arranging for rapid evaporation initially which cools, and then freezes, the bacterial suspension extremely quickly.

Survival after freeze-drying is generally recognized to be largely determined by the suspending medium. The highest survival is often obtained in a complex medium like 'Mist. Dessicans' which contains 1 vol. broth, 3 vol. serum, and 1 vol. 7·5 %, w/v, glucose solution (Fry & Greaves, 1951). Part of the protective effect is exerted during reconstitution of the dried cultures. Even with these precautions 100 % cells do not always survive; but experience suggests that freeze-drying is adequate to preserve a sizeable fraction of most cultures for several years and, more important, that selection does not occur.

Centrifugal freeze-drying requires special apparatus for which the makers supply details. Other methods using everyday materials and apparatus have been described (see Muggleton, 1963; Martin, 1964). That devised by Stamp (1947) is given here in detail as it seems adequate for many species, and is more convenient than freeze-drying because a large number of separate stocks can be prepared quickly and stored in a single tube.

Drying in gelatin–ascorbic acid broth (Stamp, 1947). An overnight culture on solid medium is suspended in a molten solution of gelatin in broth at 37°. Drops of suspension are distributed on a non-wettable surface by a Pasteur pipette and are dried *in vacuo* over P_2O_5 for at least 3 days. Each drop forms a thin disc which is stored in a dry atmosphere.

The materials required are:

(1) Suspending medium.

Broth	100 ml.
Gelatin	10 g.

Adjust to pH 7·5. Distribute in 5 ml. volumes in screw-capped bottles. Autoclave at 115° for 15 min. The original used Lemco-peptone broth (p. 48).

(2) Sterile ascorbic acid.

Ascorbic acid	1 g.
Distilled water	100 ml.

Neutralize with NaOH. Sterilize by immersion in boiling water for 1 min.

(3) Waxed filter paper. Dip circular pieces of filter paper in molten paraffin wax held in a glass Petri dish; drain; place in a sterile Petri dish; and allow to cool.

Grow the culture overnight on a slope of nutrient agar. Melt 5 ml. of suspending medium by warming to 37° and add 0·5 ml. ascorbic acid solution. The latter is not essential. Add 1–2 ml. medium to the culture and suspend the growth with a bent wire. Distribute drops of suspension on the waxed filter paper with a fine Pasteur pipette. Transfer the open dishes to a desiccator containing P_2O_5 and dry at a pressure of 100–300 mm. Hg. for 2–3 days at room temperature. After drying, the drops form thin discs which are transferred to a sterile container. A point to note with pathogens is that the discs may spring away immediately they are touched. It is safer to detach each disc from the plate with a sterile scalpel blade while holding it down by a stiff wire, and also to work in a safety cabinet or to put the Petri dish on a tray lined with absorbent paper soaked in lysol. The discs are stored in a tube closed by a cotton wool plug, usually at reduced pressure over P_2O_5. Another method of storage is to place an 8 × 35 mm.

tube on top of self-indicating silica gel in a screw-capped bottle previously sterilized, and the gel dried, by hot air in the usual way (p. 108). When cool, the bottles are closed by rubber-lined screw caps sterilized separately by autoclaving.

The culture is reconstituted by picking up a disc with the moistened tip of a straight wire and transferring it to broth. This is incubated at 37° until the gelatin dissolves when the resulting bacterial suspension is subcultured to solid medium.

Losses from contamination are avoided by drying the stock strain at high concentration (10^{10}–10^{11} bacteria/ml. suspending medium) and, on recovery, by plating out as soon as the gelatin disc has melted.

Storage in nutrient media. Dehydration is a common cause of death and often occurs in screw-capped culture bottles with chipped rims or caps lacking liners. The risk of dehydration is lessened by using soft nutrient agar (e.g. 0·6 %, w/v) into which the strains are stabbed; by closing culture tubes with bungs dipped in molten paraffin wax; or by covering the bacterial growth on a slope or in a stab culture with sterile liquid paraffin coming at least 2 cm. above the level of the medium.

Certain media are customarily used for stocks of certain genera: for example, Dorset's egg medium for Enterobacteriaceae, and Robertson's meat medium for streptococci (see Cruikshank, 1965). Liquid media have the disadvantage that a fast-growing contaminant can swamp the stock. Nutritionally exacting organisms should be kept in medium supplemented with the appropriate growth factors to remove the selective advantage otherwise possessed by non-exacting mutants.

Serology

The essential point of immunological techniques in the present context is that they offer a most powerful yet simple means for characterizing macromolecules like proteins or polysaccharides. Many macromolecules are *antigenic*, since, on injection into an animal, they cause it to form new species of serum globulin, or *antibodies*. In general, an antigen combines with its homologous antibody but not with antibody to other antigens. Similarly, antibody will combine only with antigens identical or closely related to those which provoked its formation. This specificity of reaction is the basis of serological technique, since the simple procedure of preparing an antiserum to a known antigen, like an enzyme preparation, immediately provides a means for determining its relation to other antigens. Thus, to establish the identity of the enzyme formed by an inducible wild-type strain and its constitutive mutant by purely chemical means would be a major undertaking, but is readily done by showing that they react identically with homologous and heterologous antisera.

Much of practical serology is concerned, not with bringing about the union of antigen and antibody, as this occurs within a few seconds of mixing, but with demonstrating that union has occurred. If the antigen is an enzyme or the tail fibre of a phage particle, the reaction is often manifested immediately by loss of enzymic function or of phage infectivity. In many systems, however, inactivation does not result and the reaction is then measured by one of a number of secondary phenomena like co-precipitation of antigen and antibody; agglutination of a particulate antigen like a bacterium to form macroscopic clumps; or fixation of complement (a class of protein of normal serum) to the antigen-antibody complexes. Such techniques have many degrees of sophistication, ranging from visual estimates of the reaction to precise measurement by chemical assay of the reactants. A particularly sensitive method with high powers of discrimination, which is nevertheless extremely simple to set up, involves diffusion of antigen and antibody towards each other in agar ('immunodiffusion'). Wherever they are present in a critical ratio, precipition occurs and is visible as an opaque white line (see Crowle, 1961). Quantitative precipitation tests are used for titrating the 'cross-reacting material', or defective enzyme, produced by many mutants (see Helinski & Yanofsky, 1966), as well as the β-galactosidase of *E. coli* which is not inactivated by antibody (see Cohn, 1952). The essential steps in applying serological techniques therefore resolve themselves into preparing the antigen, immunizing an experimental animal, collecting antisera and titrating antibody activity. Before commencing any animal work, care should be taken to comply with the statutory regulations governing animal experimentation (Lane-Petter *et al.* 1967).

Many antigens are used without elaborate purification, since it is usually simpler to remove unwanted antibodies by absorbing the sera than by purifying the antigen. High-titre phage stocks prepared in broth or in overlays are suitable for antiphage sera, while bacterial suspensions harvested from agar or broth can be used for antibacterial antibody. Sera to specific bacterial antigens can often be obtained directly by immunizing with a different species, the majority of whose antigens are unrelated to those of the host used in the experiment. Thus, antisera specific for sex pili can be prepared by injecting plasmid-carrying *Salmonella typhimurium*, and using the sera for experiments with plasmid-carrying *E. coli*, or *vice versa*.

The only antigen common to both strains is the sex pilus, so that although the serum will contain antibody to salmonella somatic and flagellar antigens, it will not agglutinate cultures of *E. coli* not forming sex pili. Similarly, antisera specific for salmonella somatic antigen 1 (formed by *Salm. typhimurium* following lysogenic conversion by phage P22) is prepared by injecting *Salm. senftenberg* (somatic antigens 1,3,19) as this will not agglutinate non-lysogenic *Salm. typhimurium* whose antigens are 4,5,12. Phage and bacterial suspensions not infrequently prove toxic on injection. It may be worth washing phage by centrifugation to remove soluble toxic material derived from lysed bacteria. Laboratory strains of *E. coli* are least toxic when inoculated alive. Salmonella and presumably other enterobacteria are rendered less toxic while retaining their somatic antigenic specificity by the following procedure (Roschka, quoted by Edwards & Ewing, 1962). Harvest agar-grown cells in saline to give a dense suspension. Heat for 2 hr. in a boiling water bath at 100°. Centrifuge. Resuspend the deposit in 95 %, v/v, aqueous ethanol and leave at 37° for 4 hr. Centrifuge, twice wash the deposit with acetone, suspend in a small volume of acetone and dry overnight at 37°. For use, the powder is suspended in sterile 0·85 %, w/v, saline to a density of 10^9 cells/ml. of which 0·5 ml. is injected intravenously as the first dose. The advantages of this method are that fewer animals die, higher titres are obtained and the dried antigen keeps indefinitely.

Before starting a course of immunization, a sample of blood must be taken to provide a baseline for the experiment. At least two animals should always be used for each antigen for, not only may animals be lost from intercurrent infection, but there is often such great variability in immune response that some nearly always respond better than others. Antigens are usually administered by injecting aqueous suspensions into the marginal ear vein of a rabbit. Two alternatives worth considering are the intramuscular injection of long-acting oily suspensions (e.g. in Freund's adjuvant, obtainable from Difco) and the use of smaller animals, notably mice, from which antibody is collected by inducing the intraperitoneal accumulation of ascitic fluid either chemically or with an ascites tumour (Berkovich, 1963). Whatever method is used, it should be borne in mind that the globulin fraction containing antibody activity, and also its degree of specificity, is determined by the duration of immunization, its route, and by the nature of both antigen and host (see Mayersbach, 1967; Holborow, 1967). No special apparatus is needed for rabbits which usually remain calm if the room is quiet and they are given a secure foothold on some sacking and then wrapped up firmly with only their heads exposed. It is completely unnecessary to confine rabbits in special boxes with their heads protruding through a shutter. Not only is the animal scared, but, if it jerks or bucks, will break its neck. Excessive sterile precautions are not required when collecting blood: it is sufficient to clean the ear with alcohol after shaving and to collect in a sterile 25 ml. bottle or 50 ml. centrifuge tube. Up to 35 ml. blood can be obtained from a large rabbit without harm: drinking water should be freely provided afterwards. The blood is usually allowed to clot at room temperature, the edges of the clot freed from the glass with a sterile straight wire, and the bottle put at 37° to allow clot retraction and expression of serum. After a few hours, the serum is transferred to a sterile centrifuge tube with a sterile pipette and centrifuged for 10 min. at 500 g in a swing-out head to deposit the red cells. The serum, which should now be clear and show no signs of haemolysis, is stored in a sterile bottle at 4° or at −20° or, occasionally, is freeze-dried. Preservatives like Merthiolate (p. 114) should not be added to sera to be used with phage or living bacteria. Occasionally, a fatty layer forms at the surface of stored serum but is easily removed by centrifugation. Contaminated sera can be sterilized by centrifugation followed by filtration through an asbestos-cellulose pad (p. 119). Before separating serum from clot, it may be convenient to store the blood in the refrigerator overnight, and often results in the expression of a greater volume of serum. If this is done, however, the blood must be allowed to warm up slowly to room temperature or the cells lyse on pipetting.

The most appropriate method for titrating antibody is necessarily determined by the system, and the standard techniques are described at length by Kabat & Mayer (1961) and Williams & Chase (1967). However, no amount of detail should be allowed to mask the fundamental fact that serology offers a simple tool of outstanding sensitivity for the analysis of many genetic and physiological problems.

REFERENCES

ADAMS, M. H. (1948). Surface inactivation of bacterial viruses and of proteins. *J. gen. Physiol.* **31**, 417.

ADAMS, M. H. (1959). *Bacteriophages.* New York: Interscience Publishers Inc.

ADELBERG, E. A. (1958). Selection of bacterial mutants which excrete antagonists of antimetabolites. *J. Bact.* **76**, 326.

ADELBERG, E. A., MANDEL, M. & CHEN, G. C. C. (1965). Optimal conditions for mutagenesis by *N*-methyl-*N'*-nitro-*N*-nitrosoguanidine in *Escherichia coli* K12. *Biochem. biophys. Res. Commun.* **18**, 788.

ADELBERG, E. A. & MYERS, J. W. (1953). A modification of the penicillin technique for the selection of auxotrophic bacteria. *J. Bact.* **65**, 348.

ADLER, H. I., FISHER, W. D., COHEN, A. & HARDIGREE, A. A. (1967). Miniature *Escherichia coli* cells deficient in DNA. *Proc. natn. Acad. Sci. U.S.A.* **57**, 321.

AJL, S. J. & WERKMAN, C. H. (1949). Anaerobic replacement of carbon dioxide. *Proc. Soc. exp. Biol. Med.* **70**, 522.

ALPER, T. (1961). Effects on subcellular units and free-living cells. In *Mechanisms in Radiobiology*, vol. I, p. 353. Ed. by M. Errera and A. Forssberg. New York and London: Academic Press.

ALPER, T., GILLIES, N. E. & ELKIND, M. M. (1960). The sigmoid survival curve in radiobiology. *Nature, Lond.* **186**, 1062.

ALPER, T. & STERNE, M. (1933). The measurement of the opacity of bacterial cultures with a photo-electric cell. *J. Hyg., Camb.* **33**, 497.

AMES, B. N. & MARTIN, R. G. (1964). Biochemical aspects of genetics: the operon. *A. Rev. Biochem.* **33**, 235.

ANAGNOSTOPOULOS, C. & SPIZIZEN, J. (1961). Requirements for transformation in *Bacillus subtilis.* *J. Bact.* **81**, 741.

ANDERSON, E. S. (1968). The ecology of transferable drug resistance in the enterobacteria. *A. Rev. Microbiol.* **22**, 131.

ANDERSON, E. S., ARMSTRONG, J. A. & NIVEN, J. S. F. (1959). Fluorescence microscopy: observation of virus growth with aminoacridines. *Symp. Soc. gen. Microbiol.* **9**, 224.

ANDERSON, P. A. & PETTIJOHN, D. E. (1960). Synchronization of division in *Escherichia coli. Science, N.Y.* **131**, 1098.

ARBER, W. & LINN, S. (1969). DNA modification and restriction. *A. Rev. Biochem.* **38**, 467.

ARCHAMBAULT, J. & MCCRADY, M. H. (1942). Dissolved air as a source of error in fermentation tube results. *Amer. J. publ. Hlth* **32**, 1164.

ARMITAGE, P. (1953). Statistical concepts in the theory of bacterial mutation. *J. Hyg., Camb.* **51**, 162.

ARMITAGE, P. (1957). Studies in the variability of pock counts. *J. Hyg., Camb.* **55**, 564.

ARMITAGE, P. (1959*a*). An examination of some experimental cancer data in the light of the one-hit theory of infectivity titrations. *J. natn. Cancer Inst.* **23**, 1313.

REFERENCES

ARMITAGE, P. (1959*b*). Host variability in dilution experiments. *Biometrics* **15**, 1.

ARMITAGE, P. & ALLEN, I. (1950). Methods of estimating the LD 50 in quantal response data. *J. Hyg., Camb.* **48**, 298.

ARMITAGE, P. & BARTSCH, G. E. (1960). The detection of host variability in a dilution series with single observations. *Biometrics* **16**, 582.

ARMITAGE, P., COURT BROWN, W. M. & DOLL, R. (1959). Dose-response relationship in radiation leukaemia. *Nature, Lond.* **184**, 1669.

ARMITAGE, P., MEYNELL, G. G. & WILLIAMS, T. (1965). Birth–death and other models for microbial infection. *Nature, Lond.* **207**, 570.

ARMITAGE, P. & SPICER, C. C. (1956). The detection of variation in host susceptibility in dilution counting experiments. *J. Hyg., Camb.* **54**, 401.

ASHESHOV, I. N. & HEAGY, F. C. (1951). A 'loop' method for counting viable bacteria or bacteriophage. *Canad. J. med. Sci.* **29**, 1.

ATWOOD, K. C. & NORMAN, A. (1949). On the interpretation of multi-hit survival curves. *Proc. natn. Acad. Sci., U.S.A.* **35**, 696.

ATWOOD, K. C., SCHNEIDER, L. K. & RYAN, F. J. (1951). Periodic selection in *Escherichia coli*. *Proc. natn. Acad. Sci., U.S.A.* **37**, 146.

AUBER, E. (1950). 'Cold' stain for acid-fast bacteria. *Canad. J. publ. Hlth* **41**, 31.

AUGER, W. J. (1939). A new method of culturing sputum on solid media using carbon dioxide for the isolation of pneumococci. *Brit. J. exp. Path.* **20**, 439.

BAKER, K. (1968). Low cost continuous culture apparatus. *Lab. Pract.* **17**, 817.

BALL, C. O. & OLSON, F. C. W. (1957). *Sterilization in Food Technology*. New York: McGraw-Hill Book Co. Inc.

BANNISTER, D. & GLOVER, S. W. (1968). Restriction and modification of bacteriophages by R^+ strains of *Escherichia coli* K12. *Biochem. biophys. Res. Commun.* **30**, 735.

BARER, R. (1966). Phase contrast and interference microscopy in cytology. In *Physical Techniques in Biological Research*, 2nd ed. vol. III, part A. Ed. by A. W. Pollister. New York and London: Academic Press.

BARER, R. (1968*a*). *Lecture Notes on the Use of the Microscope*, 3rd ed. Oxford: Blackwell Scientific Publications.

BARER, R. (1968*b*). Maximum-intensity fluorescence microscopy. *Nature, Lond.* **217**, 672.

BARER, R., ROSS, K. F. A. & TKACZYK, S. (1953). Refractometry of living cells. *Nature, Lond.* **171**, 720.

BARER, R. & WEINSTEIN, W. (1953). A rational microscope lamp. *Jl R. microsc. Soc.* **73**, 148.

BARNES, M. & PARKER, M. S. (1967). Use of the Coulter Counter to measure osmotic effects on the swelling of mould spores during germination. *J. gen. Microbiol.* **49**, 287.

BARRY, V. C., CONALTY, M. L., DENNENY, J. M. & WINDER, F. (1956). Peroxide formation in bacteriological media. *Nature, Lond.* **178**, 596.

BARTHOLOMEW, J. W. (1962). Variables influencing results, and the precise definition of steps in Gram staining as a means of standardizing the results obtained. *Stain Technol.* **37**, 139.

BARTHOLOMEW, J. W. & MITTWER, T. (1952). The Gram stain. *Bact. Rev.* **16**, 1.

BARTON-WRIGHT, E., CUTLER, D. W. & CRUMP, L. M. (1936). Contamination in Petri dish boxes. *Nature, Lond.* **137**, 110.

REFERENCES

Barton-Wright, E. C. (1952). *The Microbiological Assay of the Vitamin B-Complex and Amino Acids.* London: Sir Isaac Pitman and Sons, Ltd.

Barton-Wright, E. C. (1963). *Practical Methods for the Microbiological Assay of the Vitamin B-complex and Amino Acids.* London: United Trade Press Ltd.

Baten, W. D. & Stafseth, H. J. (1956). Probit analysis applied to *Salmonella pullorum* survival data. *J. Bact.* **71**, 214.

Battley, E. H. & Bartlett, E. J. (1966). A convenient pH gradient method for determination of the maximum and minimum pH for microbial growth. *Antonie van Leeuwenhoek* **32**, 245.

Bauer, D. J. (1961). The zero effect dose (E_0) as an absolute numerical index of antiviral chemotherapeutic activity in the pox virus group. *Brit. J. exp. Path.* **42**, 201.

Bauer, W. & Vinograd, V. (1968). The interaction of closed circular DNA with intercalative dyes. I. The superhelix density of SV40 DNA in the presence and absence of dye. *J. molec. Biol.* **33**, 141.

Bauman, N. & Davis, B. D. (1957). Selection of auxotrophic bacterial mutants through diaminopimelic acid or thymine deprival. *Science, N.Y.* **126**, 170.

Bazaral, M. & Helinski, D. R. (1968). Circular DNA forms of colicinogenic factors E1, E2 and E3 from *Escherichia coli. J. molec. Biol.* **36**, 185.

Beacham, I. R., Eisenstark, A., Barth, P. T. & Pritchard, R. H. (1968). Deoxynucleoside-sensitive mutants of *Salmonella typhimurium. Molec. gen. Genetics* **102**, 112.

Beam, C. A. (1955). The influence of ploidy and division stage on the anoxic protection of *Saccharomyces cerevisiae* against X-ray inactivation. *Proc. natn. Acad. Sci., U.S.A.* **41**, 857.

Beard, J. W., Sharp, D. G. & Eckert, E. A. (1955). Tumour viruses. *Adv. Virus Res.* **3**, 149.

Bennett, A. H., Osterberg, H., Jupnik, H. & Richards, O. W. (1951). *Phase Microscopy. Principles and Applications.* New York: John Wiley and Sons, Inc.

Bennett, E. O. & Williams, R. P. (1957). A comparison of methods for determining bacterial mass with particular emphasis upon the use of total phosphorus. *Appl. Microbiol.* **5**, 14.

Benzer, S. (1952). Resistance to ultraviolet light as an index to the reproduction of bacteriophage. *J. Bact.* **63**, 59.

Berkovich, S. (1963). A simple rapid method for production of viral antibody in mice. *Proc. Soc. exp. Biol. Med.* **111**, 127.

Berkson, J. (1929). A probability nomogram for estimating the significance of rate differences. *Am. J. Hyg.* **9**, 695.

Bernheim, R. (1964). The effect of pyruvate and acetate on the rate of decrease in optical density of suspensions of *Pseudomonas aeruginosa* in sodium, potassium or sodium-potassium phosphate buffers. *J. gen. Microbiol.* **43**, 327.

Bigelow, W. D. (1921). The logarithmic nature of thermal death time curves. *J. infect. Dis.* **29**, 528.

Black, R. H. & Dillon, P. (1947). A note on the Durham fermentation tube. *Med. J. Aust.* **1**, 402.

REFERENCES

Bliss, C. I. (1937). The calculation of the time-mortality curve. *Ann. appl. Biol.* **24**, 815.

Blum, H. F. (1941). *Photodynamic Action and Diseases caused by Light.* New York: Reinbold Publishing Corporation.

Boné, G. J. & Parent, G. (1963). Stearic acid, an essential growth factor for *Trypanosoma cruzi. J. gen. Microbiol.* **31**, 261.

Bouanchaud, D. H., Scavizzi, M. R. & Chabbert, Y. A. (1968). Elimination by ethidium bromide of antibiotic resistance in enterobacteria and staphylococci. *J. gen. Microbiol.* **54**, 417.

Bowers, L. E. & Williams, O. B. (1962). Factors affecting initiation of growth of *Clostridium botulinum. Antonie van Leeuwenhoek* **28**, 435.

Bowie, J. H., Kelsey, J. C. & Thompson, G. R. (1963). The Bowie and Dick autoclave test tape. *Lancet* i, 586.

Boyce, R. P. & Setlow, R. B. (1962). A simple method of increasing the incorporation of thymidine into the deoxyribonucleic acid of *Escherichia coli. Biochim. biophys. Acta* **61**, 618.

Boyd, D. M. & Casman, E. P. (1951). Inhibition of a strain of *Brucella abortus* by medium filtered through cotton. *Publ. Hlth Rep., Wash.* **66**, 44.

Bradley, D. E. (1966). The fluorescent staining of bacteriophage nucleic acids. *J. gen. Microbiol.* **44**, 383.

Brakke, M. K. (1960). Density gradient centrifugation and its application to plant viruses. *Adv. Virus Res.* **7**, 193.

Brakke, M. K. (1968). Density gradient centrifugation. In *Methods in Virology*, vol. II, p. 93. Ed. by K. Maramorosh and H. Koprowski. New York and London: Academic Press.

Braun, W. (1965). *Bacterial Genetics*, 2nd ed. Philadelphia and London: W. B. Saunders Co.

Bray, J. (1945). A method of suppressing *Proteus* and coliform bacteria on routine blood agar plates. *J. Path. Bact.* **57**, 395.

Brenner, S. & Beckwith, J. R. (1965). *Ochre* mutants, a new class of suppressible nonsense mutants. *J. molec. Biol.* **13**, 629.

Brewer, J. H. (1940). Clear liquid mediums for the 'aerobic' cultivation of anaerobes. *J. Am. med. Ass.* **115**, 598.

Brewer, J. H. (1942). A new petri dish cover and technique for use in the cultivation of anaerobes and microaerophiles. *Science, N.Y.* **95**, 587.

Bridges, B. A. (1964). Microbiological aspects of radiation sterilization. *Progr. industr. Microbiol.* **5**, 283.

Briggs, M., Tull, G., Newland, L. G. M. & Briggs, C. A. E. (1955). The preservation of lactobacilli by freeze-drying. *J. gen. Microbiol.* **12**, 503.

Brinton, C. C. (1959). Non-flagellar appendages of bacteria. *Nature, Lond.* **183**, 782.

Brinton, C. C. (1965). The structure, function, synthesis and genetic control of bacterial pili and a molecular model for DNA and RNA transport in gram-negative bacteria. *Trans. N.Y. Acad. Sci.* **27**, 1003.

Brock, T. D. (1966). Streptomycin. *Symp. Soc. gen. Microbiol.* **16**, 131.

Brodie, J. & Shepherd, W. (1950). The effect of the gas phase on differential inhibition of intestinal bacilli. *J. gen. Microbiol.* **4**, 102.

REFERENCES

BROWN, A. M. & DINSLEY, M. (1962). Homeostasis and the F_1 hybrid. *Nature, Lond.* **196**, 910.

BROWN, H. C. (1919–20). Further observations on the standardization of bacterial suspensions. *Indian J. med. Res.* **7**, 238.

BROWN, H. C. & KIRWAN, E. W. O'G. (1914–15). Standardization of bacterial suspensions by opacity. *Indian J. med. Res.* **2**, 763.

BROWN, W. R. L. & RIDOUT, C. W. (1960). An investigation of some sterilisation indicators. *Pharm. J.* **184**, 5.

BRUCH, C. W. (1961). Gaseous sterilization. *A. Rev. Microbiol.* **15**, 245.

BRUNK, C. F. & LEICK, V. (1969). Rapid equilibrium isopycnic CsCl gradients. *Biochem. biophys. Acta* **179**, 136.

BRYAN, W. R. (1957). Interpretation of host response in quantitative studies on animal viruses. *Ann. N.Y. Acad. Sci.* **69**, 698.

BRYAN, W. R. (1959). Quantitative biological experimentation in the virus and cancer fields. *J. natn. Cancer Inst.* **22**, 129.

BRYAN, W. R. & BEARD, J. W. (1940). Correlation of frequency of positive inoculations with incubation period and concentration of purified papilloma protein. *J. infect. Dis.* **66**, 245.

BULLOCH, W. (1960). *The History of Bacteriology.* London: Oxford University Press.

BURDON, K. L. (1946). Fatty material in bacteria and fungi revealed by staining dried, fixed slide preparations. *J. Bact.* **52**, 665.

BURNET, F. M. (1925). A note on the effect of dyes on bacterial growth. *Aust. J. exp. Biol. med. Sci.* **2**, 77.

BURNET, F. M. (1930). Bacteriophage activity and the antigenic structure of bacteria. *J. Path. Bact.* **33**, 647.

BURNS, V. W. (1959). Synchronised cell division and DNA synthesis in a *Lactobacillus acidophilus* mutant. *Science, N.Y.* **129**, 566.

BUSHBY, S. R. M. & WHITBY, L. E. H. (1942). Certain properties of plasma, with a new method for large scale production of a non-clotting product. *J. R. Army med. Cps* **78**, 255.

BUTTOLPH, L. J. (1955). Practical applications and sources of ultraviolet energy. In *Radiation Biology*, vol. II. Ed. by A. Hollaender. New York: McGraw-Hill Book Co. Inc.

CAHN, R. D. (1967). Detergents in membrane filters. *Science, N.Y.* **155**, 195.

CAIRNS, H. J. F. & WATSON, G. S. (1956). Multiplicity reactivation of bacteriophages. *Nature, Lond.* **177**, 131.

CALDWELL, W. J., STULBERG, C. S. & PETERSON, W. D., Jr. (1966). Somatic and flagellar immunofluorescence of *Salmonella. J. Bact.* **92**, 1177.

CAMPBELL, A. (1957). Synchronization of cell division. *Bact. Rev.* **21**, 263.

CAMPBELL, A. M. (1968). Techniques for studying defective bacteriophages. In *Methods in Virology*, vol. IV, p. 279. Ed. by K. Maramorosch and H. Koprowski. New York and London: Academic Press.

CARO, L. (1967). Progress in high-resolution autoradiography. *Prog. Biophys.* **16**, 171.

CARO, L. G. & TUBERGEN, R. P. VAN (1962). High-resolution autoradiography. 1. Methods. *J. Cell Biol.* **15**, 173.

REFERENCES

CARRIER, E. B. & McCLESKEY, C. S. (1962). Intracellular starch formation in Corynebacteria. *J. Bact.* **17**, 1029.

CASSEL, W. A. (1951). A procedure for the simultaneous demonstration of the cell walls and chromatinic bodies of bacteria. *J. Bact.* **62**, 239.

CASTER, J. H. (1967). Selection of thymine-requiring strains from *Escherichia coli* on solid medium. *J. Bact.* **94**, 1804.

CATLIN, B. W. (1956). Response of *Escherichia coli* to ferrous ions. III. Application of a method for estimating cellular division to a study of recovery and mutation. *J. Bact.* **71**, 406.

CAVALLI, L. & MAGNI, G. (1943). Quantitative Untersuchungen über die Virulenz. III. Mitteilung: Analyse der Häufigkeits-verteilung der Absterbezeiten von infizierten Mäusen. *Zentbl. Bakt. ParasitKde* (*Abt. I*), Orig. **150**, 353.

CAVALLI, L. L. (1952). Genetic analysis of drug resistance. *Bull. Wld Hlth Org.* **6**, 185.

CAVALLI-SFORZA, L. L. & LEDERBERG, J. (1956). Isolation of pre-adaptive mutants in bacteria by sib selection. *Genetics, Princeton* **41**, 367.

CERDA-OLMEDO, E., HANAWALT, P. C. & GUEROLA, N. (1968). Mutagenesis of the replication point by nitrosoguanidine: map and pattern of replication of the *Escherichia coli* chromosome. *J. molec. Biol.* **33**, 705.

CHERRY, G. B., KEMP, S. D. & PARKER, A. (1963). The sterilization of air. *Prog. industr. Microbiol.* **4**, 35.

CHERRY, W. B. & MOODY, M. D. (1965). Fluorescent-antibody techniques in diagnostic bacteriology. *Bact. Rev.* **29**, 222.

CHICK, H. (1930). The theory of disinfection. In *A System of Bacteriology in Relation to Medicine*, vol. I, p. 179. London: H.M. Stationery Office.

CHURCH, B. D., HALVORSON, H., RAMSEY, D. S. & HARTMAN, R. S. (1956). Population heterogeneity in the resistance of aerobic spores to ethylene oxide. *J. Bact.* **72**, 242.

CITRI, N. & POLLOCK, M. R. (1966). The biochemistry and function of β-lactamase (penicillinase). *Adv. Enzymol.* **28**, 237.

CLARE, N. T. (1956). Photodynamic action and its pathological effects. In *Radiation Biology*, vol. III, p. 693. Ed. by A. Hollaender. New York: McGraw-Hill Book Inc.

CLARK, A. J. (1937). General pharmacology. In *Handbuch der experimentalle Pharmakologie, Enganzung*, vol. IV. A. Heffter. Berlin: Springer.

CLARKE, P. H. & COWAN, S. T. (1952). Biochemical methods for bacteriology. *J. gen. Microbiol.* **6**, 187.

CLIFTON, C. E. (1937). A comparison of the metabolic activities of *Aerobacter aerogenes*, *Eberthella typhi* and *Escherichia coli*. *J. Bact.* **33**, 145.

CLOWES, R. C. (1964). Transfer génétique des facteurs colicinogènes. *Annls Inst. Pasteur, Paris* **107**, *suppl.* 74.

CLOWES, R. C. & HAYES, W. (1968). *Experiments in Microbial Genetics*. Oxford and Edinburgh: Blackwell Scientific Publications.

COCHRAN, W. G. (1950). Estimation of bacterial densities by means of the 'most probable number'. *Biometrics* **6**, 105.

COETZEE, J. N. (1962). Sucrose fermentation by *Proteus hauseri*. *J. gen. Microbiol.* **29**, 455.

REFERENCES

Cohen, S., Snyder, J. C. & Mueller, J. H. (1941). Factors concerned in the growth of *Corynebacterium diphtheriae* from minute inocula. *J. Bact.* **41**, 581.

Cohen-Bazire, G. & Jolit, M. (1953). Isolement par sélection de mutants d'*Escherichia coli* synthétisant spontanément l'amylomaltase et la β-galactosidase. *Annls Inst. Pasteur, Paris* **84**, 937.

Cohn, M. (1952). Techniques and analysis of the quantitative precipitin reaction. *Meth. med. Res.* **5**, 301.

Cole, R. M. (1965). Bacterial cell-wall replication followed by immunofluorescence. *Bact. Rev.* **29**, 326.

Cole, S. W. & Onslow, H. (1916). On a substitute for peptone and a standard nutrient medium for bacteriological purposes. *Lancet* ii, 9.

Collins, J. F. (1964). The distribution and formation of penicillinase in a bacterial population of *Bacillus licheniformis*. *J. gen. Microbiol.* **34**, 363.

Conn, H. J. (1961). *Biological Stains*, 7th ed. Baltimore: Williams and Wilkins Co.

Cooper, C. M., Fernstrom, G. A. & Miller, S. A. (1944). Performance of agitated gas-liquid contactors. *Ind. Engng Chem. ind Edn* **36**, 504.

Cooper, K. E. (1963). The theory of antibiotic inhibition zones. In *Analytical Microbiology*. Ed. by F. Kavanagh. London: Academic Press.

Cooper, S. & Helmstetter, C. E. (1968). Chromosome replication and the division cycle of *Escherichia coli* B/r. *J. molec. Biol.* **31**, 519.

Corman, J., Tsuchiya, H. M., Koepsell, H. J., Benedict, R. G., Kelley, S. E., Feger, V. H., Dworschack, R. G. & Jackson, R. W. (1957). Oxygen absorption rates in laboratory and pilot plant equipment. *Appl. Microbiol.* **5**, 313.

Corwin, L. M., Fanning, G. R., Feldman, F. & Margolin, P. (1966). Mutation leading to increased sensitivity to chromium in *Salmonella typhimurium*. *J. Bact.* **91**, 1509.

Cosslett, V. E. (1966). *Modern Microscopy*. London: G. Bell and Sons Ltd.

Cotton, I. M. & Lockingen, L. S. (1963). Inactivation of bacteriophage by chloroform and X-irradiation. *Proc. natn. Acad. Sci., U.S.A.* **50**, 363.

Cowan, S. T. & Steel, K. J. (1965). *Manual for the Identification of Medical Bacteria*. Cambridge University Press.

Cowles, P. B. (1939). The disinfection concentration exponent. *Yale J. Biol. Med.* **12**, 697.

Coyne, F. P. (1933). The effect of carbon dioxide on bacterial growth. *Proc. R. Soc.* B **113**, 196.

Crone, P. B. (1948). The counting of surface colonies of bacteria. *J. Hyg., Camb.* **46**, 426.

Crowle, A. J. (1961). *Immunodiffusion*. New York and London: Academic Press.

Cruikshank, R., ed. (1965). *Mackie & McCartney's Handbook of Bacteriology*, 11th ed. Edinburgh and London: E. and S. Livingstone Ltd.

Cummins, A. B. & Hale, F. B., Jr. (1956). Filtration. In *Technique of Organic Chemistry*, vol. III, pt. 1, p. 607. Ed. by A. Weissberger. New York: Interscience Publishers Inc.

Curby, W. A., Swanton, E. M. & Lind, H. E. (1963). Electrical counting characteristics of several equivolume micro-organisms. *J. gen. Microbiol.* **32**, 33.

REFERENCES

CURRAN, H. R. (1952). Resistance in bacterial spores. *Bact. Rev.* **16**, 111.

CUTLER, R. G. & EVANS, J. E. (1966). Synchronization of bacteria by a stationary-phase method. *J. Bact.* **91**, 469.

CUZIN, F. & JACOB, F. (1967). Mutations de l'épisome F. d'*Escherichia coli* K12. 1. Mutations défectives. *Annls Inst. Pasteur, Paris* **112**, 1.

DACIE, J. V. & LEWIS, S. M. (1963). *Practical Haematology*. London: J. and A. Churchill Ltd.

DADE, H. A. (1958). Simple field condensers for Köhler illumination. *J. Quekett micr. Cl.* **28**, 31.

DANNENBERG, A. M. & SCOTT, E. M. (1956). Determination of respiratory LD 50 from number of primary lesions as illustrated by meliodosis. *Proc. Soc. exp. Biol. Med.* **92**, 571.

DARMADY, E. M., HUGHES, K. E. A. & JONES, J. D. (1958). Thermal death-time of spores in dry heat. *Lancet* ii, 766.

DARMADY, E. M., HUGHES, K. E. A., JONES, J. D., PRINCE, D. & TUKE, W. (1961). Sterilization by dry heat. *J. clin. Path.* **14**, 38.

DAVIS, B. D. (1948). Absorption of bacteriostatic quantities of fatty acid from media by large inocula of tubercle bacilli. *Publ. Hlth Rep., Wash.* **63**, 455.

DAVIS, B. D. (1950). Studies on nutritionally deficient bacterial mutants isolated by means of penicillin. *Experientia* **6**, 41.

DAVIS, B. D. & DUBOS, R. J. (1947). The binding of fatty acids by serum albumin, a protective growth factor in bacteriological media. *J. exp. Med.* **86**, 215.

DAVIS, B. D. & MINGIOLI, E. S. (1950). Mutants of *Escherichia coli* requiring methionine or vitamin B_{12}. *J. Bact.* **60**, 17.

DEARMON, I. A. & LINCOLN, R. E. (1959). Number of animals required in the bio-assay of pathogens. *J. Bact.* **78**, 651.

DEINDOERFER, F. H. (1957). Calculation of heat sterilization times for fermentation media. *Appl. Microbiol.* **5**, 221.

DELBRÜCK, M. (1945). The burst size distribution in the growth of bacterial viruses (bacteriophages). *J. Bact.* **50**, 131.

DELORY, G. E. & KING, E. J. (1945). A sodium carbonate-bicarbonate buffer for alkaline phosphatases. *Biochem. J.* **39**, 245.

DEMAIN, A. L. (1958). Minimal media for quantitative studies with *Bacillus subtilis*. *J. Bact.* **75**, 517.

DEMEREC, M., ADELBERG, E. A., CLARK, A. J. & HARTMAN, P. E. (1966). A proposal for a uniform nomenclature in bacterial genetics. *Genetics* **54**, 61. Also *J. gen. Microbiol.* (1968), **50**, 1.

DEMEREC, M. & CAHN, E. (1953). Studies of mutability in nutritionally deficient strains of *Escherichia coli*. I. Genetic analysis of five auxotrophic strains. *J. Bact.* **65**, 72.

DETTORI, R. & NERI, M. G. (1966). Filamentous sex-specific bacteriophage of *E. coli* K12. II. Thermal inactivation of phage Ec9 compared with that of two other single-stranded DNA bacteriophages. *G. Microbiol.* **14**, 205.

DEWEY, W. C. & COLE, A. (1962). Effects of heterogeneous populations on radiation survival curves. *Nature, Lond.* **194**, 660.

DIENES, L. (1967). Permanent stained agar preparations of *Mycoplasma* and of *L* forms of bacteria. *J. Bact.* **93**, 689.

REFERENCES

DOBSON, A. & BULLEN, J. J. (1964). A method for the control of E_h and pH during bacterial growth. *J. gen. Microbiol.* **35**, 169.

DONALD, C., PASSEY, B. I. & SWABY, R. J. (1952). A comparison of methods for removing trace metals from microbiological media. *J. gen. Microbiol.* **7**, 211.

DOUGHERTY, R. M., MCCLOSKEY, R. V. & STEWART, R. B. (1960). Analysis of the single dilution method of titration of psittacosis virus. *J. Bact.* **79**, 899.

DRABBLE, W. T. & STOCKER, B. A. D. (1968). R (transmissible drug resistance) factors in *Salmonella typhimurium*: pattern of transduction by phage P22 and ultraviolet-protection effect. *J. gen. Microbiol.* **53**, 109.

DRENNAN, J. G. & TEAGUE, O. (1917). A selective medium for the isolation of *B. pestis* from contaminated plague lesions and observations on the growth of *B. pestis* on autoclaved nutrient agar. *J. med. Res.* **36**, 519.

DUBNAU, E. & STOCKER, B. A. D. (1964). Genetics of plasmids in *Salmonella typhimurium*. *Nature, Lond.* **204**, 1112.

DUBOS, R. (1930). The bacteriostatic action of certain components of commercial peptones as affected by conditions of oxidation and reduction. *J. exp. Med.* **52**, 331.

DUGUID, J. P. (1951). The demonstration of bacterial capsules and slime. *J. Path. Bact.* **63**, 673.

DUGUID, J. P., ANDERSON, E. S. & CAMPBELL, I. (1966). Fimbriae and adhesive properties in salmonellae. *J. Path. Bact.* **92**, 107.

DUGUID, J. P. & WILKINSON, J. F. (1961). Environmentally induced changes in bacterial morphology. *Symp. Soc. gen. Microbiol.* **11**, 69.

DULBECCO, R. & VOGT, M. (1954). Plaque formation and isolation of pure lines with poliomyelitis viruses. *J. exp. Med.* **99**, 167.

ECHOLS, H., GAREN, A., GAREN, S. & TORRIANI, A. (1961). Genetic control of repression of alkaline phosphatase in *E. coli*. *J. molec. Biol.* **3**, 425.

ECKER, R. E. & LOCKHART, W. R. (1959*a*). A rapid membrane filter method for direct counts of microorganisms from small samples. *J. Bact.* **77**, 173.

ECKER, R. E. & LOCKHART, W. R. (1959*b*). Calibration of laboratory aeration apparatus. *Appl. Microbiol.* **7**, 102.

ECKERT, E. A., BEARD, D. & BEARD, J. W. (1956). Virus of avian erythroblastosis. I. Titration of infectivity. *J. natn. Cancer Inst.* **16**, 1099.

ECKERT, E. A., WATERS, N. F., BURMEISTER, B. R., BEARD, D. & BEARD, J. W. (1954). Dose-response relations in experimental transmission of avian erythromyeloblastic leukosis. IV. Strain differences in host-response to the virus. *J. natn. Cancer Inst.* **14**, 1067.

EDDY, A. A. (1953). Death rate of populations of *Bact. lactis aerogenes*. III. Interpretation of survival curves. *Proc. R. Soc.* B **141**, 137.

EDEBO, L., HEDÉN, C.-G., HOLME, T. & ZACHARIAS, B. (1962). Laboratory equipment for growing aerobic bacteria in high yields. *Acta path. microbiol. scand.* **55**, 91.

EDWARDS, P. R. & EWING, W. H. (1962). *Identification of Enterobacteriaceae*, 2nd ed. Minneapolis, Minnesota: Burgess Publishing Co.

EDWARDS, S. & MEYNELL, G. G. (1968). General method for isolating derepressed bacterial sex factors. *Nature, Lond.* **219**, 869.

EHRLICH, R. (1955). Technique for microscopic count of microorganisms directly on membrane filters. *J. Bact.* **70**, 265.

EISENHART, C. & WILSON, P. W. (1943). Statistical methods and control in bacteriology. *Bact. Rev.* **7**, 57.

EISENSTARK, A. (1968). Bacteriophage techniques. In *Methods in Virology*, vol. I, p. 449. Ed. by K. Maramorosch and H. Koprowski. New York and London: Academic Press.

ELFORD, W. J. (1938). The sizes of viruses and bacteriophages, and methods for their determination. In *Handbuch der Virusforschung*, vol. I, p. 126. Ed. by R. Doerr and C. Hallauer. Wien: Julius Springer.

ELLAR, D., LUNDGREN, D. G., OKAMURA, K. & MARCHESSAULT, R. H. (1968). Morphology of poly-β-hydroxybutyrate granules. *J. molec. Biol.* **35**, 489.

ELLIS, E. L. & DELBRÜCK, M. (1939). The growth of bacteriophage. *J. gen. Physiol.* **22**, 365.

EMMENS, C. W. (1948). *Principles of Biological Assay.* London: Chapman and Hall Ltd.

ERCOLANI, G. L. (1967*a*). Bacterial canker of tomato. I. Analysis of some factors affecting the response of tomato to *Corynebacterium michiganense* (E. F. Sm.) Jens. *Phytopathologia mediterranea* **6**, 19.

ERCOLANI, G. L. (1967*b*). Bacterial canker of tomato. II. Interpretation of the aetiology of the quantal response of tomato to *Corynebacterium michiganense* (E.F.Sm) Jens. by the hypothesis of independent action. *Phytopathologia mediterranea* **6**, 30.

ERRINGTON, F. P., POWELL, E. O. & THOMPSON, N. (1965). Growth characteristics of some Gram-negative bacteria. *J. gen. Microbiol.* **39**, 109.

FALKOW, S., HAAPALA, D. K. & SILVER, R. P. (1969). Relationships between extrachromosomal elements. *Ciba Foundation Symposium on Bacterial Episomes and Plasmids*, p. 136. Ed. G. E. W. Wolstenholme and M. O'Connor. London: J. and A. Churchill, Ltd.

FARRER, K. T. H. (1946). The liberation of aneurin, cocarboxylase, riboflavin and nicotinic acid from autolysing yeast cells. *Aust. J. exp. Biol. med. Sci.* **24**, 213.

FAZEKAS DE ST GROTH, S. (1955). Production of non-infective particles among influenza viruses: do changes in virulence accompany the von Magnus phenomenon? *J. Hyg., Camb.* **53**, 276.

FERNELIUS, A. L., WILKES, C. E., DEARMON, I. A. & LINCOLN, R. E. (1958). A probit method to interpret thermal inactivation of bacterial spores. *J. Bact.* **75**, 300.

FERRY, J. D. (1936). Ultrafilter membranes and ultrafiltration. *Chem. Rev.* **18**, 373.

FILDES, P. (1920). A new medium for the growth of *B. influenzae*. *Brit. J. exp. Path.* **1**, 129.

FILDES, P. & GLADSTONE, G. P. (1939). Glutamine and the growth of bacteria. *Brit. J. exp. Path.* **20**, 334.

FILDES, P. & MCINTOSH, J. (1921). An improved form of McIntosh and Fildes' anaerobic jar. *Brit. J. exp. Path.* **2**, 153.

FILDES, P. & SMART, W. A. M. (1926). Volumetric measurement by drops in bacteriological technique. *Brit. J. exp. Path.* **7**, 68.

FINKELSTEIN, R. A. & LANKFORD, C. E. (1957). A bacteriotoxic substance in autoclaved culture media containing glucose and phosphate. *Appl. Microbiol.* **5**, 74.

FINN, R. K. (1955). Measurements of lag. *J. Bact.* **70**, 352.

FINN, R. K. (1967). Agitation and aeration. In *Biochemical and Biological Engineering Science*, vol. I, p. 69. Ed. by N. Blakebrough. London and New York: Academic Press.

FINNEY, D. J. (1951). The estimation of bacterial densities from dilution series. *J. Hyg., Camb.* **49**, 26.

FINNEY, D. J. (1952). *Probit Analysis*, 2nd ed. Cambridge University Press.

FINNEY, D. J. (1964). *Statistical Method in Biological Assay*, 2nd ed. London: Charles Griffin and Co. Ltd.

FINNEY, D. J., HAZLEWOOD, T. & SMITH, M. J. (1955). Logarithms to base 2. *J. gen. Microbiol.* **12**, 222.

FISHER, R. A. (1950). *Statistical Methods for Research Workers*, 11th ed. Edinburgh: Oliver and Boyd.

FISHER, R. A. & YATES, F. (1963). *Statistical Tables for Biological, Agricultural and Medical Research*, 6th ed. Edinburgh: Oliver and Boyd.

FLAMM, W. G., BOND, H. E. & BURR, H. E. (1966). Density-gradient centrifugation of DNA in a fixed-angle rotor. *Biochem. biophys. Acta* **129**, 310.

FLECK, J., MINCK, R. & KIRN, A. (1962). Etude de l'action inhibitrice spécifique des antisérums sur les cultures des bactéries. III. Localisation de l'antigène H par la technique de l'immuno-fluorescence. *Annls Inst. Pasteur, Paris* **102**, 243.

FLUEGEL, W. (1963). A shelf-type gradient incubator. *Canad. J. Microbiol.* **9**, 859.

FOLEY, M. K. & SCHAUB, I. G. (1944). Importance of the use of thioglycollate medium in diagnostic bacteriology. *J. Bact.* **47**, 455.

FORD, J. E., PERRY, K. D. & BRIGGS, C. A. E. (1958). Nutrition of lactic acid bacteria isolated from the rumen. *J. gen. Microbiol.* **18**, 273.

FORRO, F. (1965). Autoradiographic studies of bacterial chromosome replication in amino-acid deficient *Escherichia coli* $15T^-$. *Biophys. J.* **5**, 629.

FOTHERGILL, P. G. & HIDE, D. (1962). Comparative nutritional studies of *Pythium* spp. *J. gen. Microbiol.* **29**, 325.

FOWLER, J. F. (1964). Differences in survival curve shapes for formal multi-target- and multi-hit models. *Physics Med. Biol.* **9**, 177.

FOWLKS, W. L. (1959). The mechanism of the photodynamic effect. *J. invest. Dermatol.* **32**, 233.

FRANÇON, M. (1961). *Progress in Microscopy*. Oxford: Pergamon Press.

FREDERICQ, P. (1947). Eosin methyl-green sulfite agar: a modification of Levine's E.M.B. agar. *J. Bact.* **54**, 662.

FREDERICQ, P. (1965). A note on the classification of colicines. *Zentbl. Bakt. Parasitkde (Abt I)* **196**, 140.

FREDRICKSON, A. G. (1966). Stochastic models for sterilization. *Biotech. Bioengng* **8**, 167.

FREIFELDER, D. (1966). Acridine orange- and methylene blue-sensitized induction of *Escherichia coli* lysogenic for phage λ. *Virology* **30**, 567.

FRIEDMAN, M. & COWLES, P. B. (1953). The bacteriophages of *Bacillus megaterium*. I. Serological, physical and biological properties. *J. Bact.* **66**, 379.

FRY, R. M. & GREAVES, R. I. N. (1951). The survival of bacteria during and after drying. *J. Hyg., Camb.* **49**, 220.

FUERST, C. R. & STENT, G. S. (1956). Inactivation of bacteria by decay of incorporated radioactive phosphorus. *J. gen. Physiol.* **40**, 73.

REFERENCES

FULTHORPE, A. J. (1951). The variability in gel-producing properties of commercial agar and its influence on bacterial growth. *J. Hyg., Camb.* **49**, 127.

FULTON, R. W. (1962). The effect of dilution on necrotic ringspot infectivity and the enhancement of infectivity by noninfective virus. *Virology* **18**, 477.

GADEN, E. L. (1962). Improved shaken flask performance. *Biotechnol. Bioengng* **4**, 99.

GALLOWAY, I. A. & ELFORD, W. J. (1931). Filtration of the virus of foot-and-mouth disease through a new series of graded collodion membranes. *Brit. J. exp. Path.* **12**, 407.

GARBER, N., ZOHAR, D. & MICHLIN, H. (1968). Rapid chemical determination of chloramphenicol acetylation by cell extracts and by cell suspensions of *E. coli* B. *Nature, Lond.* **219**, 401.

GARD, S. (1940). Encephalomyelitis of mice. II. A method for the measurement of virus activity. *J. exp. Med.* **72**, 69.

GARD, S. (1943). Purification of poliomyelitis viruses. Experiments on murine and human strains. *Acta med. scand.*, suppl. 143.

GAREN, A. (1968). Sense and nonsense in the genetic code. *Science, N.Y.* **160**, 149.

GAREN, A., GAREN, S. & WILHELM, R. C. (1965). Suppressor genes for nonsense mutations. I. The *Su-1*, *Su-2* and *Su-3* genes of *Escherichia coli*. *J. molec. Biol.* **14**, 167.

GAREN, A. & SIDDIQI, O. (1962). Suppression of mutations in the alkaline phosphatase structural cistron of *E. coli*. *Proc. natn. Acad. Sci., U.S.A.* **48**, 1121.

GART, J. J. & WEISS, G. H. (1967). Graphically oriented tests for host variability in dilution experiments. *Biometrics* **23**, 269.

GASPAR, A. J. & LEISE, J. M. (1956). Inhibitory effect of grid imprints on growth of *Pasteurella tularensis* on membrane filters. *J. Bact.* **71**, 728.

GERHARDT, P. & GALLUP, D. M. (1963). Dialysis flask for concentrated culture of microorganisms. *J. Bact.* **86**, 919.

GERRARD, H. N., PARKER, M. S. & PORTER, G. S. (1961). The effect of pipetting on the concentration of homogeneous spore suspensions. Part II. *J. Pharm., Lond.* **13**, 405.

GERSHENFELD, L. & TICE, L. F. (1941). Gelatin for bacteriological use. *J. Bact.* **41**, 645.

GIBBS, B. M. & SKINNER, F. A., eds. (1966). *Identification Methods for Microbiologists.* Part A. London and New York: Academic Press.

GIFFORD, G. E. (1960). Occurrence of morpholine in steam and its solution during autoclaving. *J. Bact.* **80**, 278.

GILES, K. W. & MYERS, A. (1965). An improved diphenylamine method for the estimation of deoxyribonucleic acid. *Nature, Lond.* **206**, 93.

GINOZA, W. (1958). Kinetics of heat inactivation of ribonucleic acid of tobacco mosaic virus. *Nature, Lond.* **181**, 958.

GINSBERG, D. M. & JAGGER, J. (1962). Possible errors arising from the use of fritted-glass filters for bubbling of cell suspensions, especially in irradiation experiments. *J. Bact.* **83**, 1361.

GLOVER, S. W., SCHELL, J., SYMONDS, N. & STACEY, K. A. (1963). The control of host-induced modification by phage P1. *Genet. Res.* **4**, 480.

GOLDBERG, L. J., WATKINS, H. M. S., DOLMATZ, M. S. & SCHLAMM, N. A. (1954). Studies on the experimental epidemiology of respiratory infections. VI. The relationship between dose of microorganisms and subsequent infection or death of a host. *J. infect. Dis.* **94**, 9.

GOLDMAN, M. (1968). *Fluorescent Antibody Methods*. New York and London: Academic Press.

GOLDSWORTHY, N. E., STILL, J. L. & DUMARESQ, J. A. (1938). Some sources of error in the interpretation of fermentation reactions, with special reference to the effects of serum enzymes. *J. Path. Bact.* **46**, 253.

GOLUB, O. J. (1948). A single-dilution method for the estimation of LD 50 titers of the psittacosis-LGV group of viruses in chick embryos. *J. Immun.* **59**, 71.

GORDON, R. A. & MURRELL, W. G. (1967). Simple method of detecting spore septum formation and synchrony of sporulation. *J. Bact.* **93**, 495.

GORINI, L. (1961). Effect of L-cystine on initiation of anaerobic growth of *Escherichia coli* and *Aerobacter aerogenes*. *J. Bact.* **82**, 305.

GORINI, L. & BECKWITH, J. R. (1966). Suppression. *A. Rev. Microbiol.* **20**, 401.

GORINI, L. & KAUFMAN, H. (1959). Selecting bacterial mutants by the penicillin method. *Science, N.Y.* **131**, 604.

GORRILL, R. H. & MCNEIL, E. M. (1960). The effect of cold diluent on the viable count of *Pseudomonas pyocyanea*. *J. gen. Microbiol.* **22**, 437.

GOSSLING, B. S. (1959). The loss of viability of bacteria in suspension due to changing the ionic environment. *J. appl. Bact.* **21**, 220.

GRANT, C. L. & PRAMER, D. (1962). Minor element composition of yeast extract. *J. Bact.* **84**, 869.

GRIEG, M. E. & HOOGERHEIDE, J. C. (1941). The correlation of bacterial growth with oxygen consumption. *J. Bact.* **41**, 549.

GRIGG, G. W. (1958). Competitive suppression and the detection of mutations in microbial populations. *Aust. J. biol. Sci.* **11**, 69.

GROSS, J. D. & CARO, L. G. (1966). DNA transfer in bacterial conjugation. *J. molec. Biol.* **16**, 269.

GROSS, J. D., KARAMATA, D. & HEMPSTEAD, P. J. (1968). Temperature-sensitive mutants of *Bacillus subtilis* defective in DNA synthesis. *Cold Spr. Harb. symp. quant. Biol.* **33**, 307.

GROSSMAN, J. P. & POSTGATE, J. R. (1953). Cultivation of sulphate-reducing bacteria. *Nature, Lond.* **171**, 600.

GRULA, E. A. (1960). Cell division in a species of *Erwinia*. II. Inhibition of division by D-amino acids. *J. Bact.* **80**, 375.

GRUNBAUM, A. S. & HARE, E. H. (1902). Note on media for distinguishing *B. coli*, *B. typhosus* and related species. *Br. med. J.* i, 1473.

GUIRARD, B. M. (1958). Microbial nutrition. *A. Rev. Microbiol.* **12**, 247.

GUNDERSEN, W. B. (1965). Reduced streptomycin killing in *E. coli* carrying the mu-factor in its extrachromosomal state. *Acta path. microbiol. scand.* **65**, 627.

GUNSALUS, I. C. & SHUSTER, C. W. (1961). Energy-yielding metabolism in bacteria. In *The Bacteria*, vol. II, p. 1. Ed. by I. C. Gunsalus and R. Y. Stanier. New York and London: Academic Press.

GUNTER, S. E. & KOHN, H. I. (1956). The effect of incubation time on macrocolony formation after X-irradiation of microorganisms. *J. Bact.* **71**, 124.

GUZE, L. B., ed. (1968). *Microbial Protoplasts, Spheroplasts and L-forms*. Baltimore: The Williams and Wilkins Co.

HALDANE, J. B. S. (1939). Sampling errors in the determination of bacterial or virus density by the dilution method. *J. Hyg., Camb.* **39**, 289.

HALE, C. M. F. (1953). The use of phosphomolybdic acid in the mordanting of bacterial cell walls. *Lab. Pract.* **2**, 115.

HALL, I. C. (1928). Anaerobiosis. In *The Newer Knowledge of Bacteriology and Immunology*, p. 198. Ed. by E. O. Jordan and I. S. Falk. Chicago: University of Chicago Press.

HALL, I. C. (1929). A review of the development and application of physical and chemical principles in the cultivation of obligately anaerobic bacteria. *J. Bact.* **17**, 255.

HALVORSON, H. O. (1965). Sequential expression of biochemical events during intracellular differentiation. *Symp. Soc. gen. Microbiol.* **15**, 343.

HALVORSON, H. O. & ZIEGLER, N. R. (1933). Application of statistics to problems in bacteriology. I. A means of determining bacterial population by the dilution method. *J. Bact.* **25**, 101.

HANAWALT, P. C., MAALØE, O., CUMMINGS, D. J. & SCHAECHTER, M. (1961). The normal DNA replication cycle. II. *J. molec. Biol.* **3**, 156.

HANKE, M. E. & KATZ, Y. L. (1943). An electrolytic method for controlling oxidation-reduction potential and its application in the study of anaerobiosis. *Archs Biochem. Biophys.* **2**, 183.

HANKS, J. H. (1966). Host-dependent microbes. *Bact. Rev.* **30**, 114.

HANKS, J. H. & JAMES, D. F. (1940). The enumeration of bacteria by the microscopic method. *J. Bact.* **39**, 297.

HARDWICK, W. A., GUIRARD, B. & FOSTER, J. W. (1951). Antisporulation factors in complex organic media. II. Saturated fatty acids as antisporulation factors. *J. Bact.* **61**, 145.

HAROLD, F. M. (1966). Inorganic polyphosphates in biology: structure, metabolism and function. *Bact. Rev.* **30**, 772.

HARPER, G. J. & CAWSTON, W. C. (1945). The *in vitro* determination of the sulphonamide sensitivity of bacteria. *J. Path. Bact.* **57**, 59.

HARPER, G. J. & HOOD, A. M. (1962). Lung retention in mice exposed to airborne micro-organisms. *Nature, Lond.* **196**, 598.

HARRIS, N. D. (1963). The influence of the recovery medium and the incubation temperature on the survival of damaged bacteria. *J. appl. Bact.* **26**, 387.

HARTLEY, P. (1922). The value of Douglas's medium for the production of diphtheria toxin. *J. Path. Bact.* **25**, 479.

HARTLEY, W. G. (1962). *Microscopy*. London: The English Universities Press Ltd.

HARTMAN, P. E., PAYNE, J. I. & MUDD, S. (1955). Cytological analysis of ultraviolet irradiated *Escherichia coli*. I. Cytology of lysogenic *E. coli* and a non-lysogenic derivative. *J. Bact.* **70**, 531.

HARVEY, R. J. & MARR, A. G. (1966). Measurement of size distributions of bacterial cells. *J. Bact.* **92**, 805.

HAYES, W. (1957). The kinetics of the mating process in *Escherichia coli*. *J. gen. Microbiol.* **16**, 97.

HAYES, W. (1966). Genetic transformation: a retrospective appreciation. *J. gen. Microbiol.* **45**, 385.

HAYFLICK, L., ed. (1967). Biology of the mycoplasma. *Ann. N.Y. Acad. Sci.* **143**, 1.

HAYWARD, A. C. (1957). Detection of gas production from glucose by heterofermentative lactic acid bacteria. *J. gen. Microbiol.* **16**, 9.

HAYWARD, N. J. & MILES, A. A. (1943). Iron media for cultivation of anaerobic bacteria in air. *Lancet* i, 645.

HEALY, M. J. R. (1950). The planning of probit assays. *Biometrics* **6**, 424.

HEDÉN, C.-G. (1957). Pulsing aeration of microbial cultures. *Nature, Lond.* **179**, 324.

HELINSKI, D. R. & YANOFSKY, C. (1966). Genetic control of protein structure. In *The Proteins*, vol. IV, p. 1. Ed. by H. Neurath. New York and London: Academic Press.

HELLER, C. L. (1954). A simple method for producing anaerobiosis. *J. appl. Bact.* **17**, 202.

HELMSTETTER, C. E. & COOPER, S. (1968). DNA synthesis during the division cycle of rapidly growing *Escherichia coli* B/r. *J. molec. Biol.* **31**, 507.

HENDRY, C. B. (1938). The effect of serum maltase on fermentation tests with gonococci. *J. Path. Bact.* **46**, 383.

HENRICI, A. T. (1923). A statistical study of the form and growth of *Bacterium coli*. *Proc. Soc. exp. Biol. Med.* **21**, 215.

HENRY, B. S. (1933). Dissociation in the genus *Brucella*. *J. infect. Dis.* **52**, 374.

HENRY, P. S. H. (1959). Physical aspects of sterilizing cotton articles by steam. *J. appl. Bact.* **22**, 159.

HENRY, P. S. H. & SCOTT, E. (1963). Residual air in the steam sterilization of textiles with pre-vacuum. *J. appl. Bact.* **26**, 234.

HERBERT, D. (1949). Studies on the nutrition of *Pasteurella pestis*, and factors affecting the growth of isolated cells on an agar surface. *Brit. J. exp. Path.* **30**, 509.

HERBERT, D. (1959). Some principles of continuous culture. In *Recent Progress in Microbiology*. Ed. by G. Tunevall. Stockholm: Almqvist and Wiksell.

HERBERT, D. (1961 *a*). The chemical composition of micro-organisms as a function of their environment. *Symp. Soc. gen. Microbiol.* **11**, 391.

HERBERT, D. (1961 *b*). A theoretical analysis of continuous culture systems. *Soc. chem. Ind. Monograph*, no. 12, p. 21.

HERBERT, D., ELSWORTH, R. & TELLING, R. C. (1956). The continuous culture of bacteria: a theoretical and experimental study. *J. gen. Microbiol.* **14**, 601.

HERBERT, D., PHIPPS, P. J. & TEMPEST, D. W. (1965). The chemostat: design and instrumentation. *Lab. Pract.* **14**, 1150.

HERSHEY, A. D. (1955). An upper limit to the protein content of the germinal substance of bacteriophage T2. *Virology* **1**, 108.

HERSHEY, A. D. & BRONFENBRENNER, J. (1938). Factors limiting bacterial growth. III. Cell size and 'physiologic youth' in *Bacterium coli* cultures. *J. gen. Physiol.* **21**, 721.

HERSHEY, A. D., KALMANSON, G. & BRONFENBRENNER, J. (1944). Coordinate effects of electrolyte and antibody on infectivity of bacteriophage. *J. Immun.* **48**, 221.

HEWITT, L. F. (1950). *Oxidation-Reduction Potentials in Bacteriology and Biochemistry*, 6th ed. Edinburgh: E. and S. Livingstone Ltd.

HIATT, C. W. (1960). Photodynamic inactivation of viruses. *Trans. N.Y. Acad. Sci.* **23**, 66.

REFERENCES

Hiatt, C. W. (1964). Kinetics of the inactivation of viruses. *Bact. Rev.* **28**, 150.

Hicks, J. D. & Matthaei, E. (1955). Fluorescence in histology. *J. Path. Bact.* **70**, 1.

Himmelfarb, P., Read, R. B., Jr. & Litsky, W. (1961). An evaluation of β-propiolactone for the sterilization of fermentation media. *Appl. Microbiol.* **9**, 534.

Hirota, Y. (1956). Artificial elimination of the F factor in *Bact. coli* K-12. *Nature, Lond.* **178**, 92.

Hirota, Y. (1960). The effect of acridine dyes on mating type factors in *Escherichia coli. Proc. natn. Acad. Sci., U.S.A.* **46**, 57.

Hirota, Y., Jacob, F., Ryter, A., Buttin, G. & Nakai, T. (1968). On the process of cellular division in *Escherichia coli.* 1. Asymmetrical cell division and production of deoxyribonucleic acid-less bacteria. *J. molec. Biol.* **35**, 175.

Hitchins, A. D., Kahn, A. J. & Slepecky, R. A. (1968). Interference contrast and phase contrast microscopy of sporulation and germination of *Bacillus megaterium. J. Bact.* **96**, 1811.

Hitchens, A. P. (1921). Advantages of culture mediums containing small percentages of agar. *J. infect. Dis.* **29**, 390.

Hitchens, A. P. & Leikind, M. C. (1939). The introduction of agar-agar into bacteriology. *J. Bact.* **37**, 485.

Hobson, P. N. & Summers, R. (1967). The continuous culture of anerobic bacteria. *J. gen. Microbiol.* **47**, 53.

Hoffman, R. K. & Warshowsky, B. (1958). *Beta*-Propiolactone vapor as a disinfectant. *Appl. Microbiol.* **6**, 358.

Hoffmann-Berling, H., Kaerner, H. C. & Knippers, R. (1966). Small bacteriophages. *Adv. Virus Res.* **12**, 329.

Hofsten, B. von (1962). The effect of copper on the growth of *Escherichia coli. Expl Cell Res.* **26**, 606.

Holborow, E. J. (1967). An ABC of modern immunology. *Lancet* i, 833, 890, 942, 995, 1049, 1098, 1148, 1208.

Hollaender, A. (1943). Effect of long ultraviolet and short visible radiation (3500–4900 Å) on *Escherichia coli. J. Bact.* **46**, 531.

Holliday, R. (1956). A new method for the identification of biochemical mutants of micro-organisms. *Nature, Lond.* **178**, 987.

Holt, J. M., Lux, W. & Valberg, L. S. (1963). Apparatus for the preparation of highly purified water suitable for trace metal analysis. *Canad. J. Biochem.* **41**, 2029.

Holt, L. B. (1962). The culture of *Streptococcus pneumoniae. J. gen. Microbiol.* **27**, 327.

Horn, H. J. (1956). Simplified LD_{50} (or ED_{50}) calculations. *Biometrics* **12**, 311.

Hoskins, J. K. (1934). Most probable numbers for evaluation of coli-aerogenes tests by fermentation tube method. *Publ. Hlth Rep., Wash.* **49**, 393.

Hotchkiss, R. D. (1948). A microchemical reaction resulting in the staining of polysaccharide structures in fixed tissue preparations. *Archs Biochem. Biophys.* **16**, 131.

Howe, A. F., Groom, T. & Carter, R. G. (1964). Use of polyethylene glycol in the concentration of protein solutions. *Analyt. Biochem.* **9**, 443.

REFERENCES

HOYT, A., MOORE, F. J. & THOMPSON, M. A. (1962). Analysis of mixed distributions in antituberculosis vaccine assays. *Amer. Rev. resp. Dis.* **86**, 733.

HUDDLESTON, F., HASLEY, D. E. & TORREY, J. P. (1927). Further studies on the isolation and cultivation of *Bacterium abortus* (Bang). *J. infect. Dis.* **40**, 352.

HUGH, R. & LEIFSON, E. (1953). The taxonomic significance of fermentative versus oxidative metabolism of carbohydrates by various Gram negative bacteria. *J. Bact.* **66**, 24.

HUGHES, W. H. (1956). Bacterial variation to sensitivity: an example of individuality in micro-organisms. *Nature, Lond.* **177**, 1132.

HUMPHREY, A. E. (1960). Air sterilization. *Adv. appl. Microbiol.* **2**, 301.

HUMPHREY, A. E. & NICKERSON, J. T. R. (1961). Testing thermal death data for significant non-logarithmic behavior. *Appl. Microbiol.* **9**, 282.

HUNTER, C. L. F., HARBORD, P. E. & RIDDETT, D. J. (1961). Packaging papers as bacterial barriers. In *Symposium* (1961 *a*), p. 166.

HURWITZ, C., DOPPEL, H. W. & ROSANO, C. L. (1964). Correlation of the *in vivo* action of streptomycin in survival and on protein synthesis by *Mycobacterium fortuitum*. *J. gen. Microbiol.* **35**, 159.

HURWITZ, C., ROSANO, C. L. & BLATTBERG, B. (1957). A test of the validity of reactivation of bacteria. *J. Bact.* **73**, 743.

HUTNER, S. H. (1942). Some growth requirements of *Erysipelothrix* and *Listerella*. *J. Bact.* **43**, 629.

HUTNER, S. H. (1950). Anaerobic and aerobic growth of purple bacteria (*Athiorhodaceae*) in chemically defined media. *J. gen. Microbiol.* **4**, 286.

HUTNER, S. H. & BJERKNES, C. A. (1948). Volatile preservatives for culture media. *Proc. Soc. exp. Biol. Med.* **67**, 393.

HYSLOP, N. ST G. (1961). A method of treating asbestos filter pads for virus filtration. *Nature, Lond.* **191**, 305.

INGRAM, M. & EDDY, B. P. (1953). A warning note on the use of a pipette for several dilutions on viable bacterial counts. *Lab. Pract.* **2**, 1.

IPSEN, J. (1944). Quantitative Studien über mäusepathogene Fleckfieberricketsien. *Schweiz. Z. allg. Path. Bakt.* **7**, 129.

IRWIN, J. O. (1942). The distribution of the logarithm of survival times when the true law is exponential. *J. Hyg., Camb.* **42**, 328.

ISAACS, A. (1957). Particle counts and infectivity titrations for animal viruses. *Adv. Virus Res.* **4**, 112.

IYER, V. (1960). Concentration and isolation of auxotrophic mutants of sporeforming bacteria. *J. Bact.* **79**, 309.

JACOB, F. & MONOD, J. (1961). Genetic regulatory mechanisms in the synthesis of proteins. *J. molec. Biol.* **3**, 318.

JACOB, F., SCHAEFFER, P. & WOLLMAN, E. L. (1960). Episomic elements in bacteria. *Symp. Soc. gen. Microbiol.* **10**, 67.

JACOB, F., SIMINOVITCH, L. & WOLLMAN, E. L. (1952). Sur la biosynthèse d'une colicine et sur son mode d'action. *Annls Inst. Pasteur, Paris* **83**, 295.

JACOBS, S. (1965). The determination of nitrogen in biological materials. *Meth. biochem. Anal.* **13**, 241.

JAMESON, J. E. (1961). A simple method for inducing motility and phase change in *Salmonella*. *Mon. Bull. Minist. Hlth* **20**, 14.

JETER, W. S. & WYNNE, E. S. (1949). Acid fuchsin methylene blue agar: a new differential medium for enteric bacteria. *J. Bact.* **58**, 429.

JOHNSON, F. H., EYRING, H. & POLISSAR, M. J. (1954). *The Kinetic Basis of Molecular Biology*. New York: John Wiley and Sons, Inc.

JOHNSON, R. B. (1954). Factors influencing the growth of *Shigella dysenteriae*. I. In a synthetic medium. *J. Bact.* **68**, 604.

JOHNSON, R. C. & GARY, N. D. (1963). Nutrition of *Leptospira pomona*. II. Fatty acid requirements. *J. Bact.* **85**, 976.

JOYS, T. M. (1968). The structure of flagella and the genetic control of flagellation in Eubacteriales. A review. *Antonie van Leeuwenhoek* **34**, 205.

JOYS, T. M. & STOCKER, B. A. D. (1966). Isolation and serological analysis of mutant forms of flagellar antigen *i* of *Salmonella typhimurium*. *J. gen. Microbiol.* **44**, 121.

JUDGE, L. F. & PELCZAR, M. J. (1955). The sterilization of carbohydrates with liquid ethylene oxide for microbiological fermentation tests. *Appl. Microbiol.* **3**, 292.

JUHLIN, I. & ERICSON, C. (1961). A new medium for the bacteriologic examination of stools (LSU-agar). *Acta path. microbiol. scand.* **52**, 185.

KABAT, E. A. & MAYER, M. M. (1961). *Experimental Immunochemistry*, 2nd ed. Baltimore: Charles C. Thomas.

KALCKAR, H. M. (1965). Galactose metabolism and cell 'sociology'. *Science, N.Y.* **150**, 305.

KAUDEWITZ, F. (1959). Production of bacterial mutants with nitrous acid. *Nature, Lond.* **183**, 1829.

KAVANAGH, F., ed. (1963). *Analytical Microbiology*. London: Academic Press.

KAY, D. H., ed. (1965). *Techniques for Electron Microscopy*. Oxford: Blackwell Scientific Publications.

KELLENBERGER, E. (1960). The physical state of the bacterial nucleus. *Symp. Soc. gen. Microbiol.* **10**, 39.

KELSEY, J. C. (1958). The testing of sterilisers. *Lancet* ii, 306.

KELSEY, J. C. (1961 *a*). The testing of sterilizers. 2. Thermophilic spore papers. *J. clin. Path.* **14**, 313.

KELSEY, J. C. (1961 *b*). Sterilization by ethylene oxide. *J. clin. Path.* **14**, 59.

KEMPNER, W. & SCHLAYER, C. (1942). Effect of CO_2 on the growth rate of the pneumococcus. *J. Bact.* **43**, 387.

KERRIDGE, D. (1961). The effect of environment on the formation of bacterial flagella. *Symp. Soc. gen. Microbiol.* **11**, 41.

KESTON, A. S. & ROSENBERG, D. (1967). Medium for differentiation of acid-producing colonies with homogeneously suspended calcium carbonate. *J. Bact.* **93**, 1475.

KINDLER, S. H., MAGER, J. & GROSSOWICZ, N. (1956). Nutritional studies with the *Clostridium botulinum* group. *J. gen. Microbiol.* **15**, 386.

KIRKPATRICK, J. & LENDRUM, A. C. (1939). A mounting medium for microscopical preparations giving good preservation of colour. *J. Path. Bact.* **49**, 592.

KJELDGAARD, N. O. (1967). Regulation of nucleic acid and protein formation in bacteria. *Advanc. microbial Physiol.* **1**, 39.

KLECZKOWSKI, A. (1949). The transformation of local lesion counts for statistical analysis. *Ann. appl. Biol.* **36**, 139.

REFERENCES

KLECZKOWSKI, A. (1950). Interpreting relationships between the concentrations of plant viruses and numbers of local lesions. *J. gen. Microbiol.* **4**, 53.

KLECZKOWSKI, A. (1951). Effects of non-ionizing radiations on viruses. *Adv. Virus Res.* **4**, 191.

KLIGER, I. J. & GUGGENHEIM, K. (1938). The influence of vitamin C on the growth of anaerobes in the presence of air, with special reference to the relative significance of Eh and O_2 in the growth of anaerobes. *J. Bact.* **35**, 141.

KNAYSI, G. (1935). A microscopic method of distinguishing dead from living bacterial cells. *J. Bact.* **30**, 193.

KNAYSI, G. (1957). New technique of growing molds in slide culture under environmental conditions similar to those prevailing in the petri dish. *J. Bact.* **73**, 431.

KNIGHT, B. C. J. G. (1930*a*). Oxidation-reduction studies in relation to bacterial growth. I. The oxidation-reduction potential of sterile meat broth. *Biochem. J.* **24**, 1066.

KNIGHT, B. C. J. G. (1930*b*). Oxidation-reduction studies in relation to bacterial growth. II. A method of poising the oxidation-reduction potential of bacteriological culture media. *Biochem. J.* **24**, 1075.

KNIGHT, B. C. J. G. & PROOM, H. (1950). A comparative survey of the nutrition and physiology of mesophilic species in the genus *Bacillus*. *J. gen. Microbiol.* **4**, 508.

KNOX, R. & PENIKETT, E. J. K. (1958). Influence of initial vacuum on steam sterilization of dressings. *Br. med. J.* i, 680.

KNOX, R. & PICKERILL, J. K. (1964). Efficient air removal from steam sterilisers without the use of high vacuum. *Lancet* i, 1318.

KNUDSEN, L. F. (1947). Sample size of parenteral solutions for sterility testing. *J. Am. pharm. Ass.* (*Sci. Edn*) **38**, 332.

KOCH, A. L. (1961). Some calculations on the turbidity of mitochondria and bacteria. *Biochim. Biophys. Acta* **51**, 429.

KOCH, A. L. (1966*a*). Distribution of cell size in growing cultures of bacteria and the applicability of the Collins–Richmond principle. *J. gen. Microbiol.* **54**, 409.

KOCH, A. L. (1966*b*). On evidence supporting a deterministic process of bacterial growth. *J. gen. Microbiol.* **43**, 1.

KOCH, A. L. & EHRENFELD, E. (1968). The size and shape of bacteria by light scattering measurements. *Biochim. biophys. Acta* **165**, 262.

KOGA, S. & FUJITA, T. (1962). Anomalous light scattering by microbial cell suspensions. *J. gen. appl. Microbiol., Tokyo* **8**, 223.

KONOWALCHUK, J., CLUNIE, J. C., HINTON, N. A. & REED, G. B. (1954). Antibacterial action of a reaction product of cysteine and iron. *Can. J. Microbiol.* **1**, 182.

KOPPER, P. H. (1962). Effect of sodium chloride concentration on the swarming tendency of *Proteus*. *J. Bact.* **84**, 1119.

KOPRIWA, B. M. & LEBLOND, C. P. (1962). Improvements in the coating technique of radioautography. *J. Histochem. Cytochem.* **10**, 269.

KORMAN, R. Z. & BERMAN, D. T. (1958). Medium for the differentiation of acid producing colonies of staphylococci. *J. Bact.* **76**, 454.

LAIDLAW, P. P. (1915). Some simple anaërobic methods. *Br. med. J.* i, 497.

LANE, C. R., COOK, G. T., KELSEY, J. C. & BEEBY, M. M. (1964). High-temperature sterilization for surgical instruments. *Lancet* i, 358.

LANE-PETTER, W. *et al.* eds. (1967). *The UFAW Handbook on the Care and Management of Laboratory Animals*, 3rd ed. Edinburgh and London: E. and S. Livingston Ltd.

LANKFORD, C. E., KUSTOFF, T. Y. & SERGEANT, T. P. (1957). Chelating agents in growth initiation of *Bacillus globigii*. *J. Bact.* **74**, 737.

LANKFORD, C. E., RAVEL, J. M. & RAMSEY, H. H. (1957). The effect of glucose and heat sterilization on bacterial assimilation of cystine. *Appl. Microbiol.* **5**, 65.

LAPAGE, S. P. & JAYARAMAN, M. S. (1964). Beta-galactosidase and lactose fermentation in the identification of enterobacteria including salmonellae. *J. clin. Path.* **17**, 117.

LARK, K. G. (1966). Regulation of chromosome replication and segregation in bacteria. *Bact. Rev.* **30**, 3.

LARK, K. G. & ADAMS, M. H. (1953). The stability of phages as a function of the ionic environment. *Cold Spr. Harb. Symp. quant. Biol.* **18**, 171.

LARSEN, D. H. & DIMMICK, R. I. (1964). Attachment and growth of bacteria on surfaces of continuous-culture vessels. *J. Bact.* **88**, 1380.

LATARJET, R., MORENNE, P. & BERGER, R. (1953). Un appareil simple pour le dosage des rayonnements ultraviolets émis par les lampes germicides. *Annls Inst. Pasteur, Paris* **85**, 174.

LAYBOURN, R. L. (1924). A modification of Albert's stain for the diphtheria bacillus. *J. Am. med. Ass.* **83**, 121.

LEA, D. E., HAINES, R. B. & COULSON, C. A. (1936). The mechanism of the bactericidal action of radioactive radiations. I. Theoretical. *Proc. R. Soc.* B **120**, 47.

LEACH, F. R. & SNELL, E. E. (1960). The adsorption of glycine and alanine and their peptides by *Lactobacillus casei*. *J. biol. Chem.* **235**, 3523.

LECHTMAN, M. D., BARTHOLOMEW, J. W., PHILLIPS, A. & RUSSO, M. (1965). Rapid methods of staining bacterial spores at room temperature. *J. Bact.* **89**, 848.

LEDERBERG, E. M. (1951). Lysogenicity in *E. coli* K12. *Genetics, Princeton* **36**, 560.

LEDERBERG, J. (1947). The nutrition of *Salmonella*. *Archs Biochem. Biophys.* **13**, 287.

LEDERBERG, J. (1948). Detection of fermentative variants with tetrazolium. *J. Bact.* **56**, 695.

LEDERBERG, J. (1949). Aberrant heterozygotes in *Escherichia coli*. *Proc. natn. Acad. Sci., U.S.A.* **35**, 178.

LEDERBERG, J. (1950). Isolation and characterization of biochemical mutants of bacteria. *Meth. med. Res.* **3**, 5.

LEDERBERG, J. & LEDERBERG, E. M. (1952). Replica plating and indirect selection of bacterial mutants. *J. Bact.* **63**, 399.

LEDERBERG, J. & ST. CLAIR, J. (1958). Protoplasts and L-type growth of *Escherichia coli*. *J. Bact.* **75**, 143.

LEIFSON, E. (1935). New culture media based on sodium desoxycholate for the isolation of intestinal pathogens and for the enumeration of colon bacilli in milk and water. *J. Path. Bact.* **40**, 581.

REFERENCES

LEIFSON, E. (1951). Staining, shape, and arrangement of bacterial flagella. *J. Bact.* **62**, 377.

LEIFSON, E. (1958). Timing of the Leifson flagella stain. *Stain Technol.* **33**, 249.

LEIFSON, E. (1961). The effect of formaldehyde on the shape of bacterial flagella. *J. gen. Microbiol.* **25**, 131.

LELLOUCH, J. & WAMBERSIE, A. (1966). Estimation par la methode du maximum de vraisemblance des courbes de survie de microorganismes irradiés. *Biometrics* **22**, 673.

LEPPER, E. & MARTIN, C. J. (1930). The oxidation-reduction potential of cooked-meat media following the inoculation of bacteria. *Brit. J. exp. Path.* **11**, 140.

LEVINE, B. S. & BLACK, L. A. (1948). Newly proposed staining formulas for the direct microscopic examination of milk. *Am. J. publ. Hlth* **38**, 1210.

LEVINE, M. (1943). The effect of concentration of dyes on differentiation of enteric bacteria on eosin-methylene-blue agar. *J. Bact.* **45**, 471.

LEVINE, M. & SCHOENLEIN, H. W. (1930). *A Compilation of Culture Media for the Cultivation of Microorganisms.* London: Baillière, Tindall and Cox.

LEVINSON, H. S., HYATT, M. T. & HOLMES, P. K. (1967). Transition of bacterial spores into vegetative cells. *Trans. N.Y. Acad. Sci.* **30**, 81.

LEWIS, I. M. (1930). The inhibition of *Phytomonas malvaceara* in culture media containing sugars. *J. Bact.* **19**, 423.

LEY, H. L. & MUELLER, J. H. (1946). On the isolation from agar of an inhibitor for *Neisseria gonorrhoeae*. *J. Bact.* **52**, 453.

LICHSTEIN, H. C. (1960). Microbial nutrition. *A. Rev. Microbiol.* **14**, 17.

LIDWELL, O. M., NOBLE, W. C. & DOLPHIN, G. W. (1959). The use of radiation to estimate the numbers of micro-organisms in air-borne particles. *J. Hyg., Camb.* **57**, 299.

LIE, S. (1964). The mutagenic effect of hydroxylamine on *Escherichia coli. Acta path. microbiol. scand.* **62**, 575.

LIEBERT, F. & KAPER, L. (1937). Some notes on the influence of light on coloured coli-media. *Antonie van Leeuwenhoek* **4**, 164.

LIN, E. C. C., LERNER, S. A. & JORGENSEN, S. E. (1962). A method for isolating constitutive mutants for carbohydrate-catabolizing enzymes. *Biochim. biophys. Acta* **60**, 422.

LINCOLN, R. L. & DE ARMON, I. A. (1959). Homogeneity of response of mouse and guinea-pig strains to virulence tests with *Bacillus anthracis* and *Pasteurella tularensis*. *J. Bact.* **78**, 640.

LISTER, J. (1878). On the lactic fermentation and its bearings on pathology. *Trans. path. Soc., Lond.* **29**, 425.

LITCHFIELD, J. T. (1949). A method for rapid graphic solution of time–per cent effect curves. *J. Pharmac. exp. Ther.* **97**, 399.

LITCHFIELD, J. T. & WILCOXON, F. (1949). A simplified method of evaluating dose-effect experiments. *J. Pharmac. exp. Ther.* **96**, 99.

LIU, O. C. & HENLE, W. (1953). Studies on host-virus interactions in the chick embryo-influenza virus system. VII. Data concerning the significance of infectivity titration end-points and the separation of clones at limiting dilutions. *J. exp. Med.* **97**, 889.

LIU, Y. T. & TAKAHASHI, H. (1964). Screening of bacterial biochemical auxotrophs by glycine medium. *J. gen. appl. Microbiol., Tokyo* **10**, 79.

REFERENCES

LOCKHART, W. R. & SQUIRES, R. W. (1963). Aeration in the laboratory. *Adv. appl. Microbiol.* **5**, 157.

LONG, D. A., MILES, A. A. & PERRY, W. L. M. (1954). The assay of tuberculin. *Bull. World Hlth Org.* **10**, 989.

LOVELESS, A. (1966). *Genetic and Allied Effects of Alkylating Agents.* University Park and London: Pennsylvania State University Press.

LOW, B. (1968). Formation of merodiploids in matings with a class of *rec*$^{-}$ recipient strains of *Escherichia coli* K12. *Proc. natn. Acad. Sci., U.S.A.* **60**, 160.

LOWE, G. H. (1960). A study of the factors controlling lactose-fermentation in the coliform and paracolon groups. *J. gen. Microbiol.* **23**, 127.

LOWRY, O. H., ROSEBROUGH, N. J., FARR, A. L. & RANDAL, R. J. (1951). Protein measurement with the Folin phenol reagent. *J. biol. Chem.* **193**, 265.

LUBIN, M. (1959). Selection of auxotrophic bacterial mutants by tritium-labelled thymidine. *Science, N.Y.* **129**, 838.

LUBIN, M. (1962). Enrichment of auxotrophic mutant populations by recycling. *J. Bact.* **83**, 696.

LUCKE, W. H. & SARACHEK, A. (1953). X-ray inactivation of polyploid *Saccharomyces*. *Nature, Lond.* **171**, 1014.

LÜDERITZ, O., STAUB, A. M. & WESTPHAL, O. (1966). Immunochemistry of O and R antigens of *Salmonella* and related *Enterobacteriaceae*. *Bact. Rev.* **30**, 192.

LUMPKINS, E. D. & ARVESON, J. S. (1968). Improved technique for staining bacteria on membrane filters. *Appl. Microbiol.* **16**, 433.

LURIA, S. E. & DULBECCO, R. (1949). Genetic recombinations leading to production of active bacteriophage from ultraviolet inactivated bacteriophage particles. *Genetics, Princeton* **34**, 93.

LWOFF, A. & MONOD, J. (1947). Essai d'analyse du rôle de l'anhydride carbonique dans la croissance microbienne. *Annls Inst. Pasteur, Paris* **73**, 323.

LYMAN, C. M., MOSELEY, O., WOOD, S. & HALE, F. (1946). Note on the use of hydrogen peroxide-treated peptone in media for the microbiological determination of amino acids. *Archs Biochem. Biophys.* **10**, 427.

MAALØE, O. (1955). The international reference preparation for opacity. *Bull. Wld Hlth Org.* **12**, 769.

MAALØE, O. (1960). The nucleic acids and the control of bacterial growth. *Symp. Soc. gen. Microbiol.* **10**, 272.

MAALØE, O. (1962). Synchronous growth. In *The Bacteria*, vol. IV, p. 1. Ed. by I. C. Gunsalus and R. Y. Stanier. New York and London: Academic Press.

MAAS, W. K. (1961). Studies on repression of arginine biosynthesis in *Escherichia coli*. *Cold Spr. Harb. Symp. quant. Biol.* **26**, 183.

MACCACARO, G. A. & HAYES, W. (1961). The genetics of fimbriation in *Escherichia coli*. *Genet. Res.* **2**, 394.

McCARTHY, K., DOWNIE, A. W. & ARMITAGE, P. (1958). The antibody response in man following infection with viruses of the pox group. I. An evaluation of the pock counting method for measuring neutralising antibody. *J. Hyg., Camb.* **56**, 84.

McCRADY, M. H. (1918). Table for rapid interpretation of fermentation-tube results. *Can. Publ. Hlth J.* **9**, 201.

McCULLOCH, E. C. (1945). *Disinfection and Sterilization*, 2nd ed. London: Henry Kimpton.

REFERENCES

McDaniel, L. E., Bailey, E. G. & Zimmerli, A. (1965*a*). Effect of oxygen supply rates on growth of *Escherichia coli.* I. Studies in unbaffled and baffled shake flasks. *Appl. Microbiol.* **13**, 109.

McDaniel, L. E., Bailey, E. G. & Zimmerli, A. (1965*b*). Effect of oxygen supply rates on growth of *Escherichia coli.* II. Comparison of results in shake flasks and 50-liter fermentor. *Appl. Microbiol.* **13**, 115.

McDonald, I. J. (1963). Methionine requirement for growth of a species of micrococcus. *Can. J. Microbiol.* **9**, 415.

McIlwain, H. (1938). The effect of agar on the production of staphylococcal α-haemolysin. *Brit. J. exp. Path.* **19**, 411.

Mager, J., Kuczynski, M., Schatzberg, G. & Avi-Dor, Y. (1956). Turbidity changes in bacterial suspensions in relation to osmotic pressure. *J. gen. Microbiol.* **14**, 69.

Mai, K. & Bonitz, K. (1963). Statistical variation found at pock and plaque counting of vaccinia virus. *Nature, Lond.* **197**, 166.

Maizels, M. (1944). Processing of plasma with kaolin. *Lancet* ii, 205.

Mallette, M. F. (1967). A pH buffer devoid of nitrogen, sulfur and phosphorus for use in bacteriological systems. *J. Bact.* **94**, 283.

Maneval, W. E. (1934). Negative staining of micro-organisms. *Science, N.Y.* **80**, 292.

Marcus, S. & Greaves, C. (1950). Danger of false results using screw-capped tubes in diagnostic bacteriology. *J. Lab. clin. Med.* **36**, 134.

Maré, I. J., Coetzee, J. N. & Klerk, H. C. de (1964). Agar electrophoresis of colicines with an *Alcaligenes faecalis* indicator strain. *Nature, Lond.* **202**, 213.

Markham, R. (1942). A steam distillation apparatus suitable for microkjeldahl analysis. *Biochem. J.* **36**, 790.

Martin, D. S. (1932). The oxygen consumption of *Escherichia coli* during the lag and logarithmic phases of growth. *J. gen. Physiol.* **15**, 691.

Martin, L. C. (1961). *Technical Optics*, 2nd ed., vol. II. London: Sir Isaac Pitman and Sons Ltd.

Martin, L. C. (1966). *The Theory of the Microscope.* London: Blackie and Son.

Martin, S. M. (1964). Conservation of microorganisms. *A. Rev. Microbiol.* **18**, 1.

Marvin, D. A. & Hohn, B. (1969). Filamentous bacterial viruses. *Bact. Rev.* **33**, 172.

Mason, D. J. & Powelson, D. M. (1956). Nuclear division as observed in live bacteria by a new technique. *J. Bact.* **71**, 474.

Mather, K. (1949). The analysis of extinction time data in bioassay. *Biometrics* **5**, 127.

Matney, T. S. & Achenbach, N. E. (1962). New uses for membrane filters. III. Bacterial mating procedure. *J. Bact.* **84**, 874.

Maw, J. & Meynell, G. G. (1968). The true division and death rate of *Salmonella typhimurium* in the mouse spleen determined with phage P22. *Brit. J. exp. Path.* **49**, 597.

Maxon, W. D. & Johnson, M. J. (1953). Aeration studies in propagation of bakers' yeast. *Ind. Engng Chem. ind. Edn* **45**, 2554.

May, P. S., Winter, J. W., Fried, G. H. & Antopol, W. (1960). Effect of tetrazolium salts on selected bacterial species. *Proc. Soc. exp. Biol. Med., N.Y.* **105**, 364.

REFERENCES

MAYER, G. D. & TRAXLER, R. W. (1962). Action of metal chelates on growth initiation of *Bacillus subtilis*. *J. Bact.* **83**, 1281.

MAYERSBACH, H. V. (1967). Principles and limitations of immunohistochemical methods. *Jl R. microsc. Soc.* **87**, 295.

MAYOR, H. D. (1963). Biophysical studies on viruses using the fluorochrome acridine orange. *Prog. med. Virol.* **4**, 70.

MAYOR, H. D. (1963). The nucleic acids of viruses as revealed by their reactions with fluorochrome acridine orange. *Int. Rev. exper. Path.* **2**, 1.

MAYOR, H. D. & HILL, N. O. (1961). Acridine orange staining of a single-stranded DNA bacteriophage. *Virology* **14**, 264.

MAZZONE, H. M. (1968). Equilibrium centrifugation. In *Methods in Virology*, vol. II, p. 41. Ed. by K. Maramorosch and H. Koprowski. New York and London: Academic Press.

MEMORANDUM (1962). *The Sterilization, Use and Care of Syringes.* Med. Res. Counc. Memorandum, no. 41.

MERRILL, M. H. (1930). Carbohydrate metabolism of organisms of the genus *Mycobacterium*. *J. Bact.* **20**, 235.

MEYNELL, E. W. (1961). A phage, $\phi\chi$, which attacks motile bacteria. *J. gen. Microbiol.* **25**, 253.

MEYNELL, E. W. (1963). Reverting and non-reverting rough variants of *Bacillus anthracis*. *J. gen. Microbiol.* **32**, 55.

MEYNELL, E. W. & DATTA, N. (1966). The relation of resistance transfer factors to the *F* factor of *Escherichia coli* K12. *Genet. Res.* **7**, 134.

MEYNELL, E. W. & DATTA, N. (1967). Mutant drug resistance factors of high transmissibility. *Nature, Lond.* **214**, 885.

MEYNELL, E. W. & DATTA, N. (1969). Transferable drug resistance: sex factor activity. *Ciba Foundation Symposium on Bacterial Episomes and Plasmids*. Ed. G. E. W. Wolstenholme and M. O'Connor. London: J. and A. Churchill Ltd.

MEYNELL, E. W. & MEYNELL, G. G. (1964). The roles of serum and carbon dioxide in capsule formation by *Bacillus anthracis*. *J. gen. Microbiol.* **34**, 153.

MEYNELL, E. W., MEYNELL, G. G. & DATTA, N. (1968). Phylogenetic relationships of drug-resistance factors and other transmissible bacterial plasmids. *Bact. Rev.* **32**, 55.

MEYNELL, G. G. (1957). Inherently low precision of infectivity titrations using a quantal response. *Biometrics* **13**, 149.

MEYNELL, G. G. (1958). The effect of sudden chilling on *Escherichia coli*. *J. gen. Microbiol.* **19**, 380.

MEYNELL, G. G. (1959). Use of superinfecting phage for estimating the division rate of lysogenic bacteria in infected animals. *J. gen. Microbiol.* **21**, 421.

MEYNELL, G. G. (1963). Interpretation of distributions of individual response times in microbial infections. *Nature, Lond.* **198**, 970.

MEYNELL, G. G. & LAWN, A. M. (1968). Filamentous phages specific for the I sex factor. *Nature, Lond.* **217**, 1184.

MEYNELL, G. G. & MAW, J. (1968). Evidence for a two-stage model of microbial infection. *J. Hyg., Camb.* **66**, 273.

MEYNELL, G. G. & MEYNELL, E. W. (1958). The growth of micro-organisms *in vivo* with particular reference to the relation between dose and latent period. *J. Hyg., Camb.* **56**, 323.

REFERENCES

MEYNELL, G. G. & MEYNELL, E. W. (1966). The biosynthesis of poly D-glutamic acid, the capsular material of *Bacillus anthracis*. *J. gen. Microbiol.* **43**, 119.

MEYNELL, G. G. & STOCKER, B. A. D. (1957). Some hypotheses on the aetiology of fatal infections in partially resistant hosts and their application to mice challenged with *Salmonella paratyphi-B* or *Salmonella typhimurium* by intraperitoneal injection. *J. gen. Microbiol.* **16**, 38.

MEYNELL, G. G. & WILLIAMS, T. (1967). Estimating the date of infection from individual response times. *J. Hyg., Camb.* **65**, 131.

MICHEL, J. F., CAMI, B. & SCHAEFFER, P. (1968). Sélection de mutants de *Bacillus subtilis* bloqués au début de la sporulation. II. Sélection par adaptation a une nouvelle source de carbone et par vieillissement de cultures sporulées. *Annls Inst. Pasteur, Paris* **114**, 21.

MILES, A. A. & MISRA, S. S. (1938). The estimation of the bactericidal power of the blood. *J. Hyg., Camb.* **38**, 732.

MITANI, M. & IINO, T. (1968). Electron microscopy of salmonella flagella in methylcellulose solution. *J. gen. Microbiol.* **50**, 459.

MITCHELL, P. & MOYLE, J. (1956). Osmotic function and structure in bacteria. *Symp. Soc. gen. Microbiol.* **6**, 150.

MITCHISON, D. A. & SPICER, C. C. (1949). A method of estimating streptomycin in serum and other body fluids by diffusion through agar enclosed in glass tubes. *J. gen. Microbiol.* **3**, 184.

MITCHISON, J. M. (1961). The growth of single cells. III. *Streptococcus faecalis*. *Exp. Cell Res.* **22**, 208.

MOATS, W. A. (1959). Application of periodic acid-Schiff type stains to bacteria in milk. *J. Bact.* **78**, 589.

MOLINA, E. C. (1942). *Poisson's Exponential Binomial Limit*. Princeton, N.J.: Van Nostrand Co. Inc.

MONOD, J. (1949). The growth of bacterial cultures. *A. Rev. Microbiol.* **3**, 371.

MOORE, W. T. & TAYLOR, C. B. (1950). An improved colony illuminator. *J. gen. Microbiol.* **4**, 448.

MORAN, P. A. P. (1954*a*). The dilution assay of viruses. *J. Hyg., Camb.* **52**, 189.

MORAN, P. A. P. (1954*b*). The dilution assay of viruses. II. *J. Hyg., Camb.* **52**, 444.

MORAN, P. A. P. (1958). Another test for heterogeneity of host resistance in dilution assays. *J. Hyg., Camb.* **56**, 319.

MORRISON, G. A. & HINSHELWOOD, C. (1949). Nitrogen utilisation and growth of coliform bacteria. Part III. Nitrogen utilisation and lag phase. *J. chem. Soc.* p. 380.

MORSE, M. L. & ALIRE, M. L. (1958). An agar medium indicating acid production. *J. Bact.* **76**, 270.

MORTON, H. E. (1945). On the amount of carbon dioxide supplied for the primary isolation of *Neisseria gonorrhoeae*. *J. Bact.* **50**, 589.

MOSELEY, B. E. B. (1968). The repair of damaged DNA in irradiated bacteria. *Adv. microb. Physiol.* **2**, 173.

MOSS, F. J. & SAEED, M. (1967). Continuous culture. *Prog. industr. Microbiol.* **6**, 209.

MUELLER, J. H. & JOHNSON, E. R. (1941). Acid hydrolysates of casein to replace peptone in the preparation of bacteriological media. *J. Immun.* **40**, 33.

REFERENCES

MUELLER, J. H. & MILLER, P. A. (1941). Production of diphtheric toxin of high potency (100 Lf) in a reproducible medium. *J. Immun.* **40**, 21.

MUGGLETON, P. W. (1963). The preservation of cultures. *Prog. industr. Microbiol.* **4**, 189.

MULLER, H. J. (1954). The manner of production of mutations by radiation. In *Radiation Biology*, vol. I, p. 475. Ed. by A. Hollaender. New York: McGraw-Hill Book Co. Ltd.

MUNRO, H. N. & FLECK, A. (1966). The determination of nucleic acids. *Meth. biochem. Analysis* **14**, 113.

MUNSON, R. J. & BRIDGES, B. A. (1964). 'Take-over'—an unusual selection process in steady-state cultures of *Escherichia coli*. *J. gen. Microbiol.* **37**, 411.

MURRAY, R. G. E. (1960). The internal structure of the cell. In *The Bacteria*, vol. I, p. 35. Ed. by I. C. Gunsalus and R. Y. Stanier. New York and London: Academic Press.

MURRAY, R. G. E. & ROBINOW, C. F. (1952). A demonstration of the disposition of the cell wall of *Bacillus cereus*. *J. Bact.* **63**, 298.

MURRELL, W. G. (1967). The biochemistry of the bacterial endospore. *Adv. microb. Physiol.* **1**, 133.

NAIRN, R. C., ed. (1969). *Fluorescent Protein Tracing*, 3rd ed. Edinburgh: E. and S. Livingstone Ltd.

NAKAE, T. (1966). Method of temperature-gradient incubation and its application to microbiological examinations. *J. Bact.* **91**, 1730.

NEWCOMBE, H. B. & RHYNAS, P. O. W. (1958). Chloroform-induced loss of a host range mutant phenotype in virulent coliphage *lambda*. *Virology* **5**, 1.

NEWTON, B. A. (1955). A fluorescent derivative of polymyxin: its preparation and use in studying the site of action of the antibiotic. *J. gen. Microbiol.* **12**, 226.

NIEL, C. B. VAN, RUBEN, S., CARSON, S. F., KAMEN, M. D. & FOSTER, J. W. (1942). Radioactive carbon as an indicator of carbon dioxide utilization. VIII. The rôle of carbon dioxide in cellular metabolism. *Proc. natn. Acad. Sci., U.S.A.* **28**, 8.

NIEMAN, C. (1954). Influence of trace amounts of fatty acids on the growth of microorganisms. *Bact. Rev.* **18**, 147.

NIKAIDO, H. (1968). Biosynthesis of cell wall lipopolysaccharide in Gram-negative enteric bacteria. *Adv. Enzymol.* **31**, 77.

NOMURA, M. (1967). Colicins and related bacteriocins. *A. Rev. Microbiol.* **21**, 257.

NORMAN, R. L. & KEMPE, L. L. (1960). Electronic computer solution for the MPN equation used in the determination of bacterial populations. *J. biochem. microbiol. technol. Engng* **2**, 157.

NORRIS, K. P. & POWELL, E. O. (1961). Improvements in determining total counts of bacteria. *Jl R. microsc. Soc.* **80**, 107.

NOVICK, R. P. (1969). Extrachromosomal inheritance in bacteria. *Bact. Rev.* **33**.

NOVICK, R. P. & RICHMOND, M. H. (1965). Nature and interactions of the genetic elements governing penicillinase synthesis in *Staphylococcus aureus*. *J. Bact.* **90**, 467.

OKADA, T., HOMMA, J. & SONAHARA, H. (1962). Improved method for obtaining thymineless mutants of *Escherichia coli* and *Salmonella typhimurium*. *J. Bact.* **84**, 602.

REFERENCES

Okada, M. & Watanabe, T. (1968). Transduction with phage P1 in *Salmonella typhimurium*. *Nature, Lond.* **218**, 185.

Ollett, W. S. (1947). A method for staining both Gram-positive and Gram-negative bacteria in sections. *J. Path. Bact.* **59**, 357.

O'Meara, R. A. Q. & Macsween, J. C. (1937). The influence of copper in peptones on the growth of certain pathogens in peptone broth. *J. Path. Bact.* **44**, 225.

Opfell, J. B. (1965). Cold sterilization techniques. *Adv. appl. Microbiol.* **7**, 81.

Orgel, L. E. (1965). The chemical basis of mutation. *Adv. Enzymol.* **27**, 290.

Ørskov, I. & Ørskov, F. (1966). Episome-carried surface antigen K88 of *Escherichia coli*. I. Transmission of the determinant of the K88 antigen and influence on the transfer of chromosomal markers. *J. Bact.* **91**, 69.

Ørskov, J. (1922). Method for the isolation of bacteria in pure culture from single cells and procedure for the direct tracing of bacterial growth on a solid medium. *J. Bact.* **7**, 537.

Oster, G. (1960). Light scattering. In *Technique of Organic Chemistry*, 3rd ed., vol. I, pt. 3, p. 2107. Ed. by A. Weissberger. New York: Interscience Publishers Inc.

Ozeki, H. (1968). Methods for the study of colicine and colicinogeny. In *Methods in Virology*, vol. IV, p. 565. Ed. by K. Maramorosch and H. Koprowski. New York and London: Academic Press.

Ozeki, H., Stocker, B. A. D. & de Margerie, H. (1959). Production of colicine by single bacteria. *Nature, Lond.* **184**, 337.

Painter, P. R. & Marr, A. G. (1967). Inequality of mean interdivision time and doubling time. *J. gen. Microbiol.* **48**, 155.

Painter, P. R. & Marr, A. G. (1968). Mathematics of microbial populations. *A. Rev. Microbiol.* **22**, 519.

Pappenheimer, A. M. & Johnson, S. J. (1936). Studies in diphtheria toxin production. I: The effect of iron and copper. *Brit. J. exp. Path.* **17**, 335.

Pardee, A. B. (1957). An inducible mechanism for accumulation of melibiose in *Escherichia coli*. *J. Bact.* **73**, 376.

Parker, C. A. (1961). A method for the measurement of total nitrogen in microgram amounts. *Aust. J. exp. Biol. med. Sci.* **39**, 515.

Pasteur, L. (1861). Animalcules infusoires vivant sans gaz oxygène libre et determinant des fermentations. *C. r. hebd. Séanc. Acad. Sci., Paris* **52**, 344.

Pattee, P. A. (1966). Use of tetrazolium for improved resolution of bacteriophage plaques. *J. Bact.* **92**, 787.

Patton, A. R. & Hill, E. G. (1948). Inactivation of nutrients by heating with glucose. *Science, N.Y.* **107**, 68.

Payne, J. I., Hartman, P. E., Mudd, S. & Phillips, A. W. (1956). Cytological analysis of ultraviolet-irradiated *Escherichia coli*. III. Reactions of a sensitive strain and its resistant mutants. *J. Bact.* **72**, 461.

Pearce, T. W. & Powell, E. O. (1951). New techniques for the study of growing micro-organisms. *J. gen. Microbiol.* **5**, 91.

Pearse, A. G. E. (1968). *Histochemistry*, 3rd ed. London: J. and A. Churchill Ltd.

Pearson, E. S. & Hartley, H. O. (1954). *Biometrika Tables for Statisticians*, vol. I. Cambridge University Press.

PELONZE, P. S. & VITERI, L. E. (1926). A new medium for gonococcus culture. *J. Am. med. Ass.* **86**, 684.

PENNINGTON, D. (1950). A reaction of bacterial cells with formaldehyde. *J. Bact.* **59**, 617.

PERKINS, J. J. (1956). *Principles and Methods of Sterilization.* Springfield, Illinois: Charles C. Thomas.

PERRET, C. J. (1957). An apparatus for the continuous culture of bacteria at continuous population density. *J. gen. Microbiol.* **16**, 250.

PHILLIPS, J. H. & BROWN, D. M. (1967). The mutagenic action of hydroxylamine. *Progress in Nucleic Acid Research and Molecular Biology*, **7**, 349.

PIJPER, A. (1947). Methylcellulose and bacterial motility. *J. Bact.* **53**, 257.

PIRT, S. J. (1965). The maintenance energy of bacteria in growing cultures. *Proc. R. Soc.* B **163**, 224.

PIRT, S. J. & CALLOW, D. S. (1958). The relationship between maximum oxygen solution rates in a bacterial culture and in a sodium sulphite solution under comparable aeration conditions. *J. appl. Bact.* **21**, 206.

PITAL, A., JANOWITZ, S. L., HUDAK, C. E. & LEWIS, E. E. (1967). Fluorescein-labelled β-glucosidase as a bacterial stain. *Appl. Microbiol.* **15**, 1165.

PITTMAN, M. (1946). A study of fluid thioglycollate medium for the sterility test. *J. Bact.* **51**, 19.

PLUS, N. (1954). Etude de la multiplication du virus de la sensibilité au gaz carbonique chez la drosophile. *Bull. biol. Fr. Belg.* **87**, 248.

POLLARD, E. & REAUME, M. (1951). Thermal inactivation of bacterial viruses. *Archs Biochem. Biophys.* **32**, 278.

POLLARD, E. & SETLOW, J. (1953). Action of heat on the serological affinity of T-1 bacteriophage. *Archs Biochem. Biophys.* **43**, 136.

POLLARD, E. C. (1953). *The Physics of Viruses.* New York: Academic Press.

POLLARD, E. C. (1964). Thermal effects on protein, nucleic acid and viruses. *Adv. Chem. Phys.* **7**, 201.

POLLOCK, M. R. (1948). Unsaturated fatty acids in cotton wool plugs. *Nature, Lond.* **161**, 853.

PONTECORVO, G. (1949). Auxanographic techniques in biochemical genetics. *J. gen. Microbiol.* **3**, 122.

POPE, C. G. & SMITH, M. L. (1932). The routine preparation of diphtheria toxin of high value. *J. Path. Bact.* **35**, 573.

PORRO, T. J., DADIK, S. P., GREEN, M. & MORSE, H. T. (1963). Fluorescence and absorption spectra of biological dyes. *Stain Technol.* **38**, 37.

PORRO, T. J. & MORSE, H. T. (1965). Fluorescence and absorption spectra of biological dyes (II). *Stain Technol.* **40**, 173.

PORTER, K. R. & YEGIAN, D. (1945). Some artefacts encountered in stained preparations of tubercle bacilli. II. Much granules and beads. *J. Bact.* **50**, 563.

POSTGATE, J. R. (1963). The examination of sulphur auxotrophs—a warning. *J. gen. Microbiol.* **30**, 481. Corrigenda: *ibid.* **32**, pt. 1.

POSTGATE, J. R., CRUMPTON, J. E. & HUNTER, J. R. (1961). The measurement of bacterial viabilities by slide culture. *J. gen. Microbiol.* **24**, 15.

POSTGATE, J. R. & HUNTER, J. R. (1962). The survival of starved bacteria. *J. gen. Microbiol.* **29**, 233.

Postgate, J. R. & Hunter, J. R. (1963). The survival of starved bacteria. *J. appl. Bact.* **26**, 295.

Powell, E. O. (1955). Some features of the generation times of individual bacteria. *Biometrika* **42**, 16.

Powell, E. O. (1956*a*). Growth rate and generation time of bacteria with special reference to continuous culture. *J. gen. Microbiol.* **15**, 492.

Powell, E. O. (1956*b*). A rapid method for determining the proportion of viable bacteria in a culture. *J. gen. Microbiol.* **14**, 153.

Powell, E. O. (1956*c*). An improved culture chamber for the study of living bacteria. *Jl R. microsc. Soc.* **75**, 235.

Powell, E. O. (1958). An outline of the pattern of bacterial generation times. *J. gen. Microbiol.* **18**, 382.

Powell, E. O. (1963). Photometric methods in bacteriology. *J. Sci. Fd Agric.* **14**, 1.

Powell, E. O. (1965). Theory of the chemostat. *Lab. Pract.* **14**, 1145.

Powell, E. O. (1967). The growth rate of microorganisms as a function of substrate concentration. In: Powell *et al.* (1967).

Powell, E. O. & Errington, F. P. (1963*a*). Generation times of individual bacteria: some corroborative measurements. *J. gen. Microbiol.* **31**, 315.

Powell, E. O. & Errington, F. P. (1963*b*). The size of bacteria, as measured with the Dyson image-splitting eyepiece. *Jl R. microsc. Soc.* **82**, 39.

Powell, E. O., Evans, C. G. T., Strange, R. E. & Tempest, D. W., eds. (1967). *Microbial Physiology and Continuous Culture.* Proc. 3rd International Symposium on Continuous Culture. London: H.M. Stationery Office.

Powell, E. O. & Stoward, P. J. (1962). A photometric method for following changes in length of bacteria. *J. gen. Microbiol.* **27**, 489.

Prescott, S. C., Winslow, C. E. A. & McCrady, M. (1946). *Water Bacteriology*, 6th ed. London: Chapman and Hall Ltd.

Preston, N. W. & Morrell, A. (1962). Reproducible results with the Gram stain. *J. Path. Bact.* **84**, 241.

Price, S. A. & Gare, L. (1959). A source of error in microbiological assays attributable to a bacterial inhibitor in distilled water. *Nature, Lond.* **183**, 838.

Proom, H., Woiwod, A. J., Barnes, J. M. & Orbell, W. G. (1950). A growth-inhibitory effect on *Shigella dysenteriae* which occurs with some batches of nutrient agar and is associated with the production of peroxide. *J. gen. Microbiol.* **4**, 270.

Puck, T. T., Garen, A. & Cline, J. (1951). The mechanism of virus attachment to host cells. I. The role of ions in the primary reaction. *J. exp. Med.* **93**, 65.

Rahn, O. (1945). Physical methods of sterilization of microorganisms. *Bact. Rev.* **9**, 1.

Reed, G. B. & Orr, J. H. (1943). Cultivation of anaerobes and oxidation-reduction potentials. *J. Bact.* **45**, 309.

Reed, L. J. & Muench, H. (1938). A simple method of estimating fifty per cent endpoints. *Am. J. Hyg.* **27**, 493.

Reeve, E. C. R. (1968). Genetic analysis of some mutations causing resistance to tetracycline in *Escherichia coli* K12. *Genet. Res.* **11**, 303.

Report (1956). The bacteriological examination of water supplies. *Rep. publ. Hlth med. Subj., Lond.* No. 71, 3rd ed.

REPORT (1958). The practical aspects of formaldehyde fumigation. *Mon. Bull. Minist. Hlth* **17**, 270.

REPORT (1959). Sterilization by steam under increased pressure. *Lancet* i, 425.

REPORT (1960*a*). Methods for the destruction of organic matter. *Analyst, Lond.* **85**, 643.

REPORT (1960*b*). Sterilization by steam under increased pressure. *Lancet* ii, 1243.

REPORT (1965). Use of disinfectants in hospitals. *Br. med. J.* i, 408.

RICHARDS, J. W. (1961). Studies in aeration and agitation. *Prog. industr. Microbiol.* **3**, 141.

RICHARDS, O. W. & WATERS, P. (1967). A new interference exciter filter for fluorescence microscopy of fluorescein-tagged substances. *Stain Technol.* **42**, 320.

RICHARDSON, H., EMSLIE-SMITH, A. H. & SENIOR, B. W. (1968). Agar diffusion method for the assay of colicins. *Appl. Microbiol.* **16**, 1468.

RICHMOND, M. H. (1968). The plasmids of *Staphylococcus aureus* and their relation to other extrachromosomal elements in bacteria. *Adv. microbial Physiol.* **2**, 43.

RIDEAL, S. & WALKER, J. T. A. (1903). Standardisation of disinfectants. *Jl R. sanit. Inst.* **24**, 424.

RILEY, M. & PARDEE, A. B. (1962). Nutritional effects on frequencies of bacterial recombination. *J. Bact.* **83**, 1332.

ROBERTS, E. C. & SNELL, E. E. (1946). An improved medium for microbiological assays with *Lactobacillus casei*. *J. biol. Chem.* **163**, 499.

ROBINOW, C. F. (1943–4). Cytological observations on *Bact. coli*, *Proteus vulgaris* and various aerobic spore-forming bacteria with special reference to the nuclear structures. *J. Hyg., Camb.* **43**, 413.

ROBINOW, C. F. (1949). Nuclear apparatus and cell structure of rod-shaped bacteria. Addendum to *The Bacterial Cell* by R. J. Dubos. Cambridge, Mass.: Harvard University Press.

ROBINOW, C. F. (1956). The chromatin bodies of bacteria. *Symp. Soc. gen. Microbiol.* **6**, 181.

ROBINOW, C. F. (1960*a*). Outline of the visible organisation of bacteria. In *The Cell, Biochemistry, Physiology, Morphology*, vol. IV, pt. 1. Ed. by J. Brachet and A. E. Mirksy. New York and London: Academic Press.

ROBINOW, C. F. (1960*b*). Morphology of bacterial spores, their development and germination. In *The Bacteria*, vol. I, p. 207. Ed. by I. C. Gunsalus and R. Y. Stanier. New York & London: Academic Press.

ROBINOW, C. F. & MURRAY, R. G. E. (1953). The differentiation of cell wall, cytoplasmic membrane and cytoplasm of Gram positive bacteria by selective staining. *Exp. Cell Res.* **4**, 390.

ROE, A. F. (1942). Reclaiming agar for bacteriological use. *Science, N.Y.* **96**, 23.

ROGERS, A. W. (1967). *Techniques of Autoradiography*. Amsterdam, London and New York: Elsevier Publishing Co.

ROGERS, H. J. (1954). The rate of formation of hyaluronidase, coagulase and total extracellular protein by strains of *Staphylococcus aureus*. *J. gen. Microbiol.* **10**, 209.

ROMERO, E. & MEYNELL, E. (1969). Covert fi^- R factors in fi^+ R^+ strains of bacteria. *J. Bact.* **97**, 780.

REFERENCES

ROOK, J. J. & BRUCKMAN, H. W. L. (1953). On the suitability of gelatin for plate cultures. II. *Antonie van Leeuwenhoek* **19**, 354.

ROSS, K. F. A. (1957). The size of living bacteria. *Q. Jl Microsc. Sci.* **98**, 435.

ROSS, K. F. A. & GALAVAZI, G. (1965). The size of bacteria as measured by interference microscopy. *Jl R. microsc. Soc.* **84**, 13.

ROTH, J. R., ANTÓN, D. N. & HARTMAN, P. E. (1966). Histidine regulatory mutants in *Salmonella typhimurium*. I. Isolation and general properties. *J. molec. Biol.* **22**, 305.

ROTMAN, B., ZDERIC, J. A. & EDELSTEIN, M. (1963). Fluorogenic substrates for β-D-galactosidases and phosphatases derived from fluorescein (3,6-dihydroxy fluoran) and its monomethyl ether. *Proc. natn. Acad. Sci. U.S.A.* **50**, 1.

ROUF, M. A. (1964). Spectrochemical analysis of inorganic elements in bacteria. *J. Bact.* **88**, 1545.

ROWATT, E. (1957*a*). Some factors affecting the growth of *Bordetella pertussis*. *J. gen. Microbiol.* **17**, 279.

ROWATT, E. (1957*b*). The growth of *Bordetella pertussis*: a review. *J. gen. Microbiol.* **17**, 297.

ROWLEY, D. (1953). Interrelationships between amino-acids in the growth of coliform organisms. *J. gen. Microbiol.* **9**, 37.

RUDAT, K. D. (1955). Über die Infektiosität fixierter gefärbter Bakterien präparate. *Z. Hyg. Greuzgebiete* **1**, 70. Abstracted in *Bull. Hyg.* (1956), **31**, 662.

RUSSELL, W. M. S. & BURCH, R. L. (1959). *The Principles of Humane Experimental Technique*. London: Methuen and Co. Ltd.

RYTER, A., SCHAEFFER, P. & IONESCO, H. (1966). Classification cytologique, par leur stade de blocage, des mutants de sporulation de *Bacillus subtilis* Marburg. *Annls Inst. Pasteur, Paris* **110**, 305.

SABIN, A. B., HENNESSEN, W. A. & WINSSER, J. (1954). Studies on variants of poliomyelitis virus. 1. Experimental segregation and properties of avirulent variants of three immunologic types. *J. exp. Med.* **99**, 551.

SACKS, L. E. (1956). A pH gradient agar plate. *Nature, Lond.* **178**, 269.

SADOFF, H. L., HALVORSON, H. O. & FINN, R. K. (1956). Electrolysis as a means of aerating submerged cultures of microorganisms. *Appl. Microbiol.* **4**, 164.

SALTON, M. R. J. (1963). The relationship between the nature of the cell wall and the Gram stain. *J. gen. Microbiol.* **30**, 223.

SAMPFORD, M. R. (1952). The estimation of response-time distributions. II. Multi-stimulus distributions. *Biometrics* **8**, 307.

SANDERSON, K. E. (1967). Revised linkage map of *Salmonella typhimurium. Bact. Rev.* **31**, 354.

SANDIFORD, B. R. (1938). A new contrast stain for gonococci and meningococci in smears. *Br. med. J.* i, 1155.

SARTWELL, P. E. (1950). The distribution of incubation periods of infectious disease. *Am. J. Hyg.* **51**, 310.

SARTWELL, P. E. (1966). The incubation period and the dynamics of infectious disease. *Am. J. Epidem.* **83**, 204.

SAVAGE, G. M. & HALVORSON, H. O. (1941). The effect of culture environment on results obtained with the dilution method of determining bacterial population. *J. Bact.* **41**, 355.

SAVAGE, R. M. (1937). Experiments on the sterilising effects of mixtures of air and steam, and of superheated steam. *Q. Jl Pharm. Pharmac.* **10**, 451.

SCHAECHTER, M. & LAING, V. O. (1961). Direct observation of fusion of bacterial nuclei. *J. Bact.* **81**, 667.

SCHAECHTER, M., MAALØE, O. & KJELDGAARD, N. O. (1958). Dependency on medium and temperature of cell size and chemical composition during balanced growth of *Salmonella typhimurium*. *J. gen. Microbiol.* **19**, 592.

SCHAECTER, M., WILLIAMSON, J. P., HOOD, J. R., Jr. & KOCH, A. L. (1962). Growth, cell and nuclear division in some bacteria. *J. gen. Microbiol.* **29**, 421.

SCHEELE, L. S. & SHANNON, J. A. (1955). Public health implications in a programme of vaccination against poliomyelitis. *J. Amer. med. Ass.* **155**, 1249.

SCHELL, J. & GLOVER, S. W. (1966). The effect of heat on host-controlled restriction of phage λ in *Escherichia coli* K (P1). *J. gen. Microbiol.* **45**, 61.

SCHELL, J., GLOVER, S. W., STACEY, K. A., BRODA, P. M. A. & SYMONDS, N. (1963). The restriction of phage T3 by certain strains of *Escherichia coli*. *Genet. Res.* **4**, 483.

SCHNEIDER, H. A. & ZINDER, N. D. (1956). Nutrition of the host and natural resistance to infection. V. An improved assay employing genetic markers in the double strain inoculation test. *J. exp. Med.* **103**, 207.

SCHUHARDT, V. T., RODE, L. J., OGELSBY, G. & LANKFORD, C. E. (1952). Toxicity of elemental sulfur for *Brucellae*. *J. Bact.* **63**, 123.

SCHULTZ, J. S. (1964). Cotton closure as an aeration barrier in shaken flask fermentations. *Appl. Microbiol.* **12**, 305.

SCHULZE, I. T. & SCHLESINGER, R. W. (1963). Inhibition of infectious and haemagglutinating properties of Type 2 Dengue virus by aqueous agar extracts. *Virology* **19**, 49.

SCHÜTZE, H. & HASSANEIN, M. A. (1929). The oxygen requirements of *B. pestis* and *Pasteurella* strains. *Brit. J. exp. Path.* **10**, 204.

SCULLARD, G. & MEYNELL, E. (1966). Bacterial mass measured with the M.R.C. grey-wedge photometer. *J. Path. Bact.* **91**, 608.

SELIGMAN, S. J. & MICKEY, M. R. (1964). Estimation of the number of infectious bacterial or viral particles by the dilution method. *J. Bact.* **88**, 31.

SELKON, J. B. & MITCHISON, D. A. (1957). Viable counting of *Mycobacterium tuberculosis* in a silica gel medium. *J. gen. Microbiol.* **16**, 229.

SENEZ, J. C. (1962). Some considerations on the energetics of bacterial growth. *Bact. Rev.* **26**, 95.

SERRES, F. J. DE (1961). Some aspects of the influence of environment on the radiosensitivity of micro-organisms. *Symp. Soc. gen. Microbiol.* **11**, 196.

SHAW, W. V. (1967). The enzymatic acetylation of chloramphenicol by extracts of R factor-resistant *Escherichia coli*. *J. biol. Chem.* **242**, 687.

SHIPE, E. L. & FIELDS, A. (1954). A comparison of the molecular filtration technique with agar plate counting for enumeration of *Escherichia coli* in various aqueous concentrations of zinc and copper sulphates. *Appl. Microbiol.* **2**, 382.

SHULL, J. J. & ERNST, R. R. (1962). Graphical procedure for comparing thermal death of *Bacillus stearothermophilus* spores in saturated and superheated steam. *Appl. Microbiol.* **10**, 452.

SIDDIQI, O. (1965). Interallelic complementation *in vivo* and *in vitro*. *Brit. med. Bull.* **21**, 249.

REFERENCES

SILVER, H., SONNENWIRTH, A. C. & ALEX, N. (1966). Modifications in the fluorescence microscopy technique as applied to identification of acid-fast bacilli in tissue and bacteriological material. *J. clin. Path.* **19**, 583.

SILVER, I. H. (1963). Explosion in an autoclave caused by cellulose nitrate tubes. *Nature, Lond.* **199**, 102.

SILVERMAN, P. M., MOBACH, H. W. & VALENTINE, R. C. (1967). Sex hair (F-pili) mutants of *E. coli. Biochem. biophys. Res. Comm.* **27**, 412.

SKAAR, P. D., RICHTER, A. & LEDERBERG, J. (1957). Correlated selection for motility and sex-incompatibility in *Escherichia coli* K12. *Proc. natn. Acad. Sci., U.S.A.* **43**, 329.

SKERMAN, V. B. D. (1953). A chemical analysis of Brewer's medium for the aerobic culture of anaerobes. *Aust. J. biol. Sci.* **6**, 276.

SMITH, C. E. G. & WESTGARTH, D. R. (1957). The use of survival time in the analysis of neutralization tests for serum antibody surveys. *J. Hyg., Camb.* **55**, 224.

SMITH, C. G. & JOHNSON, M. J. (1954). Aeration requirements for the growth of aerobic microorganisms. *J. Bact.* **68**, 346.

SMITH, H. O. & LEVINE, M. (1967). A phage P22 gene controlling integration of prophage. *Virology* **31**, 207.

SMITH, H. W. & HALLS, S. (1968). The transmissible nature of the genetic factor in *Escherichia coli* that controls enterotoxin production. *J. gen. Microbiol.* **52**, 319.

SMITH, S. M., OZEKI, H. & STOCKER, B. A. D. (1963). Transfer of *colE1* and *colE2* during high-frequency transmission of *colI* in *Salmonella typhimurium. J. gen. Microbiol.* **33**, 231.

SMITH, P. F. (1960). Nutritional requirements of PPLO and their relation to metabolic function. *Ann. N.Y. Acad. Sci.* **79**, 508.

SNYDER, T. L. (1947). The relative errors of bacteriological plate counting methods. *J. Bact.* **54**, 641.

SØRENSEN, H. (1962). Decomposition of lignin by soil bacteria and complex formation between autoxidized lignin and organic nitrogen compounds. *J. gen. Microbiol.* **27**, 21.

SPAUN, J. (1962). Problems in standardization of turbidity determinations on bacterial suspensions. *Bull. Wld Hlth Org.* **26**, 219.

SPICER, C. C. (1956). The estimation of the parameters of an exponentially declining population. *J. Hyg., Camb.* **4**, 304.

SPIEGELMAN, S., ARONSON, A. I. & FITZ-JAMES, P. C. (1958). Isolation and characterization of nuclear bodies from protoplasts of *Bacillus megaterium. J. Bact.* **75**, 102.

SPIEGLER, K. S. & WYLLIE, M. R. J. (1956). Electrical potential differences. In *Physical Techniques in Biological Research*, vol. II, p. 301. Ed. by G. Oster and A. W. Pollister. New York: Academic Press Inc.

SPITZNAGEL, J. K. & SHARP, D. G. (1959). Magnesium and sulfate ions as determinants in the growth and reproduction of *Mycobacterium bovis. J. Bact.* **78**, 453.

STACEY, K. A. & SIMSON, E. (1965). Improved method for the isolation of thymine-requiring mutants of *Escherichia coli. J. Bact.* **90**, 554.

STAMP, LORD (1947). The preservation of bacteria by drying. *J. gen. Microbiol.* **1**, 251.

REFERENCES

STAPERT, E. M., SOKOLSKI, W. T. & NORTHAM, J. I. (1962). The factor of temperature in the better recovery of bacteria from water by filtration. *Can. J. Microbiol.* **8**, 809.

STARKS, O. B. & KOFFLER, H. (1949). Aerating liquids by agitating on a mechanical shaker. *Science, N.Y.* **109**, 495.

STEINER, R. F. & EDELHOCH, H. (1962). Fluorescent protein conjugates. *Chem. Rev.* **62**, 457.

STENT, G. S. (1958). Mating in the reproduction of bacterial viruses. *Adv. Virus Res.* **5**, 95.

STERNE, M. (1958). The growth of *Brucella abortus* strain 19 in aerated dialysed media. *J. gen. Microbiol.* **18**, 747.

STEVENS, W. L. (1958). Dilution series: a statistical test of technique. *Jl R. statist. Soc.* B **20**, 205.

STEYERMARK, A. (1961). *Quantitative Organic Microanalysis*, 2nd ed. New York and London: Academic Press.

STOCKER, B. A. D., SMITH, S. M. & OZEKI, H. (1963). High infectivity of *Salmonella typhimurium* newly infected by the *colI* factor. *J. gen. Microbiol.* **30**, 201.

STOKES, J. L. & BAYNE, H. G. (1958). Dwarf colony variants of salmonellae. *J. Bact.* **76**, 136.

STROBEL, M. & NOMURA, M. (1966). Restriction of the growth of bacteriophage BF23 by a colicine (ColI-P9) factor. *Virology* **28**, 763.

STRUGGER, S. (1949). *Fluoreszenzmikroskopie und Mikrobiologie*. Hannover: M. and H. Schaper.

SUTHERLAND, I. W. & WILKINSON, J. F. (1961). A new growth medium for virulent *Bordetella pertussis*. *J. Path. Bact.* **82**, 431.

SWAROOP, S. (1956). Estimation of bacterial density of water samples. *Bull. Wld Hlth Org.* **14**, 1089.

SYKES, G., ed. (1956). *Constituents of Bacteriological Culture Media.* Cambridge University Press.

SYKES, G. (1965). *Disinfection and Sterilization.* 2nd ed. London: E. and F. N. Spon Ltd.

SYMPOSIUM (1956). *Bacterial Anatomy. Symp. Soc. gen. Microbiol.* no. 6.

SYMPOSIUM (1961). *Recent Developments in the Sterilization of Surgical Materials.* London: Pharmaceutical Press.

SZYBALSKI, W. (1968). Use of cesium sulfate for equilibrium density gradient centrifugation. In *Methods in Enzymology*, vol. XII B, p. 330. Ed. by L. Grossman and K. Moldave. New York and London: Academic Press.

TATUM, E. L. & LEDERBERG, J. (1947). Gene recombination in the bacterium *Escherichia coli*. *J. Bact.* **53**, 673.

TAUBENECK, U. (1959). New grid-replica for precise localization in slide cultures. *J. Bact.* **77**, 506.

TAYLOR, A. L. & TROTTER, C. D. (1967). Revised linkage map of *Escherichia coli*. *Bact. Rev.* **31**, 332.

TAYLOR, C. B. (1950). An improved method for the preparation of silica gel media for microbiological purposes. *J. gen. Microbiol.* **4**, 235.

TAYLOR, J. (1962). The estimation of numbers of bacteria by tenfold dilution series. *J. appl. Bact.* **25**, 54.

REFERENCES

TAYSUM, D. H. (1956). *A priori* calculation of the number of bacteria in a turbid suspension. *J. chem. Phys.* **25**, 183.

TEMPEST, D. W. & DICKS, J. W. (1967). Inter-relationships between potassium, magnesium, phosphorus and ribonucleic acid in the growth of *Aerobacter aerogenes* in a chemostat. In: Powell *et al.* (1967).

TEMPEST, D. W., HERBERT, D. & PHIPPS, P. J. (1967). Studies on the growth of *Aerobacter aerogenes* at low dilution rates in a chemostat. In: Powell *et al.* (1967).

TESSMAN, I. (1968). Mutagenic treatment of double- and single-stranded DNA phages T4 and S13 with hydroxylamine. *Virology* **35**, 331.

THOMAS, R. (1968). Lysogeny. *Symp. Soc. gen. Microbiol.* **18**, 315.

THOMPSON, W. R. (1947). Use of moving averages and interpolation to estimate median-effective dose. *Bact. Rev.* **11**, 115.

TILL, J. E., McCULLOCH, E. A. & SIMINOVITCH, L. (1964). A stochastic model of stem cell proliferation, based on the growth of spleen colony-forming cells. *Proc. natn. Acad. Sci., U.S.A.* **51**, 29.

TILLEY, F. W. (1939). An experimental study of the relation between concentration of disinfectants and time required for disinfection. *J. Bact.* **38**, 499.

TINT, H. & GILLEN, A. (1961). A nomograph for determining fifty per cent end points. *J. appl. Bact.* **24**, 83.

TIPPETT, L. H. C. (1932). A modified method of counting particles. *Proc. R. Soc.* A **137**, 434.

TOENNIES, G., ISZARD, L., ROGERS, N. B. & SHOCKMAN, G. D. (1961). Cell multiplication studied with an electronic particle counter. *J. Bact.* **82**, 857.

TOMCSIK, J. (1956*a*). Antibodies as indicators for bacterial surface structures. *A. Rev. Microbiol.* **10**, 213.

TOMCSIK, J. (1956*b*). Bacterial capsules and their relation to the cell wall. *Symp. Soc. gen. Microbiol.* **6**, 41.

TOMCSIK, J. (1961). Immunocytologie de la surface bactérienne. *Annali Microbiol.* **9**, 91.

TOMCSIK, J. & GUEX-HOLZER, S. (1954). Demonstration of the bacterial capsule by means of a pH-dependent, salt-like combination with proteins. *J. gen. Microbiol.* **10**, 97.

TOMLINSON, A. H. (1967). Filters for use with an iodine-quartz lamp to excite immunofluorescence. *Immunology* **13**, 323.

TOMOEDA, M., INUZUKA, M., KUBO, N. & NAKAMURA, S. (1968). Effective elimination of drug resistance and sex factors in *Escherichia coli* by sodium dodecyl sulfate. *J. Bact.* **95**, 1078.

TOPLIN, I. & GADEN, E. L. (1961). The chemical sterilization of liquid media with beta-propiolactone and ethylene oxide. *J. biochem. microbiol. Technol. Engng* **3**, 311.

TORRIANI, A. & ROTHMAN, F. (1961). Mutants of *Escherichia coli* constitutive for alkaline phosphatase. *J. Bact.* **81**, 835. Erratum: *J. Bact.* **82**, 791.

TRAXLER, R. W. & LANKFORD, C. E. (1957). Observations on cystine degradation and bacteriotoxicity of peptones. *Appl. Microbiol.* **5**, 70.

TREVAN, J. W. (1927). The error of determination of toxicity. *Proc. R. Soc.* B **101**, 483.

REFERENCES

Turner, G. S. & Kaplan, C. (1965). Observations on photodynamic inactivation of vaccinia virus and its effect on immunogenicity. *J. Hyg., Camb.* **63**, 395.

Twigg, R. S. (1945). Oxidation-reduction aspects of resazurin. *Nature, Lond.* **155**, 401.

Tyndall, J. (1877). On heat as a germicide when discontinuously applied. *Proc. R. Soc.* **25**, 569.

Ulrich, J. A. & Larsen, A. M. (1948). A single solution indicator for anaerobiosis. *J. Bact.* **56**, 373.

Umbreit, W. W., Burris, R. H. & Stauffer, J. F. (1964). *Manometric Techniques*, 4th ed. Madison, Wis.: Burgess Publishing Co.

Valley, G. (1928). The effect of carbon dioxide on bacteria. *Q. Rev. Biol.* **3**, 209.

Verly, W. G., Barbason, H., Dusart, J. & Petitpas-Dewandre, A. (1967). A comparative study of the action of ethyl methane sulfonate and HNO_2 on the mutation to streptomycin resistance of *Escherichia coli* K12. *Biochem. biophys. Acta* **145**, 752.

Vidra, A. (1956). Studies on the cytochemistry of bacteria. *J. Bact.* **71**, 689.

Vinograd, J. (1963). Sedimentation equilibrium in a buoyant density gradient. In *Methods in Enzymology*, vol. vi, p. 854. Ed. by S. P. Colowick and N. O. Kaplan. New York and London: Academic Press.

Wachsman, J. T. & Mangalo, R. (1962). Use of 8-azaguanine for the isolation of auxotrophic mutants of *Bacillus megaterium. J. Bact.* **83**, 35.

Wade, H. E. (1955). Basophilia and high ribonucleic acid content of dividing *E. coli* cells. *Nature, Lond.* **176**, 310.

Wade, W. E., Smiley, K. I. & Boruff, C. S. (1946). An improved method for differentiating acid-forming from non-acid-forming bacteria. *J. Bact.* **51**, 787.

Walker, P. D. & Batty, I. (1965). Surface antigenic changes in *Bacillus cereus* during germination and sporulation as shown by fluorescent staining. *J. appl. Bact.* **28**, 194.

Wallis, C., Melnick, J. L. & Phillips, C. A. (1965). Bacterial and fungal decontamination of virus specimens by differential photosensitization. *Am. J. Epidemiol.* **81**, 222.

Ward, W. E. (1938). The apparent oxidation-reduction potentials of bright platinum electrodes in synthetic media cultures of bacteria. *J. Bact.* **36**, 337.

Waring, M. J. (1968). Drugs which affect the structure and function of DNA. *Nature, Lond.* **219**, 1320.

Warner, B. T. (1964). Method of graphical analysis of 2+2 and 3+3 biological assays with graded responses. *J. Pharm. Pharmac.* **16**, 220.

Watanabe, M. & August, J. T. (1968). Methods for selecting RNA bacteriophage. In *Methods in Virology*, vol. iii, p. 337. Ed. by K. Maramorosch & H. Koprowski. New York and London: Academic Press.

Watanabe, T. (1963). Infective heredity of multiple drug resistance in bacteria. *Bact. Rev.* **27**, 87.

Watanabe, T. (1969). Transferable drug resistance: the nature of the problem. *Ciba Foundation Symposium on Bacterial Episomes and Plasmids*. Ed. G. E. W. Wolstenholme and M. O'Connor. London: J. and A. Churchill, Ltd.

WATANABE, T., FURUSE, C. & SAKAIZUMI, S. (1968). Transduction of various R factors by phage P1 in *Escherichia coli* and by phage P22 in *Salmonella typhimurium*. *J. Bact.* **96**, 1791.

WATERWORTH, P. M. (1955). The stimulation and inhibition of the growth of *Haemophilus influenzae* on media containing blood. *Brit. J. exp. Path.* **36**, 186.

WATSON, H. E. (1908). A note on the variation of the rate of disinfection with change in the concentration of the disinfectant. *J. Hyg., Camb.* **8**, 536.

WEBB, H. B. (1946). Composition of Seitz filter pads. *Am. J. clin. Path.* **16**, 442.

WEBB, M. (1948). The influence of magnesium on cell division. 1. The growth of *Clostridium welchii* in complex media deficient in magnesium. *J. gen. Microbiol.* **2**, 275.

WEBB, R. B. (1954). A useful bacterial cell wall stain. *J. Bact.* **67**, 252.

WEDUM, A. G. (1936). Quantitative determinations of carbohydrate utilization by bacteria. *J. infect. Dis.* **58**, 234.

WEIL, C. S. (1952). Tables for convenient calculation of median-effective dose (LD 50 or ED 50) and instructions in their use. *Biometrics* **8**, 249.

WEINBERG, E. D. (1955). The effect of Mn^{++} and antimicrobial drugs on sporulation of *Bacillus subtilis* in nutrient broth. *J. Bact.* **70**, 289.

WHITE, D. O. & FAZEKAS DE ST GROTH, S. (1959). Variation of host resistance to influenza viruses in the allantois. *J. Hyg., Camb.* **57**, 123.

WHITTET, T. D., HUGO, W. B. & WILKINSON, G. R. (1965). *Sterilisation and Disinfection.* London: William Heinemann Medical Books Ltd.

WILKINSON, J. F. & DUGUID, J. P. (1960). The influence of cultural conditions on bacterial cytology. *Int. Rev. Cytol.* **9**, 1.

WILLETTS, N. S. (1967). The elimination of F*lac*$^+$ from *Escherichia coli* by mutagenic agents. *Biochem. biophys. Res. Comm.* **27**, 112.

WILLIAMS, C. A. & CHASE, M. W. (1967). *Methods in Immunology and Immunochemistry.* New York and London: Academic Press.

WILLIAMS, R. T. & BRIDGES, J. W. (1964). Fluorescence of solutions: a review. *J. clin. Path.* **17**, 371.

WILLIAMS, T. (1965*a*). The distribution of response times in a birth-death process. *Biometrika* **52**, 581.

WILLIAMS, T. (1965*b*). The basic birth-death model for microbial infection. *Jl R. statist. Soc.* **27**, 338.

WILLIAMS, T. (1968). Interval estimates for efficiency of plating. *J. gen. Virol.* **2**, 13.

WILLIAMS, T. & MEYNELL, G. G. (1967). Time-dependence and count-dependence in microbial infection. *Nature, Lond.* **214**, 473.

WILLIS, A. T. (1960). *Anaerobic Bacteriology in Clinical Medicine.* London: Butterworth and Co. (Publishers) Ltd.

WILSON, A. T. & BRUNO, P. (1950). The sterilization of bacteriological media and other fluids with ethylene oxide. *J. exp. Med.* **91**, 449.

WILSON, G. S. (1922). The proportion of viable bacteria in young cultures with especial reference to the technique employed in counting. *J. Bact.* **7**, 405.

WILSON, G. S. & MILES, A. A. (1964). *Topley and Wilson's Principles of Bacteriology and Immunity*, 5th ed. London: Edward Arnold Ltd.

WILSON, G. S. *et al.* (1958). Disinfection of fabrics with gaseous formaldehyde. *J. Hyg., Camb.* **56**, 488.

WINSLOW, C. E. A., WALKER, H. H. & SUTERMEISTER, M. (1932). The influence of aeration and of sodium chloride upon the growth curve of bacteria in various media. *J. Bact.* **24**, 185.

WITHELL, E. R. (1942*a*). The significance of the variation in shape of time-survivor curves. *J. Hyg., Camb.* **42**, 124.

WITHELL, E. R. (1942*b*). The evaluation of bactericides. *J. Hyg., Camb.* **42**, 339.

WITHROW, R. B. & PRICE, L. (1953). Filters for the isolation of narrow regions in the visible and near-visible spectrum. *Plant Physiol.* **28**, 105.

WITHROW, R. B. & WITHROW, A. P. (1956). Generation, control and measurement of visible and near-visible radiant energy. In *Radiation Biology*, vol. III, p. 125. Ed. by A. Hollaender. New York: McGraw-Hill Book, Inc.

WITKIN, E. (1947). Genetics of resistance to radiation in *Escherichia coli. Genetics, Princeton* **32**, 221.

WOESE, C. (1960). Thermal inactivation of animal viruses. *Ann. N.Y. Acad. Sci.* **83**, 741.

WOIWOD, A. J. (1954). The inhibition of bacterial growth by colloidal heavy-metal sulphides and by colloidal sulphur. *J. gen. Microbiol.* **10**, 509.

WOOD, B. T., THOMPSON, S. H. & GOLDSTEIN, G. (1965). Fluorescent antibody staining. III. Preparation of fluorescein-isothiocyanate-labeled antibodies. *J. Immun.* **95**, 225.

WRIGHT, H. D. (1933). The importance of adequate reduction of peptone in the preparation of media for the pneumococcus and other organisms. *J. Path. Bact.* **37**, 257.

WRIGHT, H. D. (1934). A substance in cotton wool inhibitory to the growth of the pneumococcus. *J. Path. Bact.* **38**, 499.

WYNNE, E. S., RODE, L. J. & HAYWARD, A. E. (1942). Mechanism of the selective action of eosin–methylene-blue agar on the enteric group. *Stain Technol.* **17**, 11.

WYSS, O., CLARK, J. B., HAAS, F. & STONE, W. S. (1948). The role of peroxide in the biological effects of irradiated broth. *J. Bact.* **56**, 51.

YAMADA, T., TIPPER, D. & DAVIES, J. (1968). Enzymatic inactivation of streptomycin by R factor-resistant *Escherichia coli. Nature, Lond.* **219**, 288.

YOUNG, E. G., BEGG, R. W. & RENTZ, E. (1944). The inorganic nutrient requirements of *Escherichia coli. Archs Biochem. Biophys.* **5**, 121.

YOUNG, M. R. & ARMSTRONG, J. A. (1967). Fluorescence microscopy with the quartz-iodine lamp. *Nature, Lond.* **213**, 649.

YOUNG, M. R. & SMITH, A. U. (1963). The use of euchrysine in staining cells and tissues for fluorescence microscopy. *Jl R. microsc. Soc.* **82**, 233.

ZAMENHOF, S. (1960). Effects of heating dry bacteria and spores on their phenotype and genotype. *Proc. natn. Acad. Sci., U.S.A.* **46**, 101.

ZAMENHOF, S. (1961). Gene unstabilization induced by heat and by nitrous acid. *J. Bact.* **81**, 111.

ZAMENHOF, S. (1967). Nucleic acids and mutability. *Progress in Nucleic Acid Research and Molecular Biology* **6**, 1.

ZELLE, M. R. & HOLLAENDER, A. (1954). Monochromatic ultraviolet action spectra and quantum yields for inactivation of T_1 and T_2 *Escherichia coli* bacteriophages. *J. Bact.* **68**, 210.

ZELLE, M. R., OGG, J. E. & HOLLAENDER, A. (1958). Photoreactivation of induced mutation and inactivation of *Escherichia coli* exposed to various wave lengths of monochromatic ultraviolet radiation. *J. Bact.* **75**, 190.

ZINDER, N. D. (1958). Lysogenization and superinfection immunity in *Salmonella*. *Virology* **5**, 291.

ZINDER, N. D. (1960). Sexuality and mating in *Salmonella*. *Science, N.Y.* **131**, 924.

INDEX

INDEX

INDEX

INDEX

INDEX